Protected Areas, Sustainable Tourism and Neo-liberal Governance Policies

T0203886

From its late nineteenth-century origins, the concept of protected areas has increased in scope and complexity. It now has to come to terms with the twenty-first century world of neoliberal politics, performance metrics and the growing and complex demands of tourism. This international collection of papers explores how this might be done, detailing the issues involved, and the value and values that protected areas have for economies, peoples and environments. Special attention is given to World Heritage Sites, tourism planning and local communities; to the growth of private protected areas; and to the health values of protected areas. Other subjects include private-sector business involvement in protected areas, concessions policy experiments and how the work of the world's largest protected area agency, the U.S. National Park Service, is adapting to changing political and market demands, and to the challenges of sustainable development. It concludes with a searching interview with a member of UNESCO's World Heritage Committee.

The chapters in this book were originally published in a special issue of the *Journal of Sustainable Tourism*.

Hubert Job is a professor of geography and regional science at the University of Würzburg, Germany, specialising in protected areas.

Susanne Becken is the director of the Griffith Institute for Tourism, based in Australia.

Bernard Lane was the founding editor of the *Journal of Sustainable Tourism* and is now a consultant, adviser, writer and speaker.

Protected Areas, Sustainable Tourism and Neo-liberal Governance Policies

Issues, management and research

Edited by
Hubert Job, Susanne Becken and Bernard Lane

LONDON AND NEW YORK

First published 2019
by Routledge
2 Park Square, Milton Park, Abingdon, Oxon, OX14 4RN, UK

and by Routledge
52 Vanderbilt Avenue, New York, NY 10017

First issued in paperback 2020

Routledge is an imprint of the Taylor & Francis Group, an informa business

British Library Cataloguing-in-Publication Data
A catalogue record for this book is available from the British Library

ISBN 13: 978-0-367-58648-5 (pbk)
ISBN 13: 978-1-138-31292-0 (hbk)

Typeset in Myriad Pro
by codeMantra

Publisher's Note
The publisher accepts responsibility for any inconsistencies that may have arisen
during the conversion of this book from journal articles to book chapters, namely
the possible inclusion of journal terminology.

Disclaimer
Every effort has been made to contact copyright holders for their permission to
reprint material in this book. The publishers would be grateful to hear from any
copyright holder who is not here acknowledged and will undertake to rectify any
errors or omissions in future editions of this book.

Contents

CONTENTS

Citation Information

The chapters in this book were originally published in the *Journal of Sustainable Tourism*, volume 25, issue 12 (December 2017). When citing this material, please use the original page numbering for each article, as follows:

Chapter 1
Protected Areas in a neoliberal world and the role of tourism in supporting conservation and sustainable development: an assessment of strategic planning, zoning, impact monitoring, and tourism management at natural World Heritage Sites
Hubert Job, Susanne Becken and Bernard Lane
Journal of Sustainable Tourism, volume 25, issue 12 (December 2017) pp. 1697–1718

Chapter 2
Values in nature conservation, tourism and UNESCO World Heritage Site stewardship
Janne J. Liburd and Susanne Becken
Journal of Sustainable Tourism, volume 25, issue 12 (December 2017) pp. 1719–1735

Chapter 3
Developing a typology of sustainable protected area tourism products
Elias Butzmann and Hubert Job
Journal of Sustainable Tourism, volume 25, issue 12 (December 2017) pp. 1736–1755

Chapter 4
The effects of local context on World Heritage Site management: the Dolomites Natural World Heritage Site, Italy
Maria Della Lucia and Mariangela Franch
Journal of Sustainable Tourism, volume 25, issue 12 (December 2017) pp. 1756–1775

Chapter 5
Estimating the value of the World Heritage Site designation: a case study from Sagarmatha (Mount Everest) National Park, Nepal
Nabin Baral, Sapna Kaul, Joel T. Heinen and Som B. Ale
Journal of Sustainable Tourism, volume 25, issue 12 (December 2017) pp. 1776–1791

For any permission-related enquiries please visit:
http://www.tandfonline.com/page/help/permissions

Notes on Contributors

Som B. Ale is a wildlife ecologist. He works on snow leopard research, education and conservation and teaches ecology at the University of Illinois at Chicago, USA.

Nabin Baral is a research associate in the School of Environmental and Forest Sciences at the University of Washington, Seattle, USA. He is interested in human dimensions of natural resource management, resilience of social–ecological systems, sustainable tourism and protected areas management. He researches the intricate linkages between nature and society. He received a master's degree in ecology from Tribhuvan University, Kirtipur, Nepal, a master's degree in environmental science from Florida International University, Miami, USA and a doctorate degree in forestry from Virginia Tech, Blacksburg, USA.

Susanne Becken is a professor of sustainable tourism and the director of the Griffith Institute for Tourism, Australia. She has published widely on sustainable tourism, climate change, energy use and greenhouse gas emissions, tourist behaviour, environmental policy and risk management. She has recently advised the Queensland Government as an expert on the Great Barrier Reef Water Science Ministerial Taskforce and is a member of Air New Zealand's Sustainability Advisory Panel.

Elias Butzmann is a research assistant in the Department of Tourism at the Munich University of Applied Sciences, Germany and PhD student at the University of Würzburg, Germany. His main research interests include nature-based tourism, sustainable tourism and economic impacts of tourism.

Catherine Cullinane Thomas is an economist with the U.S. Geological Survey, the scientific research agency for the U.S. Department of the Interior. She specialises in developing regional economic models related to protected-area management and visitation. Cathy is the lead developer for the National Park Service (NPS) visitor spending effects analysis.

Valentina Dinica is a senior lecturer in public policy at the School of Government in New Zealand, where she is teaching courses on environmental policy and governance, policy analysis and policy methods. Currently, she is also the director of the undergraduate programme in public policy at Victoria University of Wellington, New Zealand. Her research addresses several key areas of sustainable development: sustainable tourism, with particular focus on nature-based tourism and island tourism; renewable energy and energy efficiency in the context of climate change mitigation, water management and natural resource management. Valentina's teaching emphasises participatory

policy analysis; behavioural approaches to policy instruments' design, policy implementation and conflict resolution; and the operationalisation of innovative governance
principles for sustainability. She received her PhD degree from the University of Twente,
Enschede, the Netherlands and has been awarded numerous research grants by local,
regional and national European governments and European Union research funds.

Statia Elliot is an associate professor and director of the School of Hospitality, Food and
Tourism Management at the University of Guelph, Canada. She has extensive experience working with Canadian destination marketing organisations and specialises in
research of place image and branding, tourism destination planning and performance.
She teaches strategic marketing and tourism.

Barbara Engels is a scientific officer at the German Agency for Nature Conservation
(Bundesamt f€ur Naturschutz [BfN]) responsible for both Natural WHSs and BRs. She
is the executive secretary of the National MAB Committee: Germany. Barbara is also a
member of the IUCN World Heritage Panel and serves on the German Delegation to
the UNESCO Word Heritage Committee. For many years, she has been involved with
sustainable tourism in Germany and especially with the work on the evaluation of economic effects of tourism in protected areas.

Mariangela Franch is a full professor of marketing at the University of Trento, Italy. She
coordinates the university's Emasus research group, which draws together researchers
from management, environmental economics, statistics, psychology and economic
history to analyse sustainable economic development and sustainable management.
She teaches marketing and service management. Her current research interests include sustainable tourism and "4L tourism" (landscape, leisure, learning and limit),
tourist marketing as a factor in local economic development and digital marketing as
an innovative factor in communication tools. She has authored and co-authored contributions to national and international journals and books.

Joel T. Heinen is a professor in the Department of Earth and Environment at Florida International University, Miami, USA. Prior to his PhD degree at the University of Michigan,
Ann Arbor, USA, he was a Peace Corps volunteer in Nepal and focused his research
there on protected areas management and biodiversity conservation policy.

Hubert Job is a professor of geography and holds the chair for geography and regional
science at the University of Würzburg, Germany. His main areas of scientific research
are PAs, tourism and regional development. He is an elected member of the German
Academy for Spatial Research and Planning and an appointed member of the German
National Committee for the UNESCO programme on MAB.

Marion Joppe is a professor in the School of Hospitality, Food and Tourism Management
at the University of Guelph, Canada. She obtained her doctorate from the Aix-Marseille
University, Provence, France in law and economics of tourism in 1983 and specialises
in destination planning, development and marketing and the experiences upon which
destinations build. She has extensive private- and public-sector experience, having
worked for financial institutions, tour operators, consulting groups and government,
and has published in both North America and Europe.

Sapna Kaul is an assistant professor of health economics in the Department of Preventive
Medicine and Community Health at the University of Texas at Austin, USA. She received

a PhD degree from Virginia Tech, Blacksburg, USA. Her research focuses on implementing economic tools (e.g. cost–benefit analyses and contingent valuation), examining individual decision-making about health care services and environmental goods.

Anahita Khazaei received her PhD from the School of Hospitality, Food and Tourism Management at the University of Guelph, Canada. Her research interests include community engagement, tourism planning and sustainability. Anahita has taught marketing and strategic management and has years of experience in strategic planning and industry analysis.

Lynne Koontz serves as a lead economist for the NPS Social Science Program. She provides economic expertise and technical support to parks and NPS leadership on a variety of issues. She is a co-author of the NPS Visitor Spending Effects analysis.

Bernard Lane was the founding editor of the *Journal of Sustainable Tourism*, and was a co-editor of that journal for 25 years. He publishes, lectures, advises and broadcasts on sustainable tourism, rural tourism, heritage tourism and heritage conservation worldwide.

Janne J. Liburd is a professor of tourism and the director of the Centre for Tourism, Innovation and Culture at the University of Southern Denmark, Odense, Denmark. Janne is the chairman of the UNESCO World Heritage Wadden Sea National Park. By ministerial appointment, she serves on the Danish National Tourism Forum, charged with developing the first strategy for tourism in Denmark. She is a cultural anthropologist: her research interests are sustainable tourism development, innovation, national parks and tourism higher education.

Maria Della Lucia is an associate professor of tourism and business management at the University of Trento, Italy. Her current research interests include local development, culture-led regeneration, destination management and governance, sustainable tourism and mobility and economic impact analysis as investment decision-making tools. Her field research focuses primarily on fragmented and community-based areas, particularly Alpine and rural destinations and urban areas. She has authored and co-authored journal publications in leading tourism journals, in the *Journal of Information Technology and Tourism*, the *International Journal of Management Cases* and the *Journal of Agricultural Studies*, together with numerous book chapters.

Bret Meldrum serves as the chief social scientist for the NPS. His expertise is in social science methods, visitor use research and monitoring, having worked on numerous scientific projects across parks intended to inform planning and management applications.

Jeffrey Olson is a senior public affairs and communications staffer at the NPS headquarters in Washington, DC, USA. He also worked in the news media for 25 years and owned a small business where he learned real-world economics first-hand.

M. Nils Peterson is an associate professor in the Department of Forestry and Environmental Resources at North Carolina State University, Raleigh, USA. His research focuses on unravelling the drivers of human behaviour relevant to wildlife conservation.

Kati Pitkänen PhD, is a senior researcher at the Finnish Environment Institute, Environmental Policy Centre and was previously a postdoctoral researcher at the University of Eastern Finland and at Umeå University, Sweden. Her current research interests include the green economy, rural development, second-home and nature-based tourism and health and well-being impacts of nature and green infrastructure.

Riikka Puhakka PhD, is a postdoctoral researcher in the Department of Environmental Sciences at the University of Helsinki, Finland. Previously, she worked at the University of Oulu, Finland and at the University of Eastern Finland – Joensuu. Her research interests include nature-based tourism, sustainable tourism, human relationships with nature and the health and well-being impacts of nature.

Christopher Serenari is a human dimensions specialist with the North Carolina Wildlife Resources Commission, Raleigh, USA. His broad research interests fall under the nexus of conservation and development, focusing on why conservation policies succeed and fail in rural areas.

Pirkko Siikamäki PhD, adjunct professor, works as a regional manager for nature protection at Metsähallitus, Vantaa, Finland, which controls the state's agency, Parks & Wildlife Finland. Her research interests include the evolution of clutch size in passerine birds, conservation ecology and the genetics of rare plants, and most recently sustainable nature tourism and the relationships between biodiversity and nature-based tourism. Previously, she was a director of the University of Oulu's Oulanka Research Station, Finland.

Susan L. Slocum is an assistant professor in the Department of Tourism and Event Management at George Mason University, Manassas, USA. Sue has worked on regional planning and development for 15 years and worked with rural communities in Tanzania, the UK and the USA. Her primary focus is on rural sustainable development, policy implementation and food tourism, specifically working with small businesses and communities in less advantaged areas. Sue received her doctoral education from Clemson University, USA and has worked at the University of Bedfordshire, Luton, UK and at Utah State University, Logan, USA.

Paulina Stowhas is a faculty member within the Forestry and Agricultural Sciences Department at the University of Mayor, Santiago, Chile. Her primary research interests centre on the human dimensions of conservation in Chile.

Tim Wallace is an associate professor in the Department of Sociology and Anthropology at North Carolina State University (NCSU), Raleigh, USA. He is the director of the NCSU Ethnographic Field School in Lake Atitlán, Guatemala. He has published research on topics including heritage, identity and tourism in Madagascar, Hungary, Costa Rica, Guatemala and North Carolina.

Pamela Ziesler coordinates the visitor use statistics for the NPS. Her professional interests include field measurements, statistical modelling, optimisation and mathematical analysis to support land and resource management decisions.

Protected Areas in a neoliberal world and the role of tourism in supporting conservation and sustainable development: an assessment of strategic planning, zoning, impact monitoring, and tourism management at natural World Heritage Sites

Hubert Job, Susanne Becken and Bernard Lane

ABSTRACT
Societies collapse when there is an increasing natural resource scarcity and growing stratification of society into rich and poor. The neoliberal world of targets, business plans and short term economic justification in which we live exacerbates these risks to society. It is imperative to find new ways of governing natural ecosystems that protect them from these risks and allows usage that helps close the development gap. Tourism in Protected Areas (PAs) is one important vehicle to achieve sustainable conservation and development outcomes. This paper highlights that the increasing focus on promoting human activity, especially tourism, in and around PAs is increasingly enshrined in the mandate and governance structures of United Nations Educational, Scientific and Cultural Organisation natural World Heritage Sites and Biosphere Reserves. It reviews strategic planning, zoning, impact monitoring, and tourism management by analysing all 229 natural World Heritage Sites, revealing that both overall strategic planning and tourism planning in these sites need improvements, notably through more consistent monitoring systems. The paper concludes by exploring the benefits of embedding World Heritage Sites into Biosphere Reserves, with a particular focus on core zoning, regional product development, and improved monitoring standards, and suggests ways to disseminate good practice worldwide to all types of PAs.

Introduction: the wider context

Concern about planetary limits and the risks humankind is facing as a result of increasing environmental destruction is growing (Hall & Day, 2009). Preserving natural resources and environments, including through the mechanism of Protected Areas (PAs), has long been recognised as vital for societies to continue to exist and thrive. Diamond (2005), in his book on why societies collapse, proposed a five-point framework to capture the complexity of the human–environment relationship and long-term success. He suggests that large-scale decline in either human population and/or socio-economic complexity is influenced by serious environmental problems, climate change (natural or anthropogenic), conflict with trade partners or neighbouring enemies, and problems arising from society's own political, economic, and social arrangements. Building on this earlier work, and learning from patterns of societal collapse over history, Motesharrei, Rivas, and Kalnay (2014) extracted two key features that appear to influence whether societies collapse or not. One was the increasing scarcity of natural resources (the result of exceeding an ecological carrying capacity), and the other one was a heavy stratification of society into rich elites and poor commoners. Under the current conditions of an increasingly unequal society, Motesharrei and co-workers suggested that collapse is difficult to avoid; however, it is possible if we can reach an equilibrium in which the depletion rate of nature per head of population is sustainable, and if this depletion is more equally distributed.

It is clear that present and future attempts to ensure sustainability of humankind must focus on two key dimensions: the environment and the people that inhabit it. Today's challenge is of much greater magnitude compared with earlier societies (e.g. the Roman Empire, the Mayan civilisation, the monument building Polynesian society on the Easter Islands) that existed on a regional scale and not a global one. In the planetary reach of the problem also lies an opportunity, as long as the world community is cognisant of the above risks and agrees collectively on key measures to avoid collapse. In the past, flourishing societies have successfully managed natural environments through functioning economic and political institutions; this could be achievable in today's global society. Several major frameworks to develop more sustainable and equitable societies have recently been ratified, and are now being implemented. The two most important roadmaps are the 17 Sustainable Development Goals (SDG) (UNDP, 2016) and the "Paris Agreement" to combat global climate change (UNFCCC, 2015). Both are global in scale and universally applicable. These latest agreements build on a range of earlier frameworks and treaties, including some specifically dedicated to the conservation of our natural heritage, preservation of biodiversity and sustainable development. Whether these agreements are sufficient to avoid societal collapse depends amongst other factors on their implementation, but also on the purposeful integration of societal goals with environmental protection and the number and quality of PAs.

The key trends that determine the opportunity space of today's society are visualised in Figure 1. An exponentially growing global population – and a concomitant increase in tourist activity – is at the core of substantial land use changes and pressures on resources. In previous centuries, natural ecosystems largely made way for intensive agriculture, irrigation agriculture and urban spaces, with a trend towards more and more people living in bigger cities far away from nature (WBGU, 2016). Population growth, economic activity, and urbanisation have – perhaps paradoxically – been accompanied by a global expansion of PAs. Zimmerer (2006) found that "experiencing nature" has become a superior good that is in increasing demand by those with higher per capita incomes. However, the increase of the total area designated as PAs did not necessarily equate to an improvement in conservation outcomes and biodiversity protection, as it is often the "less valuable" land that is protected and management effectiveness varies (Geldmann et al., 2015; Lama & Job, 2014). Nevertheless, the rapidly declining share of natural ecosystems illustrated in Figure 1 highlights the critical importance of PAs for preserving the few remaining intact ecosystems. Further loss could be irreversible and compromise the opportunity of "closing the equity gap" in remote and potentially disadvantaged communities – thus making collapse more likely.

PAs play a pivotal role in the challenging task of preserving the environment and using it at the same time. The intensifying pressure on PAs demands new approaches to governing resources so that they are protected and their values are distributed more equitably. This is not a trivial task, as PAs are now often charged with several mandates: halting the loss of biodiversity (Geldmann et al., 2015), nature conservation, providing recreational and tourism opportunities, as well as educational and spiritual or inspirational tasks, and the demonstrable production of economic benefits. Protecting ecosystem services can in some cases be converted into "hard cash", for example through the sale of carbon credits (Dudley, Sandwith, & Belokurov, 2010), but in addition, governments look for income generation for local communities, job creation, royalties or other forms of state owned resource rent, and more recently, enhanced brand value and regional positioning in an increasingly globalised world (Bouma & van Beukering, 2015; Knaus, Ketterer Bonnelame, & Siegrist, 2017).

The acknowledgement of humans as acceptable elements in a PA socio-ecological system (Espiner & Becken, 2014; Mayer, Müller, Woltering, Arnegger, & Job, 2010) also bears risks. PAs are not immune

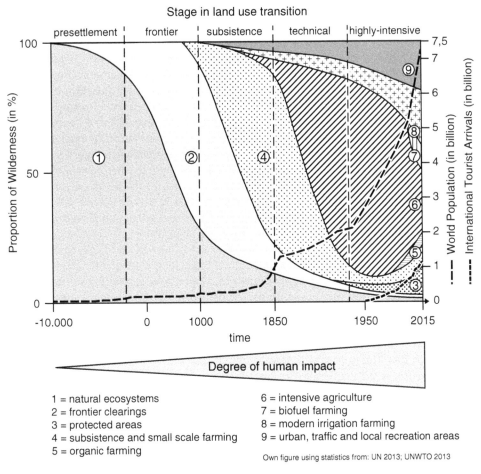

Figure 1. Land use changes, population, and tourism growth and the increasing importance of Protected Areas.

from the global "land grab" (Engels, 2015), nor from the local valorisation of nature through various forms of resource extraction. Tourism forms part of this: the commercially focused tourism sector is firmly embedded within the broader neoliberal agenda (Higgins-Desbiolles, 2010). The current Dominant Social Paradigm (Pirages & Ehrlich, 1974) is strongly shaped by anthropocentric world views and a consumerist culture, in which material abundance, perpetual economic growth, and technological progress persist and prevail (Kilbourne, Beckmann, & Thelen, 2002). One might argue that these trends are nowhere more reflected than in the hedonic and self-centred activity of leisure travel (Becken, 2017). Moreover, this dominating paradigm aligns with liberal democracies that advocate extensive private property rights and minimal intervention by the public sector, which make it increasingly difficult to protect "public goods" such as natural areas and their associated ecosystem services (Albert et al., 2017). The implications of operating within the neoliberal agenda are explored in more detail by Slocum (2017). Furthermore, the establishment of private PAs with exclusive access arrangements and tightly managed business activities may present some interesting opportunities, as discussed in the case of Chile by Serenari, Peterson, Wallace and Stowhas (2017).

At the same time, the hopes attached to tourism and its perceived potential as a vehicle for sustainable development become nowhere more apparent than in the designation by the United Nations of the 2017 International Year of Sustainable Tourism for Development. Benefits from PA tourism have long been recognised and studied in industrialized and developing countries (Carius, 2016; Job & Metzler, 2005; Job & Paesler, 2013), and the provision of visitor facilities and tourist experiences has moved to the core of management for many Parks (Butzmann & Job, 2017; Medeiros, Frickmann Young, Boniatti PAvese, & Silva Araújo, 2011; Müller, 2014). The links between tourism, conservation and rural regeneration are increasingly recognized and used by PA managements (Getzner, Lange Vik, Brendehaug, & Lane, 2014). Tourism in natural areas is also increasingly seen as a major avenue for increasing the physical and mental well-being of (often urban) visitors who recon-nect with nature (e.g. well studied in the context of PAs in Finland, see Puhakka, Pitkänen, & Sii-kamäki, 2017). Discussions about visitation and well-being are also beginning to consider new (and sometimes marginalised) members of local PA communities (Khazaei, Elliot, & Joppe, 2017). As a result, visitation can play a key role in regional development, preservation of indigenous cultures, and local identities; all of which can be compatible with the conservation mandate, if managed effec-tively (Carius, 2016). The conflicts between nature-based tourism and other development pressures (see Becken & Job, 2014; Liburd & Becken, 2017), however, adds to the complexity of PA manage-ment, especially when areas are put on the red list of heritage in danger (Engels, 2015), when they are downsized and/or partly degazetted (Ferreira et al., 2014), and/or commodified to an extent that the very naturalness of the area is eroded or destroyed (Saarinen, 2016).

In summary, at times of intensifying neoliberalism, PAs must increasingly justify their existence by providing a robust business case for public sector investment, and often tourism plays a key role in demonstrating benefits, either in the establishment of a PA or in its ongoing existence. An example from Sagarmatha, Nepal, illustrates how entry fees, in particular from international visitors, can gener-ate much needed funds for both conservation and development (Baral, Kaul, Heinen, & Ale, 2017). The Special Issue of the *Journal of Sustainable Tourism* in which this paper is published therefore examines how and to what extent sustainable tourism could help fulfil the increasing demands now being placed on PAs. The notion of widening the designation of "sacred cows" to creating "cash cows" (Müller, 2014) is at the core of the Special Issue. It aims to critically examine key issues and developments in both the management and research agendas associated with PAs, focusing on the central requirements for PAs to survive in a neo-liberal world characterised by economic imperatives, the valuation of natural capital, and maximising of returns on investment. The Special Issue also explores ways of managing for a post neo-liberal world, measuring success and introducing new pri-orities such as well-being and human health.

This opening paper sets out a series of critical reflections on the evolving conceptualisation of PAs and their ability to simultaneously deliver conservation and development benefits. The closely related United Nations Educational, Scientific and Cultural Organisation (UNESCO) concepts of natural World Heritage Sites (WHSs) and Biosphere Reserves (BRs) are studied in more detail. The findings are shown to have strong relevance for all types of PAs. This is followed by an empirical analysis of the extent of tourism planning and monitoring in natural WHSs, and an assessment of whether better planning leads to better conservation outcomes. The critical role of measurement will be highlighted, followed by a concluding discussion on ways forwards to ensure ongoing sustainability of PAs through taking advantage of a combination between the concept and practices developed in natural WHSs and those in the innovative UNESCO label of BR. Our main research questions are as follows. (1) How has the mandate changed in relation to the most prominent types of PA, namely natural WHSs and BRs? (2) What is the status of monitoring tourism in natural WHSs and is it adequate to pro-vide a robust business case for the existence of PAs? (3) Does better tourism management lead to superior natural WHS outcomes? (4) Is the combination of UNESCO's categories the "best of both worlds" to manage regional development, and especially tourism, in a sustainable manner against a further commodification of PAs? (5) Are the findings for WHSs and BRs applicable to all PAs, and how could that application take place?

The background to Protected Areas

The ecological integrity and high degree of naturalness of PAs makes it imperative to protect them to the highest standard possible. Decline of ecological quality in PAs will inevitably lead to a decline in

Table 1. Aligning major recent milestones relevant to the development and management of PAs (Sources: Becken & Job, 2014; World Bank, 2016; UNWTO, 2015).

Time-frame	Tourism	Development and population	Protected Areas
1960–1970s	1969: The 'jumbo' (Boeing 747) changes the world of travel	1968: Club of Rome founded 1972: "Limits to Growth" by the Club of Rome	1968: Man & Biosphere (MAB) Programme established at the biosphere conference in Paris 1969: First legal definition by IUCN of National Parks (in Delhi) 1972: Adoption of the UNESCO World Heritage Convention 1976: Establishment of Biosphere Reserves
1980s	1980: 278 million international arrivals	1987: Brundtland report 1987: World population exceeds 5 billion	1980: IUCN/UNEP/WWF World Conservation Strategy 1983: First World Congress of Biosphere Reserves in Minsk 1988: Establishment of the World Conservation Monitoring Centre by UNEP/IUCN/WWF
1990s	1995: 528 million international arrivals	1992: UN Earth Summit in Rio de Janeiro and Millennium Development Goals	1991: Establishment of UNEP's Global Environmental Facility (GEF) 1992: UNESCO introduces a new category of cultural landscapes 1995: 2nd World Congress on Biosphere Reserves in Seville
2000s	2002: UN announces the international Year of Ecotourism 2005: UNWTO's Sustainable Tourism – Eliminating Poverty (ST-EP) launched	2002: World Summit on Sustainable Development in Johannesburg	2008: Review of the guidelines for the IUCN PA management categories 2008: 3rd World Congress of Biosphere Reserves in 2008 in Madrid
2010+	2012: over 1 billion international arrivals 2012: Adoption of the World Heritage and Sustainable Tourism Programme by the World Heritage Committee	2014: Global Gross Domestic Product is US$74,000 billion	2010: Endorsement of the Aichi Goals as a strategic plan for the implementation of the Convention of Biological Diversity 2015: Adoption of the World Heritage Convention's "Policy on the integration of a sustainable development perspective into the processes of the Convention"
2016	United Nations Sustainable Development Goals	World population is over 7.3 billion	2016: "Lima Action Plan" as a new MAB Strategy 2015-2025

planetary ecosystem integrity and resilience, likely irreversibly as the last pools of genetic diversity and ecosystem services are eroded. However, PAs are not static and do not exist in a vacuum. They are heavily influenced by global, national and local changes in development, political systems, ideologies, and power structures. From their early conceptualisations as natural places of grandeur to be preserved, PAs now demand a complex, varied and dynamic conservation approach, moving in some cases towards the designation of carefully managed recreational parklands. The mandates of PAs have evolved to include species protection, ecosystem management, and more recently to consider an approach where humans and their needs are no longer excluded from the conservation approach, but form an integral part of it (Bender, Roth, & Job, 2017; Hammer, Mose, Siegrist, & Weixlbaumer, 2016).

The section below provides a summary of key trends and milestones that are relevant to understanding today's PA context, with a particular focus on the evolution of natural WHSs and BRs, along with the other types of PA recognised through the International Union for Conservation of Nature (IUCN) system. Both WHS and BR areas are highly relevant as they epitomise hot spots of sustainable tourism, and at the same time are unified through a transnational framework of management and governance (Job, Kraus, Merlin, & Woltering, 2013). The common approach underpinning natural WHSs is manifested in the concept of the Outstanding Universal Value (OUV). For BRs, the linking principle is one of "harmonised management and conservation of biological and cultural diversity, and economic and social development based on local community efforts and sound science" (Schaaf & Clamote Rodrigues, 2016, p. xv). WHSs and BRs have fundamentally different purposes, objectives and management principles and should, therefore, not be confused (Batisse, 1997). However, given increasing attention to market forces, BRs can illustrate how the inclusion of sustainable practices (e.g. in fisheries, agriculture, or forestry) is essential to maintain the site as a biodiversity hot spot for future generations. BRs do not require to be of OUV; they rather represent the national level of distinctive natural and cultural landscape features of a state party, and they focus on regional sustainable development. Therefore this instrument should been seen as a complementary PA category to natural WHSs.

An overview of milestones and trends

To understand the increasing demands on PAs several trends are noteworthy. Foremost, the global population has now reached over 7.3 billion (Figure 1; Table 1). The pressure on natural resources has increased accordingly, especially given that global population growth has been accompanied by substantial increases in economic activity. At the same time, and thanks to a growing middle class in many emerging economies, global tourism activity has increased substantially to reach 1.235 billion international arrivals in 2016 (UNWTO, 2017). The rising volume of international and national travel has generated growing visitor pressure at natural areas, in addition to substantial numbers of local recreationists for those areas close to major population centres (Job, Merlin, Metzler, Schamel, & Woltering, 2016; Jones & Ohsawa 2016). The tourism and development trends have been mirrored by major PA milestones that clearly indicate a shifting role of mandate and governance.

Table 1, like the discussion below, pays special attention to WHSs and BRs. They are seen as leaders within the world of PAs, both in visitor pressures and in management procedures. As part of the global UNESCO system, it is relatively easy to track key trends in WHSs and BRs, including the introduction of the concept of sustainable development, into the conservation mandate. It is also a UNESCO policy to record and publish agreed management procedures on the web. In practical terms, for empirical research on monitoring, the world's WHSs present a manageable number of sites, combining tourism hot spots, innovative good practice, benchmarked procedures, measuring impacts, and responses to the pressures of the neoliberal world. A survey of all PAs globally would entail dealing with numbers far too great to manage, with diverse legal, conceptual and operational backgrounds, as well as innumerable linguistic, national and other complications.

Integrating the sustainable development paradigm into UNESCO's World Heritage Convention

WHSs are widely recognized as the world's most special and desired PAs; as major attractions that are often commercialised, they represent an important case to understand neoliberal forces. As of June 2016, there were 229 natural WHS globally, representing about 0.1% of the total number of PAs. However, with a total coverage of 279 million hectares the natural WHSs account for over 8% of the combined surface area covered by PAs (IUCN, 2014). The World Heritage Convention's primary focus is on preservation of the world's natural and cultural heritage of OUV. Nonetheless, over time, and in a shift of the World Heritage Convention's Operational Guidelines, multiple changes have been made to broaden the integration of sustainability development into heritage protection (Engels, 2015). An interview with someone closely involved with decision making on World Heritage Sites, explains how the WHS system functions: it can be found in the same Special Issue of the *Journal of Sustainable Tourism* in which this paper is published (Engels, Becken, Job, & Lane, 2017).

Five phases in the development of the WHS system can be distinguished.

Pre-phase: translocation of Abu Simbel (1960–1971)

In Egypt's far South, the area of Abu Simbel contains dozens of archaeological sites. In 1960 with the construction of the Aswan dam, these sites were in danger of being submerged through the creation of the Lake Nasser reservoir. This plan alarmed archaeologists, who developed the idea of bringing together 50 countries to fund and get the best team to cope with the monumental rescue of Abu Simbel. The project was commissioned by UNESCO; a census of all sites at risk of flooding was taken and between 1963 and 1968 the temples were completely relocated to a higher location (Mager, 2015).

Save the heritage phase (1972–1993)

The UNESCO World Heritage Convention was adopted in 1972 and aims for the "…identification, protection, conservation, presentation and transmission to future generations of the cultural and natural heritage" (UNESCO, 1972, article 4). The convention itself has never been amended to date. But in 1993, the World Heritage Committee (WHC), as UNESCO's decision-making body, expressed the explicit need to take into account the outcome of the 1992 Rio Summit (WHC, 1993, committee decision conf. 002 VIII, 1–6)

Sustainable land-use phase (1994–2005)

In 1994, the term "sustainable land use" was first used in the Operational Guidelines (Paragraph 38), the convention's main planning instrument. The reference to sustainable land use was in connection with "cultural landscapes", a new category of WHS introduced in 1992 (UNESCO, 1994). The achievement was to protect such cultural landscapes, reflecting specific land use techniques and, therefore, maintaining the biodiversity and contributing to modern techniques of sustainable land use. Since 2002, the Operational Guidelines include the notion of sustainable development. One section specifically deals with "sustainable use" of natural and cultural heritage as contributing to the quality of life of people or "…as an instrument for the sustainable development of all societies through dialogue and mutual understanding" (WHC, 2002, article 1).

Sustainable development phase (2006–2014)

The year 2006 saw the publishing of UNESCO's Natural Heritage Strategy, which refers directly to sustainable development in its mission statement. Furthermore, the World Heritage Committee specifically added the role of communities in the implementation of the Convention in 2007 (Engels, 2015). Importantly for this present research, in 2012, the World Heritage and Sustainable Tourism Programme was adopted by the members of the committee (WHC, 2012).

UN Sustainable Development Goals (SDG) Phase (2015–today)

"World Heritage may provide a platform to develop and test new approaches that demonstrate the relevance of heritage for sustainable development with a view to its integration in the UN Post-2015 development agenda" (UNESCO, 2016a). With this statement, UNESCO acknowledges that sustainable use (i.e. implying economic transactions and market forces) of natural heritage is a key factor for its successful conservation in the long term. However, the words sustainable development are still not mentioned in the Operational Guidelines. A more systematic approach to sustainable development including strategically clear guidance is still missing (Engels, 2017).

From the above, it can be concluded that sustainable development in association with PAs is increasingly recognised, but should preferably not be realized in the protected site itself but rather in its surrounding area. Considering the wider context of the sites leads to a discussion of another PA category, namely BRs.

Evolution of UNESCO Biosphere Reserves

The global network of BRs consists of 669 reserves in 120 different countries (UNESCO, 2016b). The central proposition of BRs is one of harmonisation of biodiversity protection and sustainable development – thus, an integration of human activities with ecological systems and processes. The human element has become increasingly prominent over five phases, and accommodates opportunities for economic development – as long as it is sustainable.

Early beginnings (1968–1975)

Originating in 1968 at the biosphere conference in Paris (UNESCO, 1986), the primary role of BRs was to facilitate interdisciplinary research programs [institutionalised in the "Man and the Biosphere" (MAB) Programme] that, innovatively, consider the interplay of the natural environment with society and culture (Kammann & Möller, 2007). The goal was to develop strategies for more efficient environmental policies that could then be implemented nationally by participating countries. UNESCO (1974), as the lead agency, devised three aims that focused on (1) biodiversity protection, (2) ecological research, and (3) education.

Establishment phase (1976–1982)

The concept of BRs was implemented on the ground in 1976 with 57 newly designated reserves. An important element of the developing network was the exchange of information and insights resulting from various research projects. By 1981, the network had grown to 208 BRs. Since many newly designated BRs had already been protected through some other status (e.g. as National Parks) the focus of activity was largely restricted to nature conservation, and socio-cultural aspects remained neglected.

Pre-Seville Phase (1983–1994)

In 1983, at the first World Congress of Biosphere Reserves in Minsk, it was agreed that there was a need to better implement the development and educational mandates of BRs. Not all of the early listed sites fully addressed all three MAB objectives, as they had to "slowly grow into their role of enhancing the well-being of the local populations and contributing to the sustainable use of natural resources" (Batisse, 1997, p. 10). The shifting emphasis reflected the *Zeitgeist* of the 1987 Brundtland report and the 1992 United Nations Conference on Environment and Development (UNCED) in Rio de Janeiro. BRs were recognised as suitable vehicles to implement the UNCED outcomes, which included the Rio Declaration on Environment and Development, Agenda 21, and Forest Principles, and the beginning of three binding conventions, namely the Convention on Biological Diversity (CBD), the Framework Convention on Climate Change and the United Nations Convention to Combat Desertification.

Post-Seville phase (1995–2007)

The second World Congress on Biosphere Reserves in 1995 in Seville firmly anchored the sustainable development paradigm in the constitution of BRs. The Seville Strategy outlined strategies that help to "… reconcile conservation of biodiversity and biological resources with their sustainable use" (UNESCO, 1996, p. 3), and support the three core functions of protection, development, and knowledge generation and education. An important new element was the structuring of BRs into three spatial zones, namely a strongly protected core zone dedicated to biodiversity conservation, and a buffer zone that allows some level of human activity, as long as it is compatible with site's ecological requirements. The buffer zone is surrounded by a transition zone (which usually includes settlements), that allows cooperation between different actors and forms of use. Importantly, the buffer and transition zones provide a platform for the development of regional products, produced with the input of the local population and a diverse range of industries[1] (Carius, 2016; Kraus, Merlin, & Job, 2014; Price, 1996; Van Cuong, Dart, & Hockings, 2017). This opens the door for tourism development that, if implemented well, can support the dual mandate of conservation and development. The strategy also provides a set of proposed indicators to facilitate monitoring of progress (UNESCO, 1996), and evaluations are required every ten years to ensure compliance with BR criteria.

Rapid expansion or "Boom" phase (2008–2014)

The characteristics of this (surprisingly rarely discussed) phase were a boom in BR nominations on the one hand and the rise of the MAB programme's political recognition on the other. The key elements of the Seville Strategy were confirmed at the third World Congress of Biosphere Reserves in 2008 in the "Madrid Action Plan for Biosphere Reserves 2008–2013" (MAP), and aligned with the Millennium Development Goals[2] (MDGs). The latter was endorsed most recently through the "MAB Strategy 2015–2025" that specifically sees BRs as an integral element of the implementation of SDG (UNESCO, 2015). Again, the focus shifted towards the mandate of sustainable development facilitated through BRs. In addition to the earlier three aims, the MAP added the important role that BRs play in climate change mitigation and adaptation, as well as other processes of global environmental change (Becken & Job, 2014). The Madrid Action MAP also further strengthened BRs' role as learning sites for sustainable development. Its final evaluation revealed an ambiguous result of many sites tangibly demonstrating their capacity to promote sustainability while other sites did not provide the requested information on their related performance (http://unesdoc.unesco.org/images/0022/002280/228056E.pdf).

Delivery phase (2015–today)

The present phase focuses on the quality aspects of BRs, not on their quantity, resulting in stagnant BR numbers for the first time. The implementation of the MAB strategy is further specified by the 2016 "Lima Action Plan" (UNESCO, 2016c). Moving from rhetoric to action, this new road map is more tailored towards delivery, in contrast to the preceding MAP, which somehow lacked implementation logic. Discourses on the potential contributions to sustainable development are not sufficient anymore, evidence is needed. In response and, as part of an exit strategy, MAB governing bodies currently negotiate the delisting of sites that do not comply with their duty to periodically track and review their progress towards sustainable development. In June 2017, a total of 20 reserves withdrew from the listing. Tourism is, however, increasingly recognised as an important element of BRs and their economic valorisation (Job et al., 2013; Carius, 2016).

In their current conceptualisation, WHSs and BRs can be interpreted as a normative concept of environmental governance at multiple scales, based on non-manipulative local participation, legitimacy, and efficient conflict mitigation among stakeholders for the sustainable use of natural resources (Brenner & Job, 2012). In addition, the involvement of local populations in the BR processes was ahead of its time, compared with all other PA designations. Multi-Internationally Designated Areas, such as WHSs and BRs, are now considered not only conservation areas, but also triggers for regional development – a perspective dismissed by many conservationists (Shafer, 2015). Adequate planning and management of tourism activity in these areas is, therefore, essential, including with a focus on (co)-management, governance structures, implementation plans, community involvement, decision making, and policy development (Borrini-Feyerabend et al., 2013; Buckley, 2012; Cessford & Muhar,

2003; Eagles, 2014; Köck & Arnberger, 2017; Mayer & Job, 2014; Stoll-Kleemann, 2010; Tisdell & Wilson, 2012). This research, therefore, builds on the recommendation made in a recent IUCN report by Schaaf and Clamote Rodrigues (2016, p. xv) regarding tourism management: *"Visitor numbers should be adequately managed, and sustainable tourism strategies and plans should be developed and implemented in order to safeguard the conservation and environmental integrity ….".*

Methodology

To assess the extent of tourism planning in PAs, an analysis of natural and mixed WHSs was undertaken. Natural WHSs were selected because they are critically important for achieving global conservation goals and are magnets for tourist visitation (Conradin & Wiesmann, 2014; IUCN, 2014; Ruoss, 2016). Because of their prominence a minimum level of planning should be in place. The objective of this analysis was to determine the extent of visitor planning in natural WHS areas. In addition, and to establish the level of detail of planning and management and adherence to minimum requirements published by UNESCO, an analysis of English and Spanish language tourism management strategies and plans was undertaken via targeted content analysis. The focus was on visitor number monitoring and measurement of economic impact, and how plans address important elements of sustainable tourism management as identified by UNESCO (Becken & Wardle, 2017). Finally, subsequent analysis explored the link between planning and outcomes. The findings from this analysis may be indicative of other PAs, although it is likely that WHS represent a "best practice" scenario compared with less well-known areas, including BRs that at this point in time are less recognised globally by the general public and maybe even by decision-makers.

Data compilation and analysis

A list of all 229 natural and mixed WHSs was used to search the Internet for tourism and general management plans. Over two-thirds of the sites are located in countries where English is not an official language and Google Translate was used to search in multiple languages. As a result of the search, a database was compiled which contained web links to the relevant plans and strategies, as well as other variables including site specific data such as the year of inscription and whether the site is "in danger". Natural WHSs were also classified according to the country's development status (Human Development Index; see Becken & Wardle, 2017). In a next step, the natural WHSs were examined in terms of their level of planning. Tourism management strategies were differentiated depending on whether tourism planning was integrated into a general plan or covered in a stand-alone tourism plan. The tourism plans were then assessed with regards to their level of planning. The coding categories were as follows:

(1) management plans: "current plan exists and is publically available", "plan exists but unable to locate", "old plan is publically available", "old plan was found but a new plan is under review", "old plan exists but unable to locate", "no previous plan but a plan is under preparation", and "no mention of a plan could be located online";
(2) tourism management plans: "current plan exists and is publically available", "plan exists but unable to locate", "old plan is publically available", "old plan was found but a new plan is under review", "old plan exists but unable to locate", "no previous plan but a plan is under preparation", and "no mention of a plan could be located online";
(3) the type of tourism management strategy was assessed: stand-alone tourism management plan (if it existed) vs tourism is integrated into a general management plan; and
(4) the extent of tourism planning for those natural WHSs that addressed tourism either through a stand-alone tourism plan, or as part of, and integrated with, a general management plan: "extensive", "moderate", and "minimal".

By definition, all stand-alone tourism plans were classified as extensive. Those integrated management plans that covered tourism in great detail, usually through entire chapters or sections on tourism, or those that incorporated tourism significantly into most aspects of the plan were also classified as extensive. This was, however, slightly subjective as the amount of text or the number of times tourism was mentioned throughout the document was not always indicative of the detail or quality of the information provided. As a general rule, management plans with less than two pages of text dedicated to tourism management were classed as moderate and those with less than one page were classed as minimal. Translation programs were used to scan plans in languages other than English.

In the following stage, targeted content analysis of the extensive tourism management strategies was undertaken to examine how these address key areas of monitoring. More specifically, and informed by UNESCO's (2012) World Heritage Resource Manual, strategies were examined in relation to monitoring of visitor trends and revenue, a method for estimating economic impact, indicators, monitoring costs, and community involvement. Due to the language barrier this was performed only on plans in English and Spanish. This covered 46 (71%) of those sites with extensive tourism planning, with the remaining 29% of sites made up of eight other languages.

Finally, additional data from the World Heritage Outlook, compiled and published by the IUCN, were used to link tourism planning with an assessment of the current state of conservation and effectiveness of management (IUCN, 2014). The IUCN data consist of a collection of best-available information and the consultation of multiple "site assessors", "knowledge holders" and "experts", followed by a final review by IUCN World Heritage Panel (IUCN, 2012, pp. 9–10). Each natural WHS is rated as either "Good", "Good with some concern", "Significant concern" and "Critical." In addition to an overall rating, the Outlook also provides a rating of specific aspects of tourism and management. The scores range from "Highly Effective", through "Effective", "Some Concern", "Serious Concern", to "Data Deficient". To examine the relationship between the ordinally scaled variables of the extent of tourism planning and IUCN assessed outcomes, the non-parametric Spearman rank correlation was used. It shows the statistical dependence between the rankings of two variables, assessing how well the relationship between two variables can be described (Bühl, 2012; Job, Scheder, & Spenceley, 2017).

Limitations

The analysis presented below is based on tourism and general management plans that could be found online, and it is possible that some sites maintain adequate tourism or general management plans but these could not be located. Although Google Translate was used to search in multiple languages including languages such as Arabic that do not use the ISO basic Latin alphabet, this search approach may not have been comprehensive. Furthermore, only in-date plans were used for detailed analysis. However, leniency was given as to the time frames of management plans, with only those that provided specific end dates considered to be out-of-date. Furthermore, only natural WHS plans were included. If a particular natural WHS is made up of multiple national parks, the management plans for these individual PAs were not used in this study. It is possible that the extent of tourism planning is therefore under-estimated. Finally, it is important to note that management plans, management strategies, and sustainability plans were included. The boundaries between these types of documents are blurred and it was not the purpose of this analysis to distinguish between the different forms of planning.

Results: the status of planning and monitoring

Planning documents

The analysis identified that as of June 2016, out of the 229 natural and mixed WHSs less than half (42%) have a general management plan that is available to the public via the Internet and that is up

Table 2. Relationship between the existence of a general management plan and a tourism plan in natural and mixed WHS.

		Extent of tourism planning					
		No plan	Old plan	Minimal plan	Moderate plan	Extensive plan	Total
General management plan	No plan	36	0	0	0	0	36
	Plan	5	0	10	19	63	97
	Old plan	37	20	0	1	5	63
	Plan in preparation	6	0	0	0	3	9
	Unable to locate plan	23	0	0	0	1	24
Total		107	20	10	20	72	229

to date. A further 24 sites (10%) appear to have a plan (due to online mention of it) but it was not possible to locate it. For 36 sites (16%) no mention of a management plan could be found, and for the purpose of this analysis we assume that no comprehensive plan exists. One quarter of the sites have a plan that is out of date (Table 2).

Whilst general management plans provide an indication of the existence of planning, it was deemed important to specifically look for tourism plans or strategies. It was found that 23 natural WHSs have developed a stand-alone tourism plan; although only 11 could be located online, whilst the others were either not available or out of date. In addition, a total of 84 sites (37% of all natural WHSs) were found to have an in-date plan which incorporated tourism into a more general management plan. Thus, integrated planning seems to be relatively more common than separate, stand-alone tourism planning. The findings show that there are 105 sites for which there are no available or clearly accessible tourism plans, and, therefore, it is not possible to gain any insights into their type or level of tourism planning.

The different types of tourism management strategies were assessed in more detail. Just 63 sites (28%) have an extensive and up-to-date level of tourism planning (Table 2), either as a stand-alone plan or a substantial tourism component integrated into a general plan. The 63 sites with extensive planning exclude out-of-date plans ($N = 20$) or those that could not be located ($N = 7$ claim they have a tourism plan but it could not be found online). This means that an adequate level of planning for tourism could not be verified for 72% of sites.

A relationship between general management and tourism planning/management was found, whereby those natural WHSs with a general management plan were also much more likely to have a plan for tourism. In addition, natural WHSs with a general management plan were more likely to have extensive tourism planning (Table 2). Chi-square test results are not presented due to a large number of cells with $N < 5$. Instead, the findings are presented visually with all coding categories maintained.

Extent and types of monitoring

Based on 46 sites with extensive tourism planning, key monitoring elements within the plans were assessed. It was found that visitation data are recorded, or recording methods are discussed, in the tourism management strategies for 37 sites out of the 46 (80%). Visitor monitoring is undertaken through a variety of methods, including:

- entry fees/permits being the most commonly mentioned (50%);
- gathering data on visitor days/nights (24%); and
- using tour company/commercial data (22%).

Visitor surveys are also mentioned frequently, but it was not always clear if this meant visitors are observed and counted or if questionnaires are used to determine visitor patterns. A clear distinction can be observed between continents: far more sites in Asia, Europe/the UK, and North America

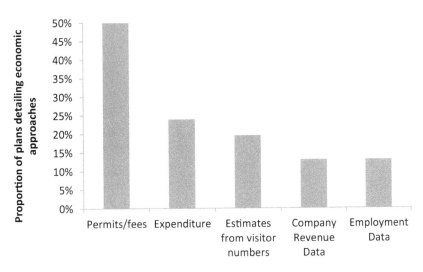

Figure 2. Methods utilised by natural and mixed WHS for estimating revenue or economic impact.

monitor visitor numbers than those in Africa and the Pacific, with slightly higher numbers in South America. Almost all sites with extensive plans record numbers.

Revenue monitoring and economic impact assessment is mentioned by 28 out of the 46 natural WHSs (61%). Several methods are discussed in the tourism management plans, including entry fees/permits (50%), expenditure data from visitors (24%), estimates from visitor numbers (20%), and company revenue and employment data (13%) (Figure 2). Differences emerged between continents, with the proportion of sites estimating revenue or economic impact being far higher in Asia, Africa and South America than those with more factually based statistics in Europe/the UK, North America, and the Pacific.

Visitor trends monitoring evident in the tourism plans includes determining visitor perceptions, visitation patterns, and demographics over multiple years. However, only 53% of sites cover visitor trends in their tourism management strategies. Rates of monitoring trends are similar across all continents, with rates slightly higher in North America (75%) and Africa (60%), and fairly low in the Pacific (25%). An interesting effect was seen in terms of country development, with rates of discussing visitor trends in the plans increasing as level of country development decreases (Figure 3). It is important to note that there are only three natural WHSs in Least Developed Countries (including the Serengeti National Park) and possibly their high level of tourism planning and monitoring is due to their iconic status that receives large scale visitation and global attention.

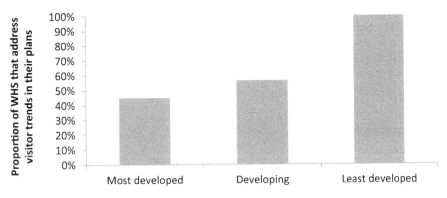

Figure 3. Proportion of natural and mixed WHSs that address visitor trends in their tourism plans, by country development.

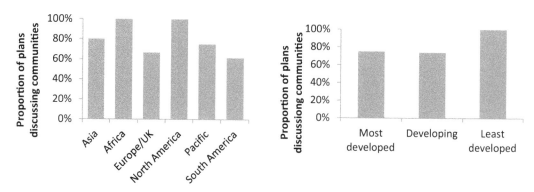

Figure 4. (a) and (b) Proportion of natural and mixed WHSs across continents and level of development that contain community engagement in their tourism plans.

Monitoring impacts is vital in ensuring that tourism does not compromise the OUV of the site. Therefore, plans were assessed in terms of information on monitoring for research, indicators and monitoring costs. Content in tourism management strategies that addresses monitoring for visitor impacts is prevalent (89%), but only 21 sites (47%) provide specific indicators to monitor. The proportion of natural WHSs that use monitoring indicators ranged from 38% to 75% across continents. Only six natural WHSs (13%) provide any costings for monitoring. None of the natural WHSs in North and South America address the issue of monitoring costs in their tourism plans.

Community engagement

Engaging with the local community and, where appropriate, facilitating their involvement in tourism ventures, is a vital component of tourism management and governance in PAs. Community engagement is generally quite high, with over three-quarters (35 sites) discussing community participation and empowerment (Figure 4(a) and 4(b)). Africa and North America have the highest level of community engagement at 100%. Least developed countries also show a 100% community engagement rate in their natural WHS tourism plans. Although community tourism is extensive in South America, this continent has the lowest rates of community engagement discussion in their tourism management strategies at just 62%.

Does tourism planning lead to better outcomes?

The results from a correlation analysis using IUCN World Heritage Outlook ratings support the hypothesis that a higher extent of tourism planning is associated with a better outcome for the WHS (Job et al., 2016). The analysis reveals a weak (i.e. with a correlation coefficient between 0.2 and 0.5, Bühl (2012), p. 420), yet significant, positive correlation between the extent of tourism planning and the overall rating in the World Heritage Outlook. A similar result can be observed for the rating in relation to "Protection and Management" (Table 3). Slightly lower correlation coefficients were found for "Management System", "Management Effectiveness", and "Education and Interpretation Programmes". The result for tourism and interpretation is not significant.

Discussion: moving forward

The analysis of UNESCO natural WHSs shows that (tourism) management is not always adequate. Indeed, only 42% of natural WHSs made a general management plan accessible on the Internet, and only 41% made either an up to date stand-alone tourism or integrated plan available. The extent of tourism planning varied, with only 28% of natural WHSs demonstrating extensive levels of planning.

Table 3. Correlation between tourism planning and ratings in the World Heritage Outlook.

Extent	Overall rating	Protection and management overall	Management system	Management effectiveness	Education and interpretation programmes	Tourism and interpretation
Correlation Coefficient	0.251**	0.238**	0.176**	0.169*	0.128	0.100
Sig. (2-tailed)	0.000	0.000	0.008	0.013	0.081	0.144
N	228	228	228	213	188	214

Monitoring of visitor numbers and trends is often carried out, but detail on methods varies and is insufficiently specific. Whilst over half of the areas measure some form of economic impact, the methods are inconsistent and a comparison across natural WHSs is difficult. In an increasingly neoliberal world that expects "returns on PA investment" this is problematic and weakens the position of natural areas. Other aspects of monitoring, including understanding trends, implementing clearly articulated indicators or measuring monitoring costs, are only addressed by a small proportion of natural WHSs. The need to engage with local communities, however, is broadly recognised and addressed in the majority of natural WHSs – possibly reflecting the increasing emphasis on the role of PAs for local development (see earlier sections). Therefore, we argue that tourism management plans should be unconditional for all natural and mixed WHS.

The analysis of IUCN World Conservation Outlook ratings indicates that better tourism planning is associated with superior conservation outcomes. However, even with the best tourism planning in some natural WHS, it is unavoidable that increasing human activity intensifies pressure on vulnerable ecosystems and natural heritage values. Today, about 10% of the UNESCO natural WHS are listed as "World Heritage in Danger", being affected by encroachment, poaching, illegal logging, mining, sale of public land and infrastructural development (Engels, 2015). For those areas, that are characterised by insufficient planning, management and monitoring, it is likely that negative impacts accumulate, and at the same time are not fully recognised or only when it is too late (Stoll-Kleemann & Job, 2008).

One approach to protect these sites is to limit or exclude human use (Serenari et al., 2017). In Kenya, for example, more and more PAs are being fenced (e.g. Aberdares, Nairobi, Nakuru, Marsabit National Park, and Mt Kenya National Park) (Job, 2014). In densely populated areas this type of "gated conservation" might be the only option. However, it poses major challenges in that local participation by the community is restricted, affecting livelihoods and the long-term acceptance of the area. Natural migrations of wildlife are also inhibited (Habel et al., 2016). The fencing off of natural areas is contradictory to the trends discussed earlier in relation to the evolution of both natural WHSs and BRs as areas that integrate nature conservation with sustainable regional development. It may also be increasingly untenable given economic imperatives and "land grabs", and other ongoing pressures (Figure 1).

As an alternative, "community-based" conservation approaches are often preferred, especially in a neoliberal paradigm that emphasises the potential of local business opportunities and favours "...the commodification of nature as it turns intrinsic or use values into exchange values..." (Saarinen, 2016, p. 6). Such a commodification is not only found in developing countries, but also in Western countries and their natural areas. Canada, for example, in its Federal Tourism Strategy focuses heavily on product development and investment for tourism in Canada's Parks (Industry Canada, 2015). Similar developments have been found in New Zealand, where neo-liberal government policies focus on facilitating economic activity in Parks, raising concerns about whether these might lead to a tipping point where tourism development irreversibly compromises conservation goals (Dinica, 2017). The examples of PAs in the United States by Slocum (2017), as well as the experience of the Great Barrier Reef in Australia (Liburd & Becken, 2017), highlight the pressures Parks face in a neoliberal world. Care must be taken about condemning commodification however. Whitney-Squire, Wright, and Alsop (2017) found that local communities do not value cultural heritage conservation unless that heritage can be commodified and shown to have financial value.

Clearly, innovative pathways are needed to bridge the traditional approach of "fences and fines" and the neoliberal ideology of "participative development" to provide robust governance structures and stewardship systems that are capable of building resilience to multiple stressors (Becken & Job, 2014; Liburd & Becken, 2017). One avenue might be to expand the mandate of natural WHSs from their primary conservation focus to a triple bottom-line approach that considers multi-facetted socio-economic outcomes. Integrating natural WHSs into the broader BR concept might just provide such a pathway. As discussed earlier, BRs explicitly consider the well-being of the local community and the regional economy outside the core zone of the area under protection. Being institutionally anchored under the same international auspices of UNESCO, the integration of the WHS and BR concepts seems promising. It may be timely to embed the 159 natural WHS that are not yet multi-designated into the broader category of BRs. How to involve stakeholders effectively and how to consider the local context as a key shaper of success is discussed in more detail for the case of the Dolomites, Italy, by Della Lucia and Franch (2017).

Following this, it is suggested to explore how the concept of the BR could lead the way to achieve complementary outcomes for conservation and sustainable development in PAs more generally. With their three core functions, namely conservation, support for science and education, and sustainable development, BRs provide a suitable template for this transformation. As mentioned above, the detailed zoning approach of BRs, with (1) core zones, (2) buffer zones[3], and (3) a so-called transition zone fostering regional development, is of particular importance; the latter providing the territory for testing sustainable development approaches and delivering ecosystem services for the people living adjacent to these Parks (Bouma & van Beukering, 2015; Plieninger, Woltering, & Job, 2016; see also Thede, Haider, & Rutherford, 2014 on PA buffer zones).

The potential for catalysing rural enterprises and stimulating the local economy in conjunction with buffer and transition zones has been demonstrated in a study of the Rhön BR (Kraus et al., 2014). The Rhön BR has been operating a regional umbrella label (Dachmarke) that sets a standard for the production and purchase of locally and sustainability produced goods. Over 200 local businesses have now subscribed to this scheme, benefitting from advice and guidelines on how to source products and manage their supply chain. Kraus et al. found that members of the Dachmarke spend more money for supply within the marketing area, leading to a lower leakage rate and a higher indirect economic impact. Such outcomes verify the principles and goals of sustainable development set through the *Seville Strategy* and the assignments of the BR. Highly networked and localised supply chains increase the region's self-sufficiency and lead to "clusters of compatible and mutually reinforcing activities" (Van der Ploeg & Renting, 2000, p. 534). Tourism then provides the injection of cash from outside the region (Knaus et al., 2017). Reinforcing circles of positive outcomes (e.g. more tourism, more local supply networks, better socio-economic outcomes, higher acceptance of the core conservation zone, etc.) have considerable educational value, which is one of the key elements of the BR concept. Therefore, intensive communication with local actors (especially business owners) is a necessary precondition to build a lasting basis of participation. At the same time it is critical to monitor economic developments and their impacts to ensure that the core of the PA is not compromised, especially as a result of increasing visitation.

The above ideas are not new, as 79 natural WHSs are already embedded in much larger BRs, whereby the core area of the BR equals the natural WHS area. In these cases, the BR buffer and development zones are designed to enable sustainable management of the wider surrounding area. A prominent example can be found in Meso-America with the Monarch Butterfly and El Vizcaino (Mexico) or the Maya Biosphere Reserve comprising Guatemala's Tikal National Park WHS. Another example is the tri-national Primeval Beech Forests of the Carpathians and the Ancient Beech Forests of Germany site, which is a Biosphere Reserve that contains multiple National Parks and Nature Reserves. These embedded models have rarely been researched but clearly demand more in-depth analysis to fully assess their potential (Engels, 2015; Shafer, 2015).

For integrating natural WHSs with BRs, it is relevant that they are both international PAs being UNESCO listed and operating in a global management environment. Therefore, lessons learned and good practices can be easily communicated through established networking platforms on a bi- or multilateral base. It is through those established platforms and through new networking methods that the application of selected aspects of WHS and BR best practice could take place. How that is best done, and how the dissemination of good practice can take place, is a future research area.

Importantly, BRs have a stricter monitoring system than natural WHSs, and a more robust reporting framework regarding conservation issues. As identified in the empirical part of this paper, at present the monitoring and reporting in natural WHSs is not fully developed or consistent. It can therefore be argued that improving reporting systems, including an evaluation of the PA's effectiveness, brings about better conservation outcomes (Geldmann et al., 2015; Hockings et al., 2008; Leverington, Hockings, Pavese, Lemos Costa, & Courrau, 2008; Stoll-Klemann, 2010). Greater focus on monitoring and reporting will also help to further advance the concept of BR systems. In particular, what is required is a standardized methodology for ecological monitoring in order to make systematic comparisons among PAs. Much more interdisciplinary research is required that includes standardized social science-based monitoring. This will enhance our conceptual and practical knowledge of people-nature relationships and true consequences of conservation and development balances in PAs.

It is also very clear that a long term view of PA management is a necessity. The need for progress in management approaches and political understanding of the issues and possibilities that PAs present has become urgent. Koontz, Cullinane, Ziesler, Olson, and Meldrum (2017) show that the USA's National Parks are receiving record visitor numbers nowadays. Those visitor numbers are making major contributions to the US national, regional and local economies. But Robbins (2017) fears that the US government proposes cutting the Park Service budget by 13% in 2018 (which would be the largest cut to the agency since Second World War), and there is already a backlog of staffing and maintenance issues. The need to showcase, or present, the positive benefits – in economic, cultural and environmental terms – is both real time and ongoing; because the tendency to cut back on conservation management is not limited to the USA.

Conclusion

Our planet may be at a tipping point towards potential collapse, due to overuse of natural resources and societal inequality. It is, therefore, urgent to consider new models for protecting natural ecosystems and at the same time ensuring sustainable development pathways. This article argued that PAs could and must play a critical role in this transformation. Sustainable tourism, in particular, may provide a vehicle to achieve regional development and support local communities. However, the sound management of tourism is a critical prerequisite for achieving sustainable outcomes. The analysis presented here shows that the majority of natural WHSs lack the type of management and tourism planning that would be required to monitor and minimise environmental impacts, and to provide a business case for governments to demonstrate tangible economic benefits. At the same time, there is evidence that for those areas that do have tourism planning, conservation outcomes are superior. Therefore our major plea is that UNESCO should make tourism management plans mandatory for all natural and mixed WHSs.

This paper also explored the potential to advance a multi-layered management approach by embedding natural WHSs into the larger BR concept. There would be several benefits, including the establishment of differentiated management zones, more robust monitoring standards, and economically competitive "local brands". Changes and adaptations to the conceptualisation of WHSs and BRs (and PAs more broadly) are not new. Since the establishment of UNESCO's MAB program there have been many significant developments. The Seville Conference in 1995, in particular, brought about significant change: the initial concept of a research program was transformed into a modern instrument for the dual mandate of conservation and sustainable regional development through BRs. For the first time, local people and their activities were explicitly integrated into the conservation goals. As a consequence, research activities and biodiversity protection tasks that were so central to the first-generation BRs now have to accommodate new demands (Köck & Arnberger, 2017). Hence, today the nature and quality of BRs is fundamentally different. Recent research has shown that post-Seville BRs are more effective in their management (Plieninger et al., 2016; Van Cuong et al., 2017), mainly due to much better implemented zoning (Price, Park, & Boumrane, 2010). This article argues that stringent PA zoning is critical for the successful integration of conservation and development, and, as such, a necessary step towards improving natural resource protection and societal equity. Both conservation and development are essential elements if we are to avoid large-scale collapse.

The array of research findings and discussions covered in this paper are further elaborated on, explained and discussed in the papers which together form *the Journal of Sustainable Tourism's* 2017 Special Issue on Protected Areas & Neo-liberal governance policies, Volume 25, Issue 11.

Notes

1. The creation of economic opportunities based on efficiency, diversity, and equity is a central element of the Seville Strategy. Key elements of success include professional management and leadership, participatory approaches, production networks, and communication between different interest groups, including public sector representatives. Developing specific eco/quality labels for BR products is one avenue of creating value and brand recognition (Kraus et al., 2014).
2. The MDG as part of the Millennium Declaration were approved by 189 countries in 2000. They specify eight goals, including poverty alleviation and environmental sustainability. The implementation of the MDGs by 2015 was extended by 15 years through the 2030 Agenda for Sustainable Development, containing 17 SDG.
3. The BR buffer zone differs from the concept of a buffer zone defined by the World Heritage Convention which is normally considerably smaller.

Acknowledgements

We would like to thank the German Federal Agency for Nature Conservation (BfN), and especially Ms Barbara Engels, Research Associate at the BfN, for funding an expert workshop on "The Economic Impacts of Protected Areas", held on the 21–25 September 2015, in Wilhelmshaven, Germany. We also thank Mr Peter Südbeck from the Wadden Sea National Park in Germany (part of the Dutch, German, and Danish Wadden Sea natural WHS) for his hospitality, and all the participants in those workshops. Our sincere thanks go to Bernard Lane for supporting this Special Issue and providing substantial intellectual and logistical input into its genesis. We also thank Peter Debrine from UNESCO, Paris, for his support and input, as well as Dr Lutz Möller, German UNESCO Commission, and Florian Carius from BfN, Bonn, for valuable feedback on a draft version of this paper.

Disclosure statement

No potential conflict of interest was reported by the authors.

References

Albert, C., von Haaren, C., Hansjürgens, B., Schröter-Schlaack, C., Dehnhardt, A., Döring, R., … Woltering, M. (2017). An economic perspective on land-use decisions in agricultural landscapes: Insights from the TEEB Germany Study. *Ecosystem Services, 25*, 69–78.

Baral, N., Kaul, S., Heinen, J. T., & Ale, S. B. (2017). Estimating the value of the World Heritage Site designation: A case study from Sagarmatha (Mt Everest) National Park, Nepal. *Journal of Sustainable Tourism*. doi:10.1080/09669582.2017.1310866

Batisse, M. (1997). Biosphere reserves: A challenge for biodiversity conservation & regional development. *Environment: Science and Policy for Sustainable Development, 39*(5), 6–33.

Becken, S. (2017). Evidence of a low-carbon tourism Paradigm? *Journal of Sustainable Tourism, 25*(6), 832–850.

Becken, S., & Job, H. (2014). Protected areas in an era of global-local change. *Journal of Sustainable Tourism, 22*(4), 507–527.

Becken, S., & Wardle, C. (2017). Tourism planning in Natural World Heritage Sites. *Griffith Institute for Tourism Research Report*. Retrieved June 7, 2017, from https://www.griffith.edu.au/__data/assets/pdf_file/0015/1009311/UNESCO-WHA-Report-Final.pdf

Bender, O., Roth, C. E., & Job, H. (2017). Protected areas and population development in the Alps. *Eco.mont Journal on Protected Mountain Areas Research and Management, 9*(SI), 5–16.

Borrini-Feyerabend, G., Dudley, N., Jaeger, T., Lassen, B., Pathak Broome, N., Phillips, A., & Sandwith, T. (2013). Governance of Protected Areas: From understanding to action (*Best Practice Protected Area Guidelines Series* No. 20). Gland: IUCN.

Bouma, J. A., & Van Beukering, P. J. H. (2015). *Ecosystem services: From concept to practice*. Cambridge: Cambridge University Press.

Brenner, L., & Job, H. (2012). Challenges to actor-oriented environmental governance: Examples from three Mexican Biosphere Reserves. *Tijdschrift voor economische en sociale geografie, 103*, 1–19.

Buckley, R. C. (2012). Tourism, conservation and the Aichi targets. *Parks, 18*(2), 12–19.

Bühl, A. (2012). *SPSS 20. Einführung in die moderne Datenanalyse*. [An introduction to modern data analysis] Munich: Pearson Deutschland.

Butzmann, E., & Job, H. (2017). Developing a typology of sustainable protected areas tourism products. *Journal of Sustainable Tourism*, doi:10.1080/09669582.2016.1206110

Carius, F. (2016). *Tourism revenue sharing with local communities* (Unpublished master's thesis). Technische Universität Kaiserslautern, Kaiserslautern.

Cessford, G., & Muhar, A. (2003). Monitoring options for visitor numbers in National Parks and natural areas. *Journal for Nature Conservation, 11*(4), 240–250.

Conradin, K., & Wiesmann, U. (2014). Does World Natural Heritage status trigger sustainable regional development efforts? *Eco.mont Journal on Protected Mountain Areas Research and Management, 6*(2), 5–12.

Della Lucia, M., & Franch, M. (2017). The effect of local context on World Heritage Site management: The Dolomites Natural World Heritage Site, Italy. *Journal of Sustainable Tourism*. doi:10.1080/09669582.2017

Diamond, J. (2005). *Collapse: How societies choose to fail or succeed*. New York, NY: Penguin.

Dinica, V. (2017). Tourism concessions in national parks: Neoliberal governance experiments for a conservation economy in New Zealand. *Journal of Sustainable Tourism*. doi:10.1080/09669582.2015.1115512

Dudley, N., Sandwith, T., & Belokurov, A. (2010). Climate change: The role of protected areas in mitigating and adapting to change. In S. Stolton & N. Dudley (Eds.), *Arguments for protected areas: Multiple benefits for conservation and use* (pp. 205–223). London: Earthscan.

Eagles, P. F. J. (2014). Research priorities in park tourism. *Journal of Sustainable Tourism, 22*(4), 528–549.

Engels, B. (2015). Natural heritage and sustainable development – A realistic option or wishful thinking? In M. Albert (Ed.), *Perceptions of sustainability in heritage studies* (pp. 49–58). Berlin: De Gruyter.

Engels, B. (2017). Natural World Heritage and the Sustainable Development Goals. In M. Albert (Ed.), *Going Beyond (Perceptions of Sustainability in Heritage Studies No. 2)* Berlin: De Gruyter (in press).

Engels, B., Becken, S., Job, H., & Lane, B. (2017). An interview with a protected area insider. *Journal of Sustainable Tourism*, doi:10.1080/09669582.2017.1380365

Espiner, S., & Becken, S. (2014). Tourist towns on the edge: Conceptualising vulnerability and resilience in a protected area tourism system. *Journal of Sustainable Tourism, 22*(4), 646–665.

Ferreira, J., Aragao, L. E. O.C., Barlow, J., Baretto, P., Berenguer, E., Bustamante, M., … Zuanon, J. (2014). Brazil's environmental leadership at risk. *Science, 346*(6210), 706–707.

Geldmann, J., Coad, L., Barnes, M., Craigie, I. D., Hockings, M., Knights, M., … Burgess, N. D. (2015). Changes in protected area management effectiveness over time: A global analysis. *Biological Conservation, 191*, 692–699.

Getzner, M., Lange Vik, M., Brendehaug, E., & Lane, B. (2014). Governance and management strategies in national parks: Implications for sustainable regional development. *International Journal of Sustainable Society, 6* (1&2), 82–101.

Habel, J. C., Teucher, M., Mulwa, R. K., Haber, W., Eggermont, H., & Lens, L. (2016). Nature conservation at the edge. *Biodiversity Conservation, 25*(4), 791–799.

Hall, C. A. S., & Day, J. W. Jr. (2009). Revisiting the Limits to Growth after Peak Oil. *American Scientist, 97*, 230–237.

Hammer, T., Mose, I., Siegrist, D., & Weixlbaumer, N. (Eds.). (2016). *Parks of the future. Protected Areas in Europe Challenging Regional and Global Change*. Munich: Oekom-Verlag.

Higgins-Desbiolles, F. (2010). The elusiveness of sustainability in tourism: The culture–ideology of consumerism and its implications. *Tourism and Hospitality Research, 10*(2), 116–129.

Hockings, M., James, R., Stolton, S., Dudley, N., Mathur, V., Makombo, J., ... Parrish, J. (2008). *Enhancing our heritage toolkit: Assessing management effectiveness of natural World Heritage Sites* (World Heritage Papers No. 23). Paris: UNESCO World Heritage Centre.

Industry Canada. (2015). *A Tourism-based comparative analysis of UNESCO World Heritage Sites in Canada, the United States and Australia*. Ottawa. Retrieved February 20, 2017, from https://www.ic.gc.ca/eic/site/034.nsf/eng/h_00492.html

IUCN. (2012). *IUCN Conservation Outlook Assessments – Guidelines for their application to natural World Heritage Sites*. Retrieved February 20, 2017, from https://www.iucn.org/sites/dev/files/import/downloads/guidelines___iucn_conser vation_outlook_assessments_08_12.pdf

IUCN. (2014). *IUCN World Heritage Outlook 2014*. Retrieved February 20, 2017, from https://portals.iucn.org/library/sites/library/files/documents/2014-039.pdf

Job, H., & Metzler, D. (2005). Regionalökonomische Effekte von Großschutzgebieten. [The regional economic effects of large Protected Areas]. *Natur und Landschaft, (80) 11*, 465–471.

Job, H., & Paesler, F. (2013). Links between nature-based tourism, protected areas, poverty alleviation and crises – The example of Wasini Island (Kenya). *Journal of Outdoor Recreation and Tourism, 1*(1/2), 18–28.

Job, H. (2014). Bevölkerungswachstum und Schutzgebiete – eine afrikanische Herausforderung [Population growth and Protected Areas – An African challenge]. *Geographische Rundschau, 66*(10), 44–47.

Job, H., Kraus, F., Merlin, C., & Woltering, M. (2013). *Wirtschaftliche Effekte des Tourismus in Biosphärenreservaten Deutschlands* [The economic effects of tourism in Biosphere Reserves in Germany]. *Naturschutz und Biologische Vielfalt*, 134. Bonn-Bad Godesberg: Bundesamt für Naturschutz.

Job, H., Merlin, C., Metzler, D., Schamel, J., & Woltering, M. (2016). *Regionalwirtschaftliche Effekte durch Naturtourismus* [The regional economic effects of nature-based tourism]. BfN-Skripten, 431. Bonn-Bad Godesberg: Bundsamt für Natur.

Job, H., Scheder, N., & Spenceley, A. (2017). *Visitation counts! Evaluation of tourism in natural World Heritage Sites* (Final Report to the UNESCO; unpublished). Paris: UNESCO.

Jones, T., & Ohsawa, T. (2016). Monitoring nature-based tourism trends in Japan's National Parks: Mixed messages from domestic and inbound visitors. *Parks, 22*(1), 25–36.

Kammann, E., & Möller, L. (2007). MAB – Der Mensch und die Biosphäre. Ein Rückblick. [Man and the Biosphere: A review]. *UNESCO Heute, 2*, 13–15.

Khazaei, A., Elliot, S., & Joppe, M. (2017). Fringe stakeholder engagement in protected area tourism planning: Inviting immigrants to the sustainability conversation. *Journal of Sustainable Tourism*. doi:10.1080/09669582.2017.1314485

Kilbourne, W. E., Beckmann, S. X., & Thelen, E. (2002). The role of the dominant social Paradigm in environmental attitudes. A multinational examination. *Journal of Business Research, 55*, 193–204.

Knaus, F., Ketterer Bonnelame, L., & Siegrist, D. (2017). The economic impact of labeled regional products: The experience of the UNESCO biosphere reserve Entlebuch. *Mountain Research and Development, 37*(1),121–130.

Köck, G., & Arnberger, A. (2017). The Austrian Biosphere Reserves in the light of changing MAB strategies. *Eco.mont Journal on Protected Mountain Areas Research and Management, 9* (Special Issue), 85–92. Retrieved June 8, 2017, from http://epub.oeaw.ac.at/0xc1aa500e_0x0034cb99.pdf

Koontz, L., Cullinane Thomas C., Ziesler, P., Olson, J., & Meldrum, B. (2017). Visitor spending effects: Assessing and showcasing America's investment in National Parks. *Journal of Sustainable Tourism*. doi:10.1080/09669582.2017.1374600

Kraus, F., Merlin, C., & Job, H. (2014). Biosphere reserves and their contribution to sustainable development. *Zeitschrift für Wirtschaftsgeographie, 5*(2–3), 164–180.

Lama, A. K., & Job, H. (2014). Protected Areas and road development: Sustainable development discourses in the Annapurna Conservation Area, Nepal. *Erdkunde, 68*(4), 229–250.

Leverington, F., Hockings, M., Pavese, H., Lemos Costa, K., & Courrau, J. (2008). *Management effectiveness evaluation in protected areas: A global study. Supplementary report no.1: Overview of approaches and methodologies*. Gatton: University of Queensland. Retrieved June 8, 2017, from https://cmsdata.iucn.org/downloads/methodologyreportdec08_fi nal.pdf

Liburd, J., & Becken, S. (2017). Values in nature conservation, tourism and UNESCO World Heritage Site stewardship. *Journal of Sustainable Tourism*. doi:10.1080/09669582.2017.1293067

Mager, T. (2015). *Schillernde Unschärfe: Der Begriff der Authentizität im architektonischen Erbe* [Iridescent Blur: The concept of authenticity in architectural heritage]. Berlin/Boston, MA: De Gruyter.

Mayer, M., & Job, H. (2014). The economics of protected areas – A European perspective. *Zeitschrift für Wirtschaftsgeographie, 58* (2–3), 73–97.

Mayer, M., Müller, M., Woltering, M., Arnegger, J., & Job, H. (2010). The economic impact of tourism in six German national parks. *Landscape and Urban Planning, 97*(2), 73–82.

Medeiros, R., Frickmann Young, C. E., Boniatti PAvese, H., & Silva Araújo, F. F. (2011). *The contribution of Brazilian conserva-tion units to the national economy*. Brasilia: UNEP–WCMC. Retrieved June 8, 2017, from http://www.ie.ufrj.br/images/gema/Gema_Artigos/2011/Medeiros_Young_2011_UNEP_Economic_Value_Protected_Areas_Brazil.pdf

Motesharrei, S., Rivas, J., & Kalnay, E. (2014). Human and nature dynamics (HANDY): Modeling inequality and use of resources in the collapse or sustainability of societies. *Ecological Economics, 101*, 90–102.

Müller, M. (2014). From sacred cow to cash cow: The shifting political ecologies of protected areas in Russia. *Zeitschrift für Wirtschaftsgeographie, 58*(2–3), 127–143.

Pirages, D. C., & Ehrlich, P. R. (1974). *Ark ii: Social response to environmental imperatives*. San Francisco, CA: Freeman.

Plieninger, T., Woltering, M., & Job, H. (2016). Implementierung des Ökosystemleistungsansatzes in deutschen Bio-sphärenreservaten [Implementing the ecosystem services approach in Germany's Biosphere Reserves]. *Raumfor-schung und Raumordnung, 74*, 541–554.

Price, M. (1996). People in biosphere reserves: An evolving concept. *Society and Natural Resources, 9*, 645–654.

Price, M. F., Park, J. J., & Boumrane, M. (2010). Reporting progress on internationally designated sites: The periodic review of biosphere reserves. *Environmental Science & Policy, 13*(6), 549–557.

Puhakka, R., Pitkänen, K., & Siikamäki, P. (2017). The health and well-being impacts of protected areas in Finland. *Journal of Sustainable Tourism*. Retrieved from https://doi.org/10.1080/09669582.2016.1243696

Robbins, J. (2017). How a surge in visitors is overwhelming America's National Parks. *Yale Environment, 360*. Retrieved August 21, 2017, from http://e360.yale.edu/features/greenlock-a-visitor-crush-is-overwhelming-americas-national-parks

Ruoss, E. (2016). Opportunities to leverage World Heritage Sites for local development in the Alps. *Journal on Protected Mountain Areas Research and Management, 8*(1), 53–61. Retrieved June 8, 2017, from http://epub.oeaw.ac.at/eco.mont-8-1/?frames=yes

Saarinen, J. (2016). Wilderness use, conservation and tourism: What do we protect and from whom? *Tourism Geographies, 18* (1), 1–8.

Schaaf, T., & Clamote Rodrigues, D. (2016). *Managing MIDAs: Harmonising the management of multiinternationally desig-nated areas: Ramsar Sites, World Heritage sites, Biosphere Reserves and UNESCO Global GeoParks*. Gland: IUCN. Retrieved January 20, 2017, from https://portals.iucn.org/library/sites/library/files/documents/2016-033.pdf

Serenari, C., Peterson, M. N., Wallace, T., & Stowhas, P. (2017). Private protected areas, ecotourism development and impacts on local people's well-being: A review from case studies in Southern Chile. *Journal of Sustainable Tourism*. doi:10.1080/09669582.2016.1178755.

Shafer, C. L. (2015). Cautionary thoughts on IUCN protected area management categories V–VI. *Global Ecology and Con-servation, 3*, 331–348.

Slocum, S. L. (2017). Operationalising both sustainability and neo-liberalism in protected areas: Implications from the USA's National Park Service's evolving experiences and challenges. *Journal of Sustainable Tourism*. doi:10.1080/09669582.2016.1260574

Stoll-Kleemann, S., & Job, H. (2008). The relevance of effective protected areas for biodiversity conservation: An introduc-tion. *Gaia, 17*(S1), 86–90.

Stoll-Kleemann, S. (2010). Evaluation of management effectiveness in protected areas: Methodologies and results. *Basic and Applied Ecology, 11*, 377–382.

Thede, A. K., Haider, W., & Rutherford, M. B. (2014). Zoning in national parks: Are Canadian zoning practices outdated? *Journal of Sustainable Tourism, 22*(4), 626–645.

Tisdell, C., & Wilson, C. (2012). *Nature-based tourism and conservation: New economic insights and case studies*. Chelten-ham: Edward Elgar Publishing.

UN (2013). *World Population Prospects: The 2012 Revision, Highlights and Advance Tables*. Department of Economic and Social Affairs, Population Division (Working Paper No. ESA/P/WP.228).

UNDP. (2016). *Sustainable Development Goals*. Retrieved December, 3, 2016, from http://www.undp.org/content/undp/en/home/sustainable-development-goals.html

UNESCO. (1972). *Convention concerning the protection of the World Cultural and Natural Heritage*. Paris: Author.

UNESCO. (1974). *Programme on Man and the Biosphere (MaB). Task Force on: Criteria and guidelines for the choice and establishment of biosphere reserves* (MaB report series 22). Paris: Author.

UNESCO. (1986). *International co-ordinating council of the programme on Man and the Biosphere (MaB report series 60) – Ninth session*. Paris: Author.

UNESCO. (1994). *Operational Guidelines for the implementation of the World Heritage Convention* (Revised edition from 1993). Paris: Author.

UNESCO. (1996). *Biosphere reserves: The Seville Strategy and the Statutory Framework of the World Network*. Paris: Author. Retrieved February, 14, 2017, from http://unesdoc.unesco.org/images/0010/001038/103849e.pdf

UNESCO. (2012). *Managing Natural World Heritage*. Resource Manual. Paris: Author.

UNESCO. (2015). *MAB Strategy 2015–2025*. Paris: Author.

UNESCO. (2016a). *World Heritage and sustainable development*. Paris: Author. Retrieved January 14, 2015, from http://worldheritagec.unesco.org/en/sustainabledevelopment/

UNESCO. (2016b). *Biosphere Reserves – Learning sites for sustainable development*. Retrieved January 14, 2015, from http://www.unesco.org/new/en/natural-sciences/environment/ecological-sciences/biosphere-reserves/

UNESCO. (2016c). *Lima Action Plan for UNESCO's Man and the Biosphere (MAB) programme and its World Network of Biosphere Reserves (2016–2025)*. Lima: Author. Retrieved June 8, 2017, from http://www.unesco.org/fileadmin/MULTIMEDIA/HQ/SC/pdf/Lima_Action_Plan_en_final.pdf

UNFCCC. (2015). *Adoption of the Paris Agreement. Conference of the Parties. Draft decision -/CP.21*. Twenty-first session, Paris. Retrieved January 14, 2015, from http://unfccc.int/resource/docs/2015/cop21/eng/l09.pdf

UNWTO. (2013). *Tourism highlights* (2013th ed.). Madrid. Retrieved 14/01/ 2015 from: http://www.e-unwto.org/doi/pdf/10.18111/9789284415427

UNWTO. (2015). Tourism highlights (2014th ed.). Madrid: Author.

UNWTO. (2017). *World tourism barometer* (*Vol. 15*). Madrid: Author. Retrieved June 11, 2017, from http://cf.cdn.unwto.org/sites/all/files/pdf/unwto_barom17_01_january_excerpt_.pdf

Van Cuong, C., Dart, P., & Hockings, M. (2017). Biosphere reserves: Attributes for success. *Journal of Environmental Management, 188*, 9–17.

Van der Ploeg, J., & Renting, H. (2000). Impact and potential: A comparative review of European rural development practices. *Sociologia Ruralis, 40*(4), 529–534.

WBGU (Wissenschaftlicher Beirat der Bundesregierung Globale Umweltveränderungen). (2016). *Der Umzug der Menschheit: Die transformative Kraft der Städte*. [The progress of mankind: The power of cities] Berlin: Author.

WHC. (2002). *Budapest Declaration on World Heritage*. UNESCO. Retrieved 8th June 2017 from: http://whc.unesco.org/en/decisions/1217/

WHC. (2012). Committee decision WHC 36 COM 5E. Retrieved June 8, 2017, from http://whc.unesco.org/en/decisions/4613

WHC (1993). Committee Decision Conf. 002 VIII, 1–6. Retrieved September 27, 2017, from http://whc.unesco.org/archive/1993/whc-93-conf002-14e.pdf

Whitney-Squire, K., Wright, P., & Alsop, J. (G.) (2017). Improving indigenous local language opportunities in community-based tourism initiatives in Haida Gwaii (British Columbia, Canada). *Journal of Sustainable Tourism*. doi:10.1080/09669582.2017.1327535

World Bank. (2016). World Development Indicators. Retrieved January 19, 2017, from http://wdi.worldbank.org/tables

Zimmerer, K. S. (2006). *Globalization and new geographies of conservation*. Chicago, IL: University of Chicago Press.

Values in nature conservation, tourism and UNESCO World Heritage Site stewardship

Janne J. Liburd and Susanne Becken

ABSTRACT

This paper seeks to understand the complex values held by those involved in Protected Area and World Heritage stewardship. Using IUCN Protected Area categories, a values framework is developed and applied to demonstrate how values guide stewardship in protected areas. In-depth interviews with key tourism operators, public sector managers and other stakeholders from the iconic World Heritage Site and tourism destination, Australia's Great Barrier Reef (GBR) reveal how shifting ideologies and government policies increased pressures on nature, resulting in new alliances between stewards from the tourism sector and national and international organisations. These alliances were built on shared nature conservation values and successfully reduced increasing development pressures. Three distinct phases in this process emerged at the GBR, which were driven by personal values held by tourism industry representatives, and their recognition of tourism's reliance on nature for business success. Changing mainstream ideologies and political values can erode World Heritage and Protected Areas, and recalibrate values – including the universal values on which World Heritage Sites depend – towards more anthropocentric interpretations. The values framework presented here could be a powerful tool for stewards involved in conservation to remind those who merely manage and govern of the original nature-focused values.

Introduction

Tourism has long been at the core of a controversial debate in protected natural areas around the commercial use of nature versus its conservation, reflecting the inherent complexity of the tourism–nature relationship, the critical influence of contextual factors, and dynamic changes in both the human and natural elements of the system. The tourism sector is not the only interest group involved in questions surrounding nature conservation, and tourism stakeholders often work alongside dedicated resource managers, community groups, non-governmental organisations (NGOs) and Indigenous people, all of whom have value-based belief systems. The values of those groups are important; they comprise the people that support protection of parks – or not (Jones & Shaw, 2012). It, therefore, becomes imperative to study the value dynamics that contribute to the tourism sector taking a more or a less symbiotic stance to nature, rather than focusing on a fixed-in-time arrangement of system attributes, interests and powers (Espiner & Becken, 2014).

How do human beings value nature? Values can be seen as determining priorities, as internal compasses or as springboards for action resembling moral imperatives that implicitly or explicitly guide action (Oyserman, 2001). Importantly, dominant values, and the paradigms in which they are embedded, change over time (Becken, 2016). And values have a geographic expression, a question discussed later in this paper.

Over the past centuries, the dominant Western view of nature has shifted substantially from a hostile view of nature to be controlled by man, followed by a nineteenth-century romantic view in awe of nature, to one of conservation or wise use (Larrère, 2008). Each of these positions reflects a philosophical separation of man from nature. Highlighting different philosophical stances, Urry (1995) divided humans into either "exploiters" or "stewards" of nature, and Harrison (1996) identified the "blue greens" (who favour a market approach), "red greens" (who mix some form of environmentalism with some form of socialism) and "really (or radical) greens" (who refuse to see nature reduced to a set of resources ready to be consumed or destroyed) (p. 69).

To gain deeper understanding about how we construct values and meanings of nature, Tribe and Liburd's (2016) conceptualisation of a knowledge force field is informative. They identify the factors creating the knowledge system as government, ideology, global capital, position and person. Governments have the power to define, select and fund areas for protection. Tribe and Liburd (2016) refer to "ideology" in a broad sense to cover the influence of both tacit and explicit value systems. Ideas of neo-liberalism and capitalism are covered under ideology, but "global capital" has a material dimension and exerts its direct power through money and its indirect power through influence. "Position" encompasses geographic location, institutional or organisational affiliation as well as language, national and cultural communities. "Person" indicates our inability to escape our embodied selves (Crouch, 2000). As human beings we carry with us autobiographies, socialisation, culture, gender, sexual orientation, instinct, our senses (Tribe & Liburd, 2016, p. 54) and thus our values. The point about the force field is that seemingly interest-free values about nature are subject to multiple forces that shape and make stewardship of nature, including both global and local ones (Becken & Job, 2014).

Natural World Heritage Sites are globally recognised as the world's most important Protected Areas (PA) (International Union for Conservation of Nature [IUCN], 2016a). There are 238 natural World Heritage Sites in the world representing about 0.1% of the total number of PAs globally and 8% of the combined surface area covered by PAs (IUCN, 2016a). The Great Barrier Reef (GBR) in Australia – the focus of this research – is one of the most iconic World Heritage Sites. The GBR World Heritage Site is a natural asset of global significance and home to the world's most diverse marine and terrestrial ecosystems. Based on a mix of different IUCN categories, the GBR is zoned to ensure an appropriate balance between ecosystem protection and economic use. It is managed under a complex governance structure that involves several Australian Federal and State Government Agencies, including the GBR Marine Park Authority (GBRMPA). In addition, a large number of natural resource management bodies, industry groups, communities and individuals are involved in GBR stewardship. Despite the protection in place, ever since its World Heritage designation in 1981, scientists have been concerned about declines in the environmental integrity of the GBR (De'ath, Fabricius, Sweatman, & Puotinen, 2012). Several Reef Water Quality Protection Plans (2003 and 2009, Queensland Government, 2014) have been implemented, and partnerships and taskforces have been formed, but the GBR continues to degrade (GBRMPA, 2016).

The GBR attracts over 2.2 million international and 1.7 million domestic visitors annually (Tourism Research Australia, 2015), generating economic benefits for Australia of AU$6.4 billion (Deloitte Access Economics, 2013). The Reef's tourism industry inherently depends on a healthy natural environment, and is also likely to benefit from the World Heritage brand value. Following the 2012 UNESCO mission to potentially add the GBR to the List of Endangered World Heritage Sites, the risk of losing the World Heritage brand triggered concerted action by the tourism sector. The perceived symbiosis between tourism and nature conservation was epitomised in the following statement by the Chief Executive Officer of the Queensland Tourism Industry Council: "A strong argument can be made that World Heritage Areas that have high visitation levels are more likely to remain protected and well managed as a result of political and commercial pressure" (Gschwind, 2013, p. 178).

The principal motivation for this research is to identify values connected to the World Heritage status and the PA itself, and understand how these values have driven and continue to drive the stewardship alliances of the GBR. Stewardship alliances can be more or less formalised, but they are likely to be strategic, as they reflect "purposive arrangements between two or more independent organisations that form part of, and are consistent with participants' overall strategies, and contribute to the achievement of their strategically significant objectives that are mutually beneficial" (Pansiri, 2008, p. 101). The distinction between stewardship and governance, the act of governing, is important. Governance denotes a "conceptual and representational role of the state in the coordination of socio-economic systems" (Hall, 2011, p. 439). The concept of governance is void of meaning without the centrality of the state, even if issues of network relationships and public–private partnerships are involved (Rhodes, 1997). Informal governance is exercised by stewards who care, display loyal devotion and identify with the conservation of PAs beyond their own and state interests. Examining the role of values in PAs is a novel way of understanding the different IUCN protection categories. The values linked with the different categories serve as moral compasses to verify how (well) an area is protected. Interpreting values as a guide for actions (Kollmuss & Agyeman, 2002) then also allows using such a value framework as a tool to ensure ongoing stewardship and conservation in the face of shifting dominant political ideologies.

The role of values is central to World Heritage Sites, which are unique PAs because they are based on the complex concept of Outstanding Universal Value (OUV) (UNESCO, 1972, 2007). Whilst many values are often local, tribal or national, the concept of universal values, by definition, should be relevant to everyone, including visitors from around the world. While "conventional" PAs may face changing values over time, such changes are more complex for World Heritage Sites. This is so because they are determined multinationally, and universality should be especially resistant to change. The relevance of understanding OUV is critical for World Heritage Sites in the day-to-day *realpolitik* of governance, where local and national stakeholders are in control, with only occasional, but potentially powerful inputs from UNESCO in Paris representing those universal values. The IUCN and UNESCO enjoy a long-standing trajectory that includes co-drafting of the 1972 World Heritage Convention text and the IUCN is explicitly recognised within the Convention as the technical Advisory Body on nature to the World Heritage Committee (IUCN, 2016b). Thus, attaching values of nature to the IUCN categories aligns well with the philosophy that underpins the OUV of World Heritage Sites.

In the following parts of this paper, a literature review unpacks the underpinning values that guide PA stewardship, and an original IUCN values framework is developed and explained. After introducing the methodology employed, the analysis of stakeholder interviews uncovers a range of values and actions that evolved in the course of three distinct phases at the GBR: pre 2012, from 2012 to 2014, and post 2014. Emphasis is given to the acute crisis in the middle phase, when the Reef was at great risk from industrial development but had reduced protection from the Government.[1] The values framework is applied to demonstrate how mainstream values changed and how, in response, GBR stewards from the tourism industry shifted alliances based on shared values to successfully fend off neo-liberal pressures. The paper concludes by arguing that the values framework can be used by anyone interested in PA management as a compass to assess and steer long-term stewardship that assures conservation of biodiversity and site integrity beyond parochial interests and changing governance regimes.

Protected areas, tourism and World Heritage

The following presents a classification of PAs and focuses on the values of nature that guide stewardship. An original value framework derived from the IUCN protection categories is presented.

Protected area categories and values

The designation of PAs is not a neutral undertaking. Framing a particular area deemed to be of national or international significance in itself is influenced by government, power, ideology, capital and ultimately by somebody in a position able to sanction and legally uphold nature conservation. Endorsing specific spaces, events, accounts and ecologies is a powerful display of selective renderings of history by government, a display that may be read as symbols used in contemporary identity creation. It is selective because the historic events chosen include but a fraction of all happenings that have taken place over a particular time span. Therefore, protected area designation can be understood as resulting from an accumulation of power, ideology and meanings ascribed to a space. Place construction of PAs is both a cultural concept and a practical project. It is cultural because shared meanings and values are continuously attributed to specific environments. The processual approach signifies that these are open to multiple interpretations, management and values, which may be contested and adapted over time.

The widely accepted definition of a PA by the EUROPARC Federation (see www.europarc.org/) and the IUCN (2000, p. 11) states: "A clearly defined geographical space, recognised, dedicated and managed, through legal or other effective means, to achieve the long-term conservation of nature with associated ecosystem services and cultural values." The IUCN (2008) identifies seven management categories to distinguish the specific aims, objectives and concerns of PAs. Detailed definitions and management actions typically involved can be found at https://www.iucn.org/theme/protected-areas/about/protected-areas-categories.

Table 1 captures the definitions and primary objectives, and lists related law and policy tools. It also adds and names six anthropocentric values of nature, developed uniquely for this research. For example, IUCN Protected Area Categories 1a and 1b represent strict preservation, which means that nature is perceived as having intrinsic value. This value of nature is referred to as "man of nature". If nature is a resource to be used by man, as seen in IUCN Category 6, it is named "man above nature".

Category 1a: areas must conserve outstanding ecosystems, species or geodiversity through strict management. Nature has intrinsic value. Termed "man of nature", humanity is only one of many other living organisms. In practical PA management terms, human intrusion is mainly for scientific purposes to inform management. Category 1b aims to preserve wilderness areas undisturbed by significant human activity.

Category 2: national or marine parks protect biodiversity, ecological processes and support recreational activity. Through management policies and strict monitoring, the term "man in nature" values education and controlled activities in nature.

Category 3: areas protect specific natural features through legislation and monitoring. "Man with nature" values the relationship between human beings and the environment but is largely alienated from nature aside from planned or occasional visits.

Category 4: protection of specific habitats or species through active intervention is valued by "man for nature". Management measures are taken to protect flora and fauna from invasive species, and various forms of animal population control may be in place.

Category 5: protection of distinct ecological, biological or cultural importance is founded on man's longitudinal interaction with nature. "Man and nature" represents co-existence that makes use of monitoring and traditional management techniques.

Category 6: conservation of ecosystems and habitats is linked to cultural practices that allow for balanced, low-level non-industrial utilisation, including "traditional" agriculture and forestry. Nature is valued, using the term "man above nature", where nature is a resource for human use although not for "industrial" exploitation or intensive "modern" farming.

Table 1 draws attention to PA categories and management tools, in combination with the anthropocentric values that guide nature conservation. In all of the IUCN categories, nature is seen as manageable and predictable by man (Plummer & Fennell, 2009), represented in particular by "man and nature" and "man above nature". Supported by well-established concepts such as carrying capacity and sustainable yield, law and policies are predicated in a hierarchical, technologically based and

Table 1. Protected area values framework.

IUCN Category	Definition	Primary objective	Law and Policy	Values of Nature
1a. Strict Nature Reserve and 1b. Wilderness Areas	Strictly protected areas set aside to protect biodiversity and possibly geological/ geomorphical features.	To conserve regionally, nationally or globally outstanding ecosystems, species and/ or geodiversity features.	Legislation and Treaties. National and international policies and treaties. Management mainly for science.	*man of nature* Human use and impacts are strictly controlled and limited. Nature has intrinsic value. No recreational visitation is allowed.
2. National Park, including Marine Reserves	Large natural or near natural areas set aside to protect large-scale ecological processes and recreation.	To protect natural biodiversity, ecological structure, supporting environmental processes.	Policy development: Setting legal standards, strict monitoring.	*man in nature* A foundation to protect and promote education. Controlled environmentally and culturally compatible recreation is possible.
3. Natural Monument or Feature	Conservation of specific natural features.	To protect specific outstanding natural features. Many enjoy high visitor value.	Compliance and Watchdog. Legislation and monitoring.	*man with nature* Recreational visits to pay tribute to specific features. These may involve education.
4. Habitat/Species Management	Protection of particular species or habitats.	To maintain, conserve and restore species and habitats.	Compliance and Intervention. Policy options, litigation, prosecution.	*man for nature* Bans on killing specific species, habitat management. Recreational activities encouraged.
5. Protected Landscape/ Seascape	The interaction of people and nature over time has produced a distinct area with significant ecological, biological, cultural and scenic value.	To protect and sustain important landscapes/ seascapes and other values created by interactions with humans.	Devolution of control. Monitoring. Traditional management practices, civil suits.	*man and nature* Co-existence. Protection of distinct areas with a pronounced mandate for sightseeing and recreational activities.
6. Protected Area with sustainable use of natural resources.	Conservation of ecosystems and habitats together with associated cultural values and traditional natural resource management systems.	To protect natural ecosystems and sustainable use when mutually beneficial.	Natural resource management.	*man above nature* Low-level, non- industrial use of natural resources. Tourism compatible with nature conservation.

Source: Adapted from Dudley (2008) and Brockington, Duffy, and Igoe (2008, p. 11).

linear fashion (Holling, Gunderson, & Ludwig, 2002; Plummer & Fennell, 2009). However, behaviours are not predictable and the real world does not operate in a mechanistic way (McDonald, 2009). Without recognition of the dynamic complexities of both nature and shifting values, such reductionist views of the world will fail to deliver the agreed conservation goals. Arguably, a holistic understanding of PA management is needed that is accountable against moral imperatives that lie beyond mere governance of those in charge at the time. This discussion continues below in relation to PA stewardship, tourism and UNESCO World Heritage.

Stewardship

The concept of stewardship differs from stakeholder and agency theories, both of which find their justification in self-preservation, economic motives and a pragmatist, rational approach to management (Bernstein, Buse, & Bilimoira, 2016; Donaldson & Davis, 1991; Freeman, Wicks, & Pamar, 2004). In other words, stakeholder and agency theories have a strong individualistic focus, which can jeopardise the greater environmental and societal good. Stewardship theory does not reject individual motivations, but suggests that those involved gain benefit by putting the interests of others above their own and pursuing actions that generate their own intrinsic rewards (Neubaum, 2013). Neubaum (2013) defines stewardship as "caring and loyal devotion to an organization, institution, or social group" (p. 2). The concept of stewardship thus puts emphasis on the people involved in conservation efforts, their personal values and dynamic interrelations.

The formal governance arrangements at UNESCO World Heritage Sites are often complemented by stewardship that involves alliances across local, national and international levels. Appreciating the dynamic nature of tourism in World Heritage Areas, these alliances are constituted in and through shifting ideologies, government, global capital, position and the persons influencing policies. Stewardship alliances between those holding similar values of nature may be central to ensuring PA resilience and integrity over time (Becken, 2013; Plummer & Fennell, 2009; Scharin et al., 2016). We hypothesise that the nature values of those who act as stewards are less vulnerable to change and erosion, as they are deeply anchored in personal eco-centric ideologies. Changes in alliances are then evidence of adaptive processes in response to the increasing vulnerabilities of exposed systems and the potential for disruption and crises (Espiner & Becken, 2014).

Tourism and World Heritage

Two interdependent dimensions of biodiversity preservation and levels of visitation underpin the IUCN classification system for PAs (Whitelaw, King, & Tolkach, 2014). Numerous insightful studies have analysed the meanings, management and use of PAs (e.g. Day et al., 2012; Liburd, 2006; McCool, 2009; Plummer & Fennell, 2009). Common to all these studies are the importance PAs have acquired as tourist destinations, against a background of growing popularity of "eco-" or "nature-based" tourism (Holden, 2015). However, the negative impacts sometimes associated with visitation and profit-seeking tourism businesses may be incompatible with nature conservation. Such a "use–conservation gap" (Jamal & Stronza, 2009, p. 171) can be illustrated by a Category 1 "man of nature" protected ecosystem, which over time opens up for human use by "man above nature", where the same ecosystem becomes a recreational resource.

The overarching goal of the UNESCO's 1972 *Convention Concerning the Protection of the World Cultural and Natural Heritage* was the protection and preservation of cultural and natural properties of "OUV". The OUV emphasis on valuing distinctive places refers to sites that are sufficiently exceptional to transcend national boundaries now *and in the future*. The permanent protection of the World Heritage Sites is of "the highest importance to the international community as a whole" (UNESCO World Heritage Centre, 2012, para. 49). The nomination, OUV classification and inscription processes for new World Heritage Sites that exist between the nation states and UNESCO speak of largely non-contested influence by government, ideology and global capital. However, World Heritage Areas, and the people who govern or protect them, do not exist in a value-free vacuum. Indeed, and as this paper will demonstrate, it seems implausible that the inscription and management of a World Heritage Site escapes the usual pressures arising from changing political systems, commercial demands and conflicts, and evolving governance arrangements. If previously agreed values function as imperatives that guide moral action, attention needs to be cultivated to the multilevel dynamics of PA management, associated values and the hitherto neglected role of (local) stewards.

This paper argues that a more holistic approach to PA management can emerge through recognition of the values held by people who devotedly care, beyond individualistic or commercial gain, while not excluding the latter. These stewardship dynamics are best understood through a value-based research approach.

Methodology

This research explores the perceptions of the values that tourism operators and managers hold in relation to the GBR World Heritage Site. To reduce the risk of being overly prescriptive, and as a result missing potentially important emerging issues, a qualitative approach was adopted. The relativist ontology underpinning the work assumes that reality is socially constructed, even if particular elements of this reality (e.g. the environmental quality of the Reef) may be measurable by objective approaches. What is important here is to understand how GBR stakeholders perceive its status as a World Heritage Site, and to identify the different values and ideologies that underpin such perceptions, including their ability to engage in day-to-day *real-politik* of PA governance. The analysis will also seek to uncover whether stakeholders see themselves as stewards and whether such self-proclaimed role is linked to strongly nature-based values.

Informant interviews

The data collection for this research was conducted between December 2014 and March 2015. To begin with, leading tourism representatives involved in the GBR were identified and contacted for an in-depth interview. A list of members of the GBRMPA's Tourism Reef Advisory Committee was used to identify key individuals. Further interviewees emerged using snowball sampling from recommendations provided by the key stakeholders first interviewed. This process also brought in managers involved in GBR governance, and ultimately resulted in 13 in-depth interviews, each ranging between 30 and 120 minutes. Whilst this is a relatively small number, it is important to note that the interviewees were leaders in their field (typically Chief Executive Officers) who collectively had accumulated considerable expertise, experience and insights, including into historic changes in the political environment, governance structures, partnerships, organisational and personal relationships, and who had been through previous crises. Table 2 provides an overview of key informants.

Interviews started out by broadly discussing what World Heritage status meant to the interviewee. This was followed by a discussion of stakeholder networks, and the extent to which these were influenced by the GBR's status as a World Heritage Site. Interviewees were asked to explore the idea of a "worst" case scenario for the Reef. This method has been found useful in previous research (Liburd, 2007). Scenarios are stories, which describe an imaginary sequence of actions and events (Rosson & Carroll, 2002). Asking for a worst case scenario prompted interviewees to first reflect upon their personal and professional motivations, on the many actors involved, the various preservation tools available, the possible loss of World Heritage status, and ultimately on the Reef itself. Having reflected upon what could go wrong, the recent (2012–2014) crisis of increasing development pressure by the resources sector, effectively opened up new imaginations of Reef stewardship.

Interviewing continued until a point of saturation was reached (Dwyer, Gill, & Seetaram 2012), and interviews were transcribed verbatim, except for three interviews where notes were taken. Interviews were then coded for emergent themes using content analysis, a commonly employed tool that is useful for uncovering knowledge and new insights from the participants' perspective (Jennings, 2010, pp. 211–213). Based on repeated readings of the textual transcriptions, UNESCO World

Table 2. Informants.

Type of organisation	Position	Place of interview	Recorded
Marine Tour Operator	CEO	Cairns	Tape, 43 minutes
Marine Tour Operator	Marine Biologist and educator	Cairns	Tape, 54 minutes
Marine Tour Operator	CEO	Port Douglas	Tape, 66 minutes
Island Resort	CEO	GBR	Tape, 58 minutes
Industry Association	CEO	Brisbane	Tape, 45 minutes
Industry Association	CEO	Cairns	Notes
State-level tourism organisation	Manager	Brisbane	Notes
GBRMPA	Manager	Cairns	Phone/notes
Ecotourism Consultant	CEO	Brisbane	Notes
Certification programme	Manager	Brisbane	Phone/notes
Destination Marketing Organisation	CEO	Cairns	Notes
University	Academic/marine tourism expert	Gold Coast	Informal interview, notes
University	Academic/protected area expert	Brisbane	Informal interview, notes

Heritage, person, position, government, ideology, shifts, crisis and temporal dimensions imposed coding themes based on theory. Secondary materials were sourced from newspapers, online news-letters, campaigns and GBR information brochures to further contextualise findings. Numerous meetings were held between the researchers in person in Australia and Denmark, and via Skype, during the entire research process. There was ongoing discussion of the material and critical evaluation of counter evidence between the authors.

The reflective research process was also influenced and assisted by the involvement of one of the researchers in the State of Queensland's Ministerial Taskforce on GBR Water Quality. The Taskforce worked from May 2015 until June 2016 to provide advice and recommendations to the Minister for Environment and Heritage Protection and Minister for National Parks and the Great Barrier Reef and the Queensland Government on issues related to water quality targets, management strategies and pro-grammes, investment priorities, and monitoring activities. The researcher's particular role was to engage with the tourism sector and facilitate communication between scientists, the Queensland Government and tourism organisations. The interaction with other Taskforce members, as well as with consulted tourism stakeholders, further deepened the researchers' understanding of key issues, relationships between people and organisations, and values held by those involved in the Reef stewardship.

Results

The researchers' engagement with the GBR over several years, and the insights gained from the interviews, helped identify three distinct phases, which emerged as changes in formal and informal value perceptions of GBR conservation and UNESCO World Heritage (see below). Changes in government, ideology and global capital, notably in the form of resource extraction and port expansion, were instrumental in defining the shifting phases. Over the years, the tourism industry has publicly expressed concern about the GBR's health and potential negative impacts of its declining health on tourism during neo-liberal governance in 2012–2014 (McLennan, Becken, & Moyle, 2015).

Based on historical accounts of the informants, and underpinned by previous publications, reports and media coverage, it became clear that the force field shaping and making stewardship of the GBR was substantially different before 2012, compared with the years thereafter. The force field as defined through government, ideology, global capital, position and personality, seemed to change substantially again in 2015, with a change in government, a declining mining industry and a booming tourism sector. The focus of analysis is on the acute crisis in 2012–2014, but the other phases are considered as well, as they critically highlight the dynamic nature of PA stewardship. The crisis prompted action at multiple levels of stewardship. We illustrate how the shifting stewardship alliances at local, national and international levels, and over time, were informed by the values assigned to nature that were strategically mobilised in GBR conservation efforts.

Phase I (pre 2012)

Prior to 2012, tourism to the GBR had been in a state of decline and the industry was hampered by the global financial crisis, lack of business confidence and reduced strategic "position" prompted by the boom in mining activities. The weak position of tourism during this phase has been analysed from various perspectives. For example, McLennan et al. (2015) examined media reporting on the "two-speed economy", finding that tourism was often framed as a (likable) underdog compared with the more powerful and successful resources sector. The Australian mining boom and the "Dutch Disease" that affected tourism as a result of increased exchange rates and labour mobility were discussed in detail by Pham, Jago, Spurr, and Marshall (2015). During the Labour-led Queensland government (Premier Anna Bligh from September 2007 to March 2012), the decline of the environmental integrity of the GBR appeared to be tolerated. One tourism consultant interviewed suggested that the southern part of the Reef was in such poor condition that the added pressure from urban land use and resource extraction "might result in sacrificing the southern end". In this quote, the IUCN protection categories appeared void of meaning, but his "position" as tourism consultant was obvious when he swiftly added: "So you better see it before it disappears." This "man above nature" valuation was reiterated by the tourism consultant when asked about the meaning of the World Heritage status, which he deemed "very important for the Reef and tourism benefit from the World Heritage listing". During Phase I, the tourism sector was able to maintain business as usual at the World Heritage Site by adapting to the gradually declining quality of the GRB.

Industry leaders at Tourism Events Queensland and the Queensland Tourism Industry Council also appreciated that World Heritage was a strong brand, which "enhanced the unparalleled natural attraction and recognition of the GBR internationally". They agreed that World Heritage had limited significance within Australia, and that the tourism industry ought to use the brand to champion the conservation of nature: "We simply haven't made enough of it. Perhaps it's because conservation is the driver of World Heritage and that tourism is dominated by marketeers?" This quote reveals how the tourism sector may have perceived World Heritage protection to be at odds with business interests, but now identified an opportunity to align these mandates based on "man and nature" values. This is supported by the Tourism Events Queensland manager who speculated that insufficient innovation and the low profit margins of Reef operators could widen the gap between conservation and use: "The tourism industry rests on their laurels. They need to be ahead of the market." The underpinning "man and nature" value here aligns industry use of the WH brand, not only for marketing and business purposes, but also with conservation itself. The Queensland Tourism Industry Council expert further explained: "brand promotion sounds like gimmicky marketing, that's not what it is. It is to give value to the proposition (…) more buy-in, more political support (…) and educating the visitors of the broader implications of GBR conservation". Here, the knowledge force field "position" (Tribe & Liburd, 2016) is helpful in understanding how an industry CEO promotes business that can also be underpinned by strong "personal" "man with nature" values to protect the Reef, as a universally distinct area with a pronounced mandate for sightseeing and educational activities.

A small family tourism business operator using the GBR, positioning himself as "man in nature", reflected on his more than 20 years of professional experience:

> I'm a skipper, I'm a dive instructor and committed to environmental best practices both working within it as a means of making a living and showing off the natural environment for the opportunity to educate and for others to experience it who have a feel and an interaction with nature, which is quite powerful.

Dedicated to "continuing a path of sustainability which makes good business sense", he actively promoted education and controlled, environmentally compatible recreation. Redirecting stewardship responsibilities towards government, he added: "But I feel disappointed that, in a political environment, sustainability is pretty low on the agenda." Other interviewees similarly suggested how government had failed to mitigate Reef deterioration: "So they preach to us that we have to sustainably manage our sites, but they're not sustainably managing, and they need to come to terms with it, the fact that they need to."

The citations by industry operators and CEOs show how the declining health and lagging conservation of the GBR appeared to be tolerated by the tourism sector during Phase I, but also that formal GRB governance is subject to a mounting critique. The urgency heralded by the small family business operator reflects Giddens' (2009) paradox of climate change. Giddens argued that since we were not unduly affected by the outcomes of climate change, we failed to act, but when we would be pressed into action by its consequences, it would be too late to do anything about them. A parallel can be drawn with the deterioration of the GBR. The tourism industry and government were aware of the declining health of the Reef, but instead of acting, the deterioration was allowed and vindicated by "man above nature".

Phase II (2012–2014)

The second phase identified from the interviews was one of the crises. The key reason was international intervention into Australia's governance of the GBR. In addition to ongoing declines of environmental parameters, such as coral cover and nitrogen loads, the catalyst for increasing problems was increasing industrial activity from the resources sector in proximity to the GBR World Heritage Site. Large-scale port development and expansions, and concomitant increases in dredging and shipping activities, became centre stage in global media coverage that raised questions about Australia's efforts to protect the GBR. The port developments were to happen in fragile marine areas and the dredging activity was feared to substantially increase turbidity and sedimentation with major negative impacts on marine life and, in particular, on coral (Becken, McLennan, & Moyle, 2014).

Notably, the failure to inform the UNESCO World Heritage Committee of several proposed industrial developments prompted a UNESCO–IUCN reactive monitoring mission in March 2012 (Brodie & Waterhouse, 2012). The mission led to the process of considering adding the GBR to the UNESCO List of World Heritage in Danger. The number and extent of the development proposals presented a high risk to the integrity and conservation of the GBR's OUVs. These developments were also received with great concern by the tourism sector, which feared repercussions from negative publicity at a global scale.

The 2012 UNESCO mission coincided with a change of state government, in which the conservative National Party (March 2012 to February 2015) took over; thus aligning ideologically with the conservative National Party Federal Government. Explicit support of major industrial and resource sector projects and simultaneously lowered priorities for environmental protection prompted environmental NGOs to launch major campaigns for the protection of the GBR. The Green Party supported those campaigns by providing so-called fact sheets, strongly criticising both the Federal Prime Minister (Tony Abbott) and the Queensland Premier (Campbell Newman) at the time (The Greens, 2015). Media coverage (e.g. Peatling, 2012) centred on quotes such as the Queensland Premier Newman's "We will protect the environment but we are not going to see the economic future of Queensland shut down." This ideological shift in "government" of the value of nature as a resource to be exploited, rather than conserved by man, included the GBR.

To the surprise of several tourism stakeholders, GBRMPA aligned itself with government when allowing dredging activities at Abbot Point. Abbot Point is a coal export port close to the Whitsunday Islands, which is a major GBR tourist destination. Criticising GBRMPA, a family business operator lamented:

> We have to have integrity otherwise everything else you've been doing doesn't mean anything. They are managing it; they are responsible for it. So to put a stamp of approval on dredging - it's totally unhelpful. (…) An environmental agency should never give endorsements for something that's politically incorrect.

The operator accused GBRMPA for failing to meet their obligations to adequately govern an IUCN category 2 (i.e. "man in nature") marine reserve. The operator's reference to "politically incorrect" (possibly meaning 'morally incorrect') uncovers the conflicting values that guide neoliberal governance of PAs (i.e. "man above nature") as opposed to stewardship by "man in nature" that lies beyond state and self-interests.

Still, a GBRMPA manager maintained that "World Heritage is our core obligation to UNESCO." She welcomed the long-standing partnership with the tourism industry, including the newly appointed Tourism Reef Advisory Committee, effective zoning, licensing and monitoring programmes. When prompted about a worst case scenario, the GBRMPA manager's response was unequivocal: "Thankfully I can't really imagine one! We have zoning plans, 30% are green zones, even the high use areas have limits to growth. Fishing is regulated. I have full confidence in our management arrangements moving forward." The underpinning rationality based on cause and effect in PA management represents a "man above nature" value. It is interesting how the "position" of the GBRMPA manager prevented her from taking a critical stance, just as the rhetoric reflects her organisational affiliation. Different to some of the other interviewees, she did not disclose "personal" values that could potentially challenge the official, organisational "position" and government "ideology". The acute crisis exposed how rational PA management and governance differ from stewardship as a values-based ethic, the latter of which is deeply anchored in eco-centric values. These values are perhaps less prone to change, which will be explored next.

In response to the erosion of the original IUCN protection categories, lacking preservation of the OUVs and clashing values of nature, the GBR operators abandoned their long-standing alliance with GBRMPA. One of the largest marine tour operators, annually carrying about 400,000 tourists to the Reef, explained how the Association of the *Australian Marine Park Tourism Operators* (AMPTO) lobbied ministers, the Federal Government and affirmed: "We're probably the people who got the government to change their position. (…) Admittedly that attracted a lot of the attention of some of the green, greener groups." Referring to the industry alliance with green organisations, such as the World Wildlife Foundation, the marine tourism operator emphasised how they had "similar concerns to some of the greener groups without necessarily sharing all their concerns or exaggeration to achieve an outcome."

Despite different "ideologies" and "positions", the tour operators and environmental organisations strategically aligned values of "man of nature", "man in nature" and "man and nature" against the natural resource depletion represented by "government" and the "global capital" of the mining industry. The goal-driven alliance between the green organisations and the GBR tourism operators strategically invoked the IUCN–UNESCO protection values (i.e. category 2 and the OUVs) during the crisis of GBR stewardship. Often the proactive support was driven by personal values, rather than official "stakeholder" roles.

The director of the large-scale marine tour company, for example, was personally proud of the World Heritage status, but made a clear separation between the values of a private "person" to the commercial "position" and the World Heritage brand value in overseas markets. Similar positions were held by industry representatives and the person responsible for coordinating "Reef community education" at a smaller tour operator. She stated:

> Personally, World Heritage to me opens up that we live on a planet and it's a global community. For the business perspective obviously, it's very important because it's a drawcard that the Reef is World Heritage listed and the tourists know that.

According to several informants, pride in the World Heritage brand or nature conservation value do not appear to be part of the dominant Australian "ideology", and the educator went on to explain that their tour operation's education programmes was orientated towards marginalised Aboriginal communities. She passionately conveyed "man in nature" values, where nature has both intrinsic and spiritual value.

Here it becomes obvious how personal and professional values may or may not align, and that these values can be strategically utilised depending on the context and situation at stake. The fluidity between personal versus professional values was surprisingly revealed by an industry association representative. He initially claimed not to be proud of the World Heritage status, emphasising that "it meant nothing to the tourism operators". However, when prompted to imagine a worst case scenario, he referred to the neo-liberal government as "environmental vandals!" He acknowledged "the Reef as an iconic type World Heritage Area, like the Galapagos Islands" and went on to argue how the GBR is endangered, but responsibly protected by the local tourism operators. The first citation alludes to World Heritage as having no commercial value. This allusion shifts radically in the subsequent statement, which become more personal and recognised outstanding ecosystems, species or geodiversity, underpinned by a "man of nature" value and a sense of tourism stewardship.

Recognising the importance of stewardship, another industry representative explained: "Mooring sites is the best stewardship model for us and the Reef, especially as the damage from anchoring is eliminated. Shared moorings don't offer the same felt responsibility" (because shared sites bear a tragedy of the commons risk, where individual operators feel less responsible). The "felt responsibility" and referral to stewardship by the tourism operators represent values of "man in nature", "man with nature" and "man for nature". These values of nature range from conserving and restoring specific outstanding natural features and species. His understanding of stewardship was clearly underpinned by care and loyal devotion to the GBR by the GBR operators. The industry association representative expressed his faith in national legislation that adheres to the original IUCN designation and the role of the tourism industry as watchdog, but not in the GBRMPA or the neo-liberal government as stewards of the GBR.

The crisis culminated when local GBR stewards from the tourism industry shifted alliances away from Government to green advocacy groups, based on shared values across national and international contexts to successfully fend off neo-liberal pressures in court. In 2014, UNESCO delayed their decision about adding the GBR to the List of World Heritage in Danger. In follow-up discussions, as part of the GBR Water Quality Taskforce consultation, several of the original tourism industry interviewees passionately reiterated the value of the World Heritage status as a "man in nature", and the critical role that tourism played in battling port development decisions.

Phase III (post 2014–2016)

In 2015, UNESCO formally recognised the noticeable increase in Government commitment to Reef protection and an observation period was established. More specifically, UNESCO requested an update to be sent to their advisory body about progress, documented in the *Reef 2050 Long Term Sustainability Plan* (2016), and that the full Committee be informed of the state of the conservation of the GBR in 2020 (Department of Environment, 2015). This new phase coincided with a change in the Queensland Government. Labour regained leadership from the National Party, which brought an augmented focus on environmental conservation. This included a restructuring of Government with the formation of Queensland's first Office of the Great Barrier Reef based on the Department of Environment and Heritage Protection. The Office reports to the Minister for the Environment and Heritage Protection, who is also the Minister for National Parks and the Great Barrier Reef. Aligned with preservation of the integrity of the GBR OUVs, the new Government committed to a ban on sea dumping of capital dredge spoil within the GBR World Heritage Site, and provided an additional AU$100 million over five years to address the issue of deteriorating water quality. The GBR Water Quality Taskforce was asked to advise the Minister on how to best invest this additional funding. The "ideological" shift in "government" conservation of PAs and the GBR reflects "man and nature" and "man for nature" values, which promote coexistence and habitat management based on a conservation commitment to maintain species and natural environments.

In parallel with the change of Government, substantial economic restructuring could be observed. Export of resources declined substantially and the end of the mining boom were visible in trade statistics. In November 2015, the *Sydney Morning Herald* asked whether "services exports fill the mining hole?" pointing to substantial "growth in net exports of services, with tourism and education playing a 'starring role'" (Cauchi, 2015, p. 1). Indeed, tourism reached record growth rates compared to the previous year, for example, with an increase of 14.8% in arrivals to North Queensland, the main hub for GBR tourism (Tourism Events Queensland, 2015). With tourism strategically repositioned on the economic agenda, the political importance of protecting both the GBR and the (growing) tourism industry were purposefully connected.

The new "man and nature", "man for nature" alliance became apparent at domestic as well as international events, including *World Environment Day 2015*, where the Queensland Tourism Industry Council organised an event focused on the GBR that specifically linked up with the Minister for Environment and Heritage Protection and the GBR Water Quality Taskforce. The event featured a short preview of a documentary on the GBR, commissioned by the BBC's Natural History Unit and with Tourism Australia's sponsorship of the BBC's world renowned presenter, Sir David Attenborough. The documentary was part of a major advertisement campaign by the tourism industry, and was also used as a vehicle to attract more support for saving the Reef, evidenced by its pre-screening at the Paris Climate Summit (COP 21) (see Scott, Hall, & Gössling, 2016).

These activities reflect a call by two tourism stakeholders for a joint communication strategy that not only brings together all the tourism operators "up and down the Reef so that they sing from the same sheet", but strategically coordinates with non-tourism bodies, such as the Department of Environment and Heritage Protection, which share similar conservation values. The ability of the tourism sector to tap into global networks, social media and tell "success stories of Reef protection", according to one tourism manager, highlights the importance of value-based collaborations that strengthen PA stewardship.

Against the background of UNESCO's decision to maintain the GBR's World Heritage listing, and increased global attention to protect the GBR, tourism was able to regain momentum in charging a new "position" for the industry and its symbiotic relationship with the GBR. The new position comfortably accommodated the prevailing underpinning (personal) values of "man in nature" and "man and nature" held by many tourism stakeholders. Furthermore, the conservation-focused position provided flexible space for those who value World Heritage in "person", and those in "positions" to value World Heritage as a brand with combined commercial and conservation interests. The consultation with tourism stakeholders as part of the Taskforce outreach re-emphasised perceptions that tourism operators see themselves as the key stewards of the GBR, working proactively to protect it, beyond state or own commercial interests while not excluding these.

At the same time, and whilst a realignment of the partnership alliance between the tourism industry and the GBRMPA could be observed, tourism stakeholders began to distance themselves from environmental lobby groups whose strong campaigns were perceived as "damaging to our tourism industry", according to an industry representative. The strategic alliance between the green organisations and the tourism operators, which had been necessary during the acute UNESCO crises in Phase 2, had become redundant. A regrouping with "government" based on realigned values appeared more effective in the immediate, post-crisis context.

Conclusion

First and foremost, the contribution of this article is the development of an original value framework based on the IUCN's PA categories and management tools. The Protected Area Values Framework identifies the PA categories and management tools and corresponding anthropocentric perspectives of nature that guide conservation. The values framework offers an important tool for ensuring greater understanding of, and reflexivity in, nature conservation, tourism and UNESCO World Heritage stewardship.

Second, the analysis demonstrates the importance of a dynamic and holistic understanding of PA stewardship, which lies beyond state governance and formal site management. The importance of governance in implementing sustainable tourism and heritage conservation has long been recognised (Bramwell, 2011; Bramwell & Lane, 2011). Better stakeholder involvement has frequently been introduced as a key to success and implementation of more sustainable practices (e.g. Waligo, Clarke, & Hawkins, 2015). Hitherto, the resilience and strategic importance of local stewards in PA conservation, and their ability to engage in day-to-day *real-politik* of governance have been neglected. Stakeholders may have different stakes over time, but extant research has failed to understand how they care and show loyal devotion to a conservation task beyond individual gains. A deeper, more holistic and also dynamic understanding of PA and World Heritage Site governance can be qualified through the notion of stewardship and underpinning anthropocentric values of nature, which may be strategically mobilised at multiple levels.

Third, to gain deeper understandings about how humans construct the values and meanings of nature which influence stewardship of PAs, a knowledge force field was applied. The force field of the tourism knowledge system is determined by government, ideology, global capital, position and person (Tribe & Liburd, 2016). The combination of the force field and values framework made it possible to exemplify how seemingly interest-free values about nature are subject to various forces that shape and make PA stewardship. The systemic recognition of shifting alliances based on shared values of nature was particularly appropriate in the face of crisis, where rational management approaches by "man above nature" proved insufficient to counter the continued deterioration of the World Heritage GBR.

By identifying values of nature among tourism operators in one of the world's most iconic destinations, the UNESCO World Heritage GBR, it became evident why new stewardship alliances emerged during a period of system crisis from 2012 to 2014. The pre- and post-crisis framed the strategic changes, notably how local tourism operators allied with international environmental conservation organisations and UNESCO against formal government (state, federal and GBRMPA) and industrial activity from the resources sector in, or in close proximity to, the GBR. The 2012–2014 crises prompted the GBR tourism operators to shift stewardship "position" from a more pragmatic one of tolerating decline (e.g. "man and nature") to one of open dissent, mainly based on personal values of "man of nature" or "man in nature". The potential power of mobilising tourism stewards is reflected by the Water Quality Taskforce (State of Queensland, 2015) interim report that states: "The Taskforce's vision for the Reef's future is that it will be healthy and resilient and continue to support an iconic and wondrous ecosystem, world class tourism, viable industries and sustainable communities" (p. 33). This quote also indicates a renewed alignment between Government and the tourism industry in the third phase identified in this research.

A future system crisis is likely to demand new stewardship alliances. This is to be expected in a complex adaptive system, which may, once again, be prompted by the continued degradation of the iconic World Heritage Site. In May 2016, concerted action from 175 tourism operators in the *Brisbane Times* (2016) criticised "government" for favouring "global capital" from the resources sector and failing to act on what they labelled a "disaster needing urgent action" (Branco, 2016, p. 1). The tourism representatives called on the Federal government:

> to rule out any financing, investment or help with associated infrastructure for the Abbot Point coal terminal expansion and Adani's controversial Carmichael mine, the largest in Australia. They pointed the finger at climate change, calling for investment in renewable energy projects, particularly in regional Queensland and a ban on any new coal mines. (Branco, 2016, p. 1)

The wider implications of this research for the governance systems for World Heritage Sites and PAs are several. Understanding the shifting values of nature, tourism and UNESCO World Heritage stewardship reveals potential for advancing PA management mechanisms, and the possibilities for real-world engagement. These insights are of pivotal importance to other UNESCO World Heritage Sites to proactively engage in systemic changes at multiple levels, whether triggered by climate change, site management or competing interests in PAs. Local stewards and PA managers may utilise the values framework as a compass to help steer and guide action if the original designation category, including OUVs, are compromised by neo-liberal (or other) ideologies, for instance, to increase visitation or use of nature. In particular, UNESCO and World Heritage Site managers should be warned against compromising conservation values.

This article helps us better understand how to make progress in PA management. Researchers are encouraged to more deeply explore the changing political, commercial and conservation *real-politik* of management and governance to help advance holistic PA stewardship. Future research is needed on how and why values change over time, about the possible need to recognise both immutable and adaptable values, and equally about how to reconcile universal versus national and local values. The Protected Area Values Framework reveals philosophical values of nature, and offers guidance for real-world engagement, but also the radical possibilities of values-based research in tourism and nature conservation towards more sustainable futures (Liburd, 2010). Future research also needs to consider philosophical values other than the Western one from which this particular research was conducted, and there might be scope to look at stewardship values in the many urban World Heritage Sites around the world.

Note

1. The declining quality of the GBR and port developments approved by both the Federal and State Governments triggered an acute (environmental and public relations) crisis beginning in 2012. The crisis not only involved an

intervention by UNESCO, but also led to global media coverage on the Reef and its "death by a thousand cuts" (Shafy, 2013). The Port of Gladstone, where substantial dredging has occurred, received particular attention (Becken et al., 2014).

Disclosure statement

No potential conflict of interest was reported by the authors.

References

Becken, S. (2013). Developing a framework for assessing resilience of tourism sub-systems to climatic factors. *Annals of Tourism Research, 43*, 506–528.

Becken, S. (2016). Evidence of a low-carbon tourism paradigm? *Journal of Sustainable Tourism*, doi 10.1080/09669582.2016.1251446

Becken, S., & Job, H. (2014). Protected areas in an era of global-local change. *Journal of Sustainable Tourism, 22*(4), 507–527.

Becken, S., McLennan, C., & Moyle, B. (2014). *World Heritage Area at risk? Resident and stakeholder perceptions of the Great Barrier Reef in Gladstone, Australia* (Griffith Institute for Tourism Research Report Series No. 2). Retrieved 25 January 2017 from https://www.griffith.edu.au/__data/assets/pdf_file/0004/614596/Report-2-World-Heritage-Area-at-Risk-GBR.pdf

Bernstein, R., Buse, K., & Bilimoria, D. (2016). Revisiting agency and stewardship theories. *Nonprofit Management & Leadership, 26*(4), 489–498.

Bramwell, B. (2011). Governance, the state and sustainable tourism: A political economy approach. *Journal of Sustainable Tourism, 19* (4&5) 459–477.

Bramwell, B., & Lane, B. (2011). Critical research on the governance of tourism and sustainability. *Journal of Sustainable Tourism, 19* (4&5) 411–421.

Branco, J. (2016). Great barrier reef tourism operators beg for change. *Brisbane Times*. Retrieved 13 May 2016 from http://www.brisbanetimes.com.au/queensland/great-barrier-reef-tourism-operators-beg-for-action-on-bleaching-20160506-goom2s.html

Brockington, D., Duffy, R., & Igoe, J. (2008). *Nature unbound. Conservation, capitalism and the future of protected areas.* London: Earthscan.

Brodie, J., & Waterhouse, J. (2012). A critical review of environmental management of the 'not so Great' Barrier Reef. *Estuarine, Coastal and Shelf Science, 104–105*(0), 1–22.

Cauchi, S. (2015, August 2017). Australian dollar still too high for post-mining economy. *Sydney Morning Herald*. Retrieved 20 February 2016 from http://www.smh.com.au/business/markets/currencies/australian-dollar-still-too-high-for-post mining-economy-20150817-gj0ilt.html

Crouch, D. (2000). Places around us: Embodied lay geographies in leisure and tourism. *Leisure Studies, 19*(2), 63–76.

Day, J., Dudley, N., Hockings, M., Holmes, G., Laffoley, D., Stolton, S., & Wells, S. (2012). *Guidelines for applying the IUCN protected area management categories to marine protected areas.* Gland: IUCN.

De'ath, G., Fabricius, K.E., Sweatman, H., & Puotinen, M. (2012). The 27-year decline of coral cover on the Great Barrier Reef and its causes. *Proceedings of the National Academy of Sciences, 109*(44), 17995–17999.

Deloitte Access Economics. (2013). *Economic contribution of the Great Barrier Reef.* Townsville: GBRMPA.

Department of Environment. (2015). *Reef 2050 long-term sustainability plan.* Canberra: Commonwealth of Australia.

Donaldson, L., & Davis, J.H. (1991). Stewardship theory or agency theory: CEO governance and shareholder returns. *Australian Journal of Management, 16*(1), 49–65.

Dwyer, L., Gill, A., & Seetaram, N. (Eds.). (2012). *Handbook of research methods in tourism: Quantitative and qualitative approaches*. Cheltenham: Edward Elgar.

Espiner, S., & Becken, S. (2014). Tourist towns on the edge: Conceptualising vulnerability and resilience in a protected area tourism system. *Journal of Sustainable Tourism, 22*(4), 646–665.

EUROPARC Federation and the International Union for the Conservation of Nature. (2000). IUCN Protected Areas Categories System: EUROPARC and WCPA, Grafenau, Germany. Retrieved 13 January 2015 from http://www.iucn.org/about/work/programmes/gpap_home/gpap_quality/gpap_pacategories/

Freeman, E.R., Wicks, A.C., & Pamar, B. (2004). Stakeholder theory and "the corporate objective revisited". *Organization Science, 15*(3), 364–369.

Giddens, A. (2009). *The politics of climate change*. London: Polity Press.

Great Barrier Reef Marine Park Authority. (2016). *Great barrier reef outlook report 2016*. Townsville: Author.

Gschwind, D. (2013). Valuing our iconic heritage areas: How the industry can support keeping the outstanding exceptional. In P. Figgis, A. Leverington, R. Mackay, A. Maclean, & P. Valentine (Eds.), (2012). *Keeping the outstanding exceptional: The future of world heritage in Australia*. Sydney: Australian Committee for IUCN.

Hall, C.M. (2011). A typology of governance and its implications for tourism policy analysis. *Journal of Sustainable Tourism, 19*(4–5), 437–457.

Harrison, D. (1996). Sustainability and tourism: Reflections from a muddy pool. In L. Briguglio, B. Archer, J. Jafari, & G. Wall (Eds.), *Sustainable tourism in islands and small states* (pp. 69–88). London: Pinter Press.

Holden, A. (2015). Evolving perspectives on tourism's interaction with nature during the last 40 years. *Tourism Recreation Research, 4*(2), 133–143.

Holling, C.S., Gunderson, L.H., & Ludwig, D. (2002). In quest of a theory of adaptive change. In L.H. Gunderson & C.S. Holling (Eds.), *Panarchy* (pp. 3–24). Washington, DC: Island Press.

International Union for the Conservation of Nature. (2008). IUCN protected areas categories system. Retrieved 13 January 2015 from http://www.iucn.org/about/work/programmes/gpap_home/gpap_quality/gpap_pacategories/

International Union for the Conservation of Nature. (2016a). Natural World Heritage: Facts and figures 2016. Retrieved 28 October 2016 from https://www.iucn.org/sites/dev/files/fact_sheet_wh_2016_desert.pdf

International Union for the Conservation of Nature. (2016b). Terms of reference for the IUCN World Heritage Panel. Retrieved 28 October 2016 from File://sdu-data0.c.sdu.dk/staff/liburd/SDU/Publikationer/JoST%202017%20GBR/IUCN%202015%20tor_for_iucn_world_heritage_panel.pdf

Jamal, T., & Stronza, A. (2009). Collaboration theory and tourism practice in protected areas: Stakeholders, structuring and sustainability. *Journal of Sustainable Tourism, 17*(2), 169–189.

Jennings, G. (2010). *Tourism research* (2nd ed.). Milton: John Wiley & Sons.

Jones, R., & Shaw, B. (2012). Thinking locally, acting globally? Stakeholder conflicts over UNESCO World Heritage inscription in Western Australia. *Journal of Heritage Tourism, 7*(1), 83–96.

Kollmuss, A., & Agyeman, J. (2002). Mind the gap: Why do people act environmentally and what are the barriers to pro-environmental behavior? *Environmental Education Research, 8*(3), 239–260.

Larrère, C. (2008). Scientific models for the protection of nature. In L. Garnier (Ed.), *Man and nature-making the relationship last* (Biosphere Reserves – Technical Notes 3) (pp. 28–31). Paris: UNESCO.

Liburd, J.J. (2006). Sustainable tourism and national park development in St. Lucia. In P.M. Burns & M. Novelli (Eds.), *Advances in tourism research: Tourism and social identities: Global frameworks and local realities* (pp. 155–176). London: Elsevier.

Liburd, J.J. (2007). Sustainable tourism, cultural practice and competence development for hotels and inns in Denmark. *Tourism Recreation Research, 32*(1), 41–49.

Liburd, J.J. (2010). Sustainable tourism development. In J.J. Liburd & D. Edwards (Eds.), *Understanding the sustainable development of tourism* (pp. 1–18). Oxford: Goodfellow Publishers.

McDonald, J.R. (2009). Complexity science: An alternative world view for understanding sustainable tourism development. *Journal of Sustainable Tourism, 17*(4), 455–471.

McCool, S.F. (2009). Constructing partnerships for protected area tourism planning in an era of change and messiness. *Journal of Sustainable Tourism, 17*(2), 133–148.

McLennan, C.L., Becken, S., & Moyle, B.D. (2015). Framing in a contested space: An analysis of media reporting on mining and tourism. *Current Issues in Tourism*, doi: 10.1080/13683500.2014.946893

Neubaum, D.O. (2013). Stewardship theory. In E. Kessler (Ed.), *Encyclopaedia of management theory* (pp. 768–769). Thousand Oaks, CA: Sage.

Oyserman, D. (2001). Values: Psychological perspectives. In N.J. Smelser, J. Wright, & B. Baltes (Eds.), *International encyclopaedia of the social and behavioural sciences* (pp. 16150–16153). Oxford: Pergamon.

Pansiri, J. (2008). The effects of characteristics of partners on strategic alliance performance in the SME dominated travel sector. *Tourism Management, 29*, 101–115.

Peatling, S. (2012). UNESCO report scathing of Great Barrier Reef management. *The Sydney Morning Herald*. Retrieved 5 March 2016 from http://www.smh.com.au/environment/conservation/unesco-report-scathing-of-great-barrier-reef-management-20120601-1zo0m.html#ixzz42CwJ1oA2

Pham, T., Jago, L., Spurr, R., & Marshall, J. (2015). The Dutch disease effects on tourism – the case of Australia. *Tourism Management, 46,* 610–622.

Plummer, R., & Fennell, D.A. (2009). Managing protected areas for sustainable tourism: Prospects for adaptive co-management. *Journal of Sustainable Tourism, 17*(2), 149–168

Queensland Government. (2014). Reef water quality protection plan – History. Retrieved 1 March 2016 from http://www.reefplan.qld.gov.au/about/history/

Rosson, M.B., & Carroll, J.M. (2002). *Usability engineering: Scenario-based development of human-computer interaction.* San Francisco, CA: Morgan Kaufmann.

Rhodes, R.A.W. (1997). *Understanding governance: Policy networks, governance, reflexivity and accountability.* Buckingham: Open University Press.

Scharin, H., Ericsdotter, S., Elliott, M., Turner, R.K., Niiranen, S., Blenckner, T., … Rockström, J. (2016). Processes for the sustainable stewardship of marine environments. *Ecological Economics, 128,* 55–67.

Scott, D., Hall, C.M., & Gössling, S. (2016). A report on the Paris Climate Change Agreement and its implications for tourism: Why we will always have Paris. *Journal of Sustainable Tourism, 24*(7), 933–948.

Shafy, S. (2013). "Death by a thousand cuts": Coal boom could destroy Great Barrier Reef. *Spiegel International.* Retrieved 13 March 2014 from http://www.spiegel.de/international/world/australia-debates-how-to-protect-the-great-barrier-reef-a-900911.html

State of Queensland. (2015). *Full interim report, December 2015.* Prepared by The Great Barrier Reef Water Science Taskforce, and the Office of the Great Barrier Reef, Brisbane: Department of Environment and Heritage Protection.

The Greens. (2015). Abbott vs. Newman in race to destroy Queensland's environment. Retrieved 5 March 2016 from http://christine-milne.greensmps.org.au/sites/default/files/greens_-_abbott_vs_newman_on_qlds_environment.pdf

Tourism and Events Queensland. (2015). Tropical Nth Qld regional snapshot. Retrieved 20 April 2016 from http://cdn.queensland.com/~/media/DB2D0BCFC9C8484DBD2E2305A33E65A6.ashx?vs=1&d=00010101T000000

Tourism Research Australia. (2015). Research. Retrieved 20 February 2017 from www.tra.gov.au

Tribe, J., & Liburd, J.J. (2016). The tourism knowledge system. *Annals of Tourism Research, 57,* 44–61.

UNESCO. (1972). *Convention concerning the protection of the world cultural and natural heritage.* Paris: UNESCO.

UNESCO. (2007). *Discussion on the outstanding universal value.* Retrieved 12 January 2016 from http://whc.unesco.org/archive/2007/whc07-31com-9e.pdf

UNESCO World Heritage Centre. (2012). Operational guidelines for the implementation of the World Heritage Convention. Retrieved 1 January 2016 from http://whc.unesco.org/archive/opguide12-en.pdf

Urry, J. (1995). *Consuming places.* London: Routledge.

Waligo, V., Clarke, J., & Hawkins, R. (2015). Embedding stakeholders in sustainable tourism strategies. *Annals of Tourism Research, 55,* 90–93.

Whitelaw, P.A., King, B.E.M., & Tolkach, D. (2014). Protected areas, conservation and tourism – financing the sustainable dream. *Journal of Sustainable Tourism, 22*(4), 584–603.

Developing a typology of sustainable protected area tourism products

Elias Butzmann and Hubert Job

ABSTRACT
Political, socio-economic and environmental changes are creating demands for protected areas (PAs) to fulfill a double mandate of both "protection" and "use". An appropriate mix of tourism products in PAs could help fulfill those demands. The conceptual framework of the Product-based Typology for Nature-based Tourism (PTNT) was developed and tested to identify and monitor suitable tourism products and users. The typology was developed in a deductive approach and empirically tested for the first time in this study of Berchtesgaden National Park (Germany). Two methodological approaches are used: first, a demand-sided approach to the motives and activities of 1092 overnight visitors in a latent class analysis to identify six tourism product clusters. Second, several common sense supply-side-defined tourism products are identified and profiled. All products are described by the motivations and attitudes of their users towards the environment and to sustainable tourism. One product category of "structured ecotourism" is identified, which seems to have the highest potential to help PAs fulfill their double mandate. The results are used to discuss an adaptation of the PTNT for sustainable protected area tourism products. Greater market knowledge, and its skilled use, could help PA managements fulfill the double mandate of PAs.

Introduction

Protected areas (PAs), as defined by the International Union for Conservation of Nature (IUCN), especially categories II and V, have a double mandate of both "protection" and "use" (Dudley, 2008; Mayer, Müller, Woltering, Arnegger, & Job, 2010). Due to political, socio-economic and environmental changes, they are increasingly required to fulfill those double mandates (Becken & Job, 2014). Under the protection mandate, PAs are regarded as a major instrument to stop the loss of global biodiversity. And they are also asked to play an important role in the adaption and mitigation processes combating climate change (CBD, 2010a, 2010b). At the same time, PAs in IUCN categories II and V are meant to fulfill their "use" mandate by offering spaces for recreation and tourism (Dudley, 2008; Job & Paesler, 2013; Mayer & Job, 2014). Tourism to PAs is becoming a global "major tourism activity" (Eagles, 2014, p. 529) or even a mass phenomenon (Weaver, 2001, p. 104). PAs worldwide receive about 8 billion visits per year, accounting for direct tourist expenditure of US $600 billion as extrapolated by Balmford et al. (2015).

Supplemental data for this article can be accessed 🔗 here.

40

While fulfilling their "use" mandate through tourism, PAs need to ensure that this is in line with their "protection" objectives, by managing visitor use at a level which will not cause "significant biological or ecological degradation to the natural resources" (Dudley, 2008, p. 16). Furthermore, PAs (especially National Parks (NPs) of IUCN category II) are often designated in peripheral areas, and subsidized with public funds to promote tourism in order to contribute to regional economic development. In times of international debt crisis, putting public investments under higher scrutiny, parks are often (at least indirectly) required to justify their returns on investment. Due to these economic pressures, the development of commodifying nature is forced upon many PAs. One way to fulfill these necessities, without betraying the protection mandate, is to develop sustainable tourism products. However, tourism products in PAs have many different forms and types (Arnegger, Woltering, & Job, 2010). As claimed for tourism destinations in general (McKercher, 2005, p. 100), PA destinations can be seen as a product class, consisting of different product lines, groups or clusters. And these tourism products can differ concerning their economic and ecological impacts. From the perspective of tourism, one possible way to fulfill the double mandate is to focus on groups of products (Weaver, 2000), which in combination can create positive tourism experiences for tourism markets, support regional economic development and are in line with the conservation of nature and "traditional" landscapes. To achieve this, sustainable tourism products in PA destinations need to be identified and monitored accordingly.

In this study, the Product-based Typology for Nature-based Tourism (PTNT) is used as a conceptual framework for the development of such monitoring work (Arnegger et al., 2010). The approach of the PTNT is unique in the context of nature-based tourism as it combines the demand- and supply-side view of tourism products in one framework to segment ideal product types. The PTNT has been developed using a deductive approach and has not been empirically applied so far (Butzmann & Schamel, 2014).

In this study, the PTNT is applied in a case study in Berchtesgaden National Park (Germany). Currently, in Germany, there are 16 NPs seeking to fulfill the double mandate. They play an essential role in achieving the National Strategy on Biodiversity of the German Government, and, reacting to that challenge, it is planned to have 20 NPs by 2020 (BMU, 2007, p. 28; Job, 2010). Furthermore, they are required to contribute to regional economic development. Studies show that German NPs annually count over 50 million visitor-days generating a gross revenue of about 2.1 billion Euros (Job et al., 2008; Mayer et al., 2010; Metzler, Woltering, & Scheder, 2016; Woltering, 2012). But these studies do not differentiate between tourism products offered in park destinations that are in line with the protection objective or if these are products that are especially interesting from a regional economic perspective. Furthermore, the international debate about ecotourism segments and products (Fennell & Weaver, 2005; Perkins & Grace, 2009; Weaver, 2005) is not taken into account.

The following goals will be pursued in this study:

1. The overall aim is to develop and empirically apply and operationalize the PTNT, by finding distinct products that can be matched with the typology framework. For this purpose, in the first section of this paper, the PTNT framework will briefly be presented and definitions of the tourism product and concepts for its operationalization from the demand side and the supply side will be discussed. Building on that work, a methodological framework for applying the typology will be presented, as tested in the case study of Germany's Berchtesgaden National Park.
2. A further aim of the study is to test the derived operationalization of the PTNT framework, to find if it is able to identify sustainable tourism and ecotourism products that can foster the fulfillment of the double mandate of PAs. For this purpose, three factors measuring attitudes towards sustainable tourism and ecotourism are derived from the literature, and used to differentiate the identified tourism products. Furthermore, tourist's gross expenditure will be explored to differentiate tourism products concerning their potential to contribute to regional economic development.

3. Based on the results, their implications for further developing the PTNT to create a typology for sustainable protected area tourism products will be considered. Practical implications for park management are then discussed.

Conceptual background

The PTNT framework and possible operationalizations

In the PTNT, nature-based tourism products are differentiated in a matrix using two dimensions. The first (supply-side) dimension of the matrix, "individuality" describes the form of tourism, defined by the service arrangement of the products. The dimension differentiates four ideal type categories: "independent", "à la carte", "customized" and "fully standardized". The second (demand) dimension, namely "nature as a point of attraction", looks on tourism products by travel motivations and activities. It consists of four ideal type categories: "nature protection", "nature experience", "sports and adventure" and "hedonistic". So, as this dimension implies, tourism products can be lined up according to the role nature has played in user's destination choice (Arnegger et al., 2010).

Hence, with four categories in each dimension, 16 ideal types of nature-based tourism products can be defined. However, empirically derived products can be "blurred" and not necessarily be located in only one category (Arnegger et al., 2010, p. 922). Furthermore, different products of the ideal types can be offered simultaneously in any one PA destination. And the portfolio of products can vary, depending on the natural setting, size and location of the PA.

The approach of the PTNT is unique in the context of nature-based tourism, as it combines the demand- and supply-side view on tourism products in one framework to segment ideal product types. Earlier typology and segmentation approaches predominantly focus on the demand side (Beh & Bruyere, 2007; Kerstetter, Hou, & Lin, 2004; Marques, Reis, & Menezes, 2010; Mehmetoglu, 2007; Palacio & McCool, 1997; Smith, Tuffin, Taplin, Moore, & Tonge, 2014) or just on the supply side (Perkins & Grace, 2009; Sirakaya, Sasidharan, & Sonmez, 1999). There is only one typology framework in the context of nature-based tourism that combines demand- and supply-side aspects, namely the model of comprehensive and minimalist manifestations of ecotourism (Weaver, 2005). The model is based on prior conceptualizations of hard, structured and soft ecotourism (Weaver & Lawton, 2002). But compared to the conceptual framework of the PTNT, it only defines the bipolar ends of two continuums (comprehensive vs. minimalist; hard vs. soft) and not ideal-type categories. Furthermore, it does not explicitly focus on tourism products.

In the two-dimensional PTNT framework, the demand-side (total view) and supply-side perspectives (specific view) on tourism products are brought together (Smith, 1994). The "à la carte", "customized" and "fully standardized" categories of the service arrangement dimension resemble the specific view of tourism products, as they can be defined via intermediate outputs sold to a tourist by tourism companies or destinations as specific tourism products (Middleton, 1989, p. 573; Smith, 1994). The "independent" category, on the other side, needs to be defined from a demand perspective, using travel motivations and activities that create the "tourist experience" (Smith, 1994, p. 591), the final output of a tourism product or the "mental construct" of the tourism product in the eyes of the traveler (Middleton, 1989, p. 573). Middleton (1988, p. 79) further emphasizes the role of activities and defines the total tourism product as a bundle of "tangible and intangible components, based on activity at a destination".

A methodological distinction between empirical segmentation frameworks in tourism is one of common sense (a priori) and data-driven (a posteriori) segmentations, or combinations of both (Dolnicar, 2004). Common sense segmentations are often found in combination with the supply-side view on tourism products and data-driven segmentations with the demand-side view. Examples for the first are Eagles (1992), Wight (1996) or Perkins and Grace (2009). Examples for the latter are studies using data-driven segmentations to derive visitor segments from a demand-side view. Even though authors of these studies do not explicitly define tourism products, they implicitly use the demand-side approach to do so. Several studies use the Recreation Experience Preference Scales (REP scales) (Manfredo, Driver, & Tarrant, 1996) as a segmentation base (i.e. Beh & Bruyere, 2007; Kerstetter et al., 2004; Marques et al., 2010; Palacio & McCool, 1997). Other studies use activities (Mehmetoglu, 2007) or a combination of both (Smith et al., 2014).

Referring to the demand-side definitions describing the tourism product (Middleton, 1988, p. 79; Smith, 1994, p. 591), both motivations and activities need to be taken into account. But there is evidence that activities are better suited for that purpose. Moscardo, Morrison, Pearce, Lang, and O'Leary (1996) claim that potential activities in a destination are easier to match with experience preferences in the destination choice process of tourists than motivations. Therefore, segmentations by activities are better suited for destination marketing, as they give evidence of the available activity-products in the destination. Further studies conclude that activity-based segmentations produce more distinct patterns (Hvenegaard, 2002, p. 15) and are more appropriate for the tourism industry than motivation-based approaches (Mehmetoglu, 2007, p. 658).

However, for further describing tourism products, motivations and experience preferences are also needed, especially to operationalize the demand dimension "nature as a point of attraction". There are several studies categorizing nature-based tourism activities or products by the increasing interest in nature. For example, Juric, Cornwell, and Mather (2002) designed an Ecotourism-Interests Scale (EIS) consisting of seven "social" and "attraction" motivation items (Eagles, 1992), which could explain participation in ecotourism activities. Perkins and Grace (2009) used the same scale to differentiate specific tourism products defined from a supply-side view. Mehmetoglu (2010) designed a Nature-Based, Eco- and Sustainable Tourists (NES) scale, which consists of three distinct scales, to differentiate market segments.

Besides studies measuring interest in nature using scales, there are also attempts to cover this by studying single items. One example is the construct of "national-park affinity". The first operationalization of this construct was developed by Küpfer (2000). Further studies applied or slightly modified the concept (Arnberger, Eder, Allex, Sterl, & Burns, 2012; Job et al., 2009; Mayer et al., 2010; Müller & Job, 2009). For example, Mayer et al. (2010) attest visitors of PAs a high "national-park affinity", if they answer affirmatively to the following three successive questions: "Do you know whether this area enjoys any special protection?", "Do you know whether this region is a national park?" and "How important was the existence of the national park in your decision to come to this region?" with the response "important" or "very important".

The concept of "national-park affinity" and items and methodologies of the described scales give impetus to develop a scale operationalizing the dimension "nature as a point of attraction".

Sustainable nature-based-tourism (ecotourism)

Sustainable tourism products can also be distinguished by supply-side and demand-side approaches. With the latter, tourism products can be described by the attitudes and behaviors of its users, leading to their definition as sustainable tourists. Building on that definition, selective marketing strategies can be developed for favorable products to foster the sustainable development of destinations (Dolnicar & Leisch, 2008). This study follows that approach, as it is consistent with the demand-side description of tourism products with "nature as a point of attraction".

Ecotourists can be described by three core criteria. First, they visit nature-based attractions to appreciate nature or to be in nature. Second, they are interested to learn about nature and, third, they are more environmentally conscious and want their holiday to be ecologically and socio-culturally sustainable (Fennell & Weaver, 2005, p. 373). The first two criteria can be measured by the motivations and activities outlined above. To differentiate what essentially defines a sustainable tourist or the sustainability criteria of ecotourists, several theoretical concepts can be used; for example, by looking at the tourist's general environmental attitude and beliefs (Zografos & Allcroft, 2007), moral obligations and personal environmental norms and values (Dolnicar & Leisch, 2007; Perkins & Brown, 2012), environmental concern and intended behavior (Fairweather, Maslin, & Simmons, 2005;

Kerstetter et al., 2004), and their actual environmental behavior in the tourism context (Dolnicar, 2010). The literature reveals that the above concepts are strongly associated: tourist's general environmental attitudes and beliefs, as well as moral obligations and personal norms and values, influence their environmental concern and intended behavior, which finally determine actual environmental behavior (Ajzen, 2011; Mehmetoglu, 2010).

The concept of environmental attitudes and beliefs has typically been operationalized by the New Environmental Paradigm (NEP) scale (Dunlap, Van Liere, Mertig, & Jones, 2000), which is strongly associated with pro-environmental intended and actual behavior. The scale consists of 15 items and has been used to position people along a continuum of anthropocentrism and eco-centrism. Several studies evaluate this relationship in the context of NPs and biodiversity (i.e. Floyd, Jang, & Noe, 1997; Müller & Job, 2009). Furthermore, there are studies in more general tourism contexts successfully applying the NEP scale to identify sustainable tourists (Fairweather et al., 2005; Mair, 2011). While some studies show that the NEP scale has its limitations (Beaumont, 2011; Zografos & Allcroft, 2007), most studies confirm the high validity of the NEP scale to measure environmental attitudes and beliefs and their relationship to behavioral intentions and actual behavior.

Different from general environmental attitudes, there does not exist one prominent scale to measure attitudes towards sustainable tourism or intended sustainable behaviors in the tourism context. But several operationalizations have been applied that reveal certain congruities (Fairweather et al., 2005; FUR, 2012; Mehmetoglu, 2010; Perkins & Brown, 2012). Commonly, the general concern about the environmental impacts of travel is queried (Mehmetoglu, 2010). Similarities also exist concerning items used for measuring the social pillar of sustainability. Other important aspects are the perceived importance of eco-labels and environmental accreditation systems (Perkins & Brown, 2012) and the willingness to pay more for a sustainable holiday (Fairweather et al., 2005; Mehmetoglu, 2010).

Methods

Study area

To pursue the aims of this study, Berchtesgaden National Park was chosen as a case study area. This IUCN category II park is the only NP in the German Alps. It is located in southeastern Germany, extends over 20,000 ha and offers a huge variety of activity opportunities covering the spectrum of the PTNT. Historical attractions close to the NP and natural attractions within its boundaries like the Watzmann peak (2713 m) and its surrounding mountains attract hiker and mountaineers and general national and international leisure tourists. The flora and fauna of the NP is appealing to nature experience tourists and volunteers to engage in wildlife viewing or nature protection activities. Berchtesgaden National Park had nearly 1.6 million visitor-days in 2014 generating a gross revenue of more than 90 million Euros; in 2002, the number of visitor-days was 1.1 million (Job, Merlin, Metzler, Schamel, & Woltering, 2016; Job, Metzler, & Vogt, 2003). As for the other German NPs – so far – no information concerning tourism products or visitor segments has been gathered in a comprehensive way. Furthermore, it is not known, if tourism, in total or of specific segments/products, is in line with the protection objective or ecotourism criteria.

Framework for empirically testing and questionnaire design

In this study, the PTNT framework will be operationalized in the context of park tourism by two approaches. First, tourism products of the total level will be derived by data-driven segmentation based on activities and nature travel motivations of PA visitors. The identified products will be further divided by a common sense segmentation into "independent", "customized" and "fully standardized" products. Second, products of the "à la carte", "customized" and "fully standardized" category will be defined from the supply side on a common sense basis as specific products offered by tourism companies or PA management. All derived products will then be characterized by the demand dimension "nature as a point of attraction" as well as further travel motivations, sociodemographic indicators, environmental attitudes and attitudes towards sustainable tourism.

The data-driven segmentation of the demand side will be done by activities and nature-related travel motivations. The activity-item list consisted of 20 activities, derived from Weaver and Lawton (2002), Palacio and McCool (1997), Mehmetoglu (2007), Marques et al. (2010) and FUR (2012). The items can be theoretically matched with the four discrete categories of the typology (see Table 1). Tourists (n = 1410) were then asked which activities they were pursuing in the PA destination and how important these activities were for them (using a five-point importance scale). Furthermore, motivation items were derived from Eagles (1992), Juric et al. (2002) and Mehmetoglu (2010), measured on a five-point scale from "I totally agree" to "I do not agree at all" for further describing the

Table 1. Items for empirical testing the product-based typology for nature-based tourism.

Motivation/activity categories of the product-based typolgy for nature-based tourism (Arnegger et al., 2010)			
Nature protection	Nature experience	Sports and adventure	Hedonistic
Activity items (used for clustering)			
- Taking part in nature protection activities (erosion control, wildlife monitoring, etc.) - Using offers for environmental education (i.e. excursions, visitor information centers, etc.) - Taking part in guided hikes for environmental education	- Visiting the national park - Taking photos of wildlife and landscape - Visiting natural attractions	- Trekking - Mountaineering - Cycling - Guided hikes with a mountain guide - Doing intensive sports (cardio, trailrunning, etc.) - Mountainbiking - Hiking	- Using wellness facilities - Making a shopping trip - Going for a walk - Visiting recreational/leisure facilities - Visiting historic and cultural attractions - Visiting special events (cultural events, concerts, events, etc.) - Doing light sporting activities (nordic walking, swimming, running)
Nature-based motivation items (used for clustering)			
- Studying nature - Learning about nature	- Seeing the wilderness and the undisturbed nature - Choosing the destination because of nature - Informing about nature and landscape of the destination before travel		
Recreation experience preference items (used for description/validation of clusters)			
- ...to study nature - ...to learn about nature	- ...to be close to nature - ...to watch the scenery	- ...to increase my endurance/my sporting skills - ...to stay physically fit - ...to have something to talk about - ...to experience an adventure - ...to experience something new	- ...to recover physically - ...to recover mentally - ...to escape the daily routine - ...to get well entertained/to just have fun - ...to get to know new people/to make new contacts - ...to relax and to laze around - ...to be with family and friends
À la carte products (examples)			
- Guided hikes by park rangers	- National park information center "Haus der Berge"	- Guided climbing/mountaineer tours - Guided mountainbike tour	- Königssee-shipping - Watzmann thermal spring - Eagle's Nest
Customized and fully standardized products			
- Excursions of the national park for nature study/protection	- Excursions of the national park for wildlife/nature watching	- Package mountaineer holidays	- Package bus trips or general leisure package holidays

"nature as a point of attraction" dimension of the defined products. Additionally, the construct of "national-park affinity" was integrated (Job et al., 2009; Mayer et al., 2010).

Supply-side-defined products had been selected on a common sense basis. Three standardized/ customized products had been defined: two types of "excursions of the national park" for environmental protection and experience, "package mountaineer holidays", "package bus trips" and "general leisure package holidays". The selection was based on an open-ended question, capturing the name of the travel agency and by an a priori selection through sampling procedure. Additionally, several à la carte products were listed in the questionnaire. So, if the assumptions of the typology framework apply for the case study, activities and benefit items of the respective category should be able to discriminate the supply-side-defined products hypothetically assigned to the respective category (see Table 1).

For further validations of the demand- and supply-side-defined products, additional background variables as items of the REP scale were included, thus capturing a more complete picture of motivations aligning with theoretical motivation categories (Crompton, 1979; Fodness, 1994) (see Table 1). Additionally, socio-demographic variables and variables concerning the travel arrangements were included in the questionnaire.

To pursue research question number two, six items of the NEP scale were used. The reduced set was chosen to capture all five theoretical facets of the NEP scale (Dunlap et al., 2000). Furthermore, three items measuring attitudes towards sustainable tourism were adopted from FUR (2012) and Mehmetoglu (2010), namely, "my holiday should be ecologically compatible, resource-efficient and environmentally friendly", "my holiday/visit should be as socially compatible as possible" and "as a tourist it is very important to me that tourism products and services are certified with an eco-label". And three items measuring behavioral intentions inquiring about the willingness to pay more for a sustainable holiday were derived from the literature (Fairweather et al., 2005; Mehmetoglu, 2010; Perkins & Brown, 2012): "I am willing to pay more money for an ecologically compatible holiday", "I am willing to pay more money for an ecological accommodation" and "I am willing to spend more money on an ecologically friendly or rather climate friendly journey (e.g. CO_2-compensation payment)". Additionally, tourist spending data for accommodation and gastronomy were captured.

Sampling procedure

Data collection was based on a combined approach of personal interviews and self-administered questionnaires. A total of 1410 personal interviews were conducted at seven locations at the main entrances to the NP on 20 sampling days (weekdays and weekends) between May and October 2014. Of the total respondents, 84% were overnight visitors, and were used as the basis for the following analysis. After the face-to-face questionnaire had been completed, the respondents were handed out a self-administered questionnaire with additional questions, including the REP, the NEP scale as well as the sustainability items. This questionnaire could be filled out at site or mailed back. Five hundred and thirty-eight self-administered questionnaires were collected from overnight visitors. To further improve the database for the analysis of supply-side-defined products as excursions and guided hikes by the NP and "package mountaineer holidays", 178 additional self-administered questionnaires had been distributed in a stratified sampling procedure with a response rate of 28%. These additional questionnaires were used to describe the a priori defined products.

Segmentation and scaling methods

For data-driven segmentation, a latent class model (LC-model) was applied (Magidson & Vermunt, 2001; Vermunt & Magidson, 2013). LC models analyze the relationships between observed variables to create an unobserved categorical (latent) variable with several discrete categories, which are basically representing latent classes. The latent variable stratifies the observed dataset so that the observed variables in each latent class become statistically independent. Twenty activity and five

Table 2. Nature interest scales (NI scales).

	Nature interest scale (five motives)	Nature interest scale (five motives and four activities)
Indicators (n = 1092)		
Motives	Factor loadings	
To learn something about nature	0.714	0.684
To study nature	0.700	0.653
To see the wilderness and the undisturbed nature	0.703	0.617
When I chose my holiday destination, nature did play a very important role	0.676	0.616
Before my vacation/stay in the region, I informed myself intensely about nature and landscape of the region	0.658	0.610
Activities		
Visiting the national park		0.676
Taking photos of wildlife and landscape		0.502
Offers for environmental education (i.e. visitor information centers)		0.474
Visiting natural attractions		0.437
Cronbach's alpha	0.724	0.757

nature-related motivation items were included in the clustering procedure (see Table 1). Cases with missing values were excluded, so a total of 1092 valid cases of overnight visitors were used for clustering. Different latent class analyses were run (one to ten clusters) based on the activity and motivation items. To ensure the detection of a global optimum, the analysis was repeated 25 times with different random initializations. A six-cluster solution was selected based on the lowest Bayesian Information Criterion (BIC) and Consistent Akaike Information Criterion (CAIC). In each repetition, the six-cluster solution showed the lowest BIC and CAIC. The classification error of this cluster model is 16%. The derived pseudo R-squared statistics (reductions of error (lambda), entropy R-squared and standard R-squared) are between 0.72 and 0.78.

For modeling the demand dimension "nature as a point of attraction", two nature interest scales (NI scales) were computed. The NI scales are based on the EIS by Juric et al. (2002) and the NES scale by Mehmetoglu (2010). The first NI scale consists of the five nature-related motives and the second of the five nature-related motives and additional nature-related activity preferences. All items were included in a principal components analysis (PCA) and load high on one factor (see Table 2).

For further describing the data-driven and common sense-defined products, three additional PCAs were conducted separately with the REP items, the NEP items as well as with the sustainability items. In each analysis, cases with missing values and items with low MSA values (measure of sampling adequacy < 0.5) indicating low correlations with other items were excluded from the analysis. The remaining items showed a statistically significant Bartlett's test of sphericity and an adequate overall MSA value. Factors were extracted based on the eigenvalue criteria (>1.0) and varimax rotation was applied. Cronbach's alpha was calculated as a reliability parameter for each factor.

Items from the REP scales load on five factors, explaining 64% of total variance (see Table 3). As hypothesized, the REP factors resemble the typology framework. The six NEP items are loading on one single factor. So, the NEP items were summated to one single scale. The sustainability items load on two factors, explaining 66% of the total variance. The first factor consists of the three items measuring the willingness to pay more for a sustainable holiday (Cronbach's alpha = 0.730), and the second factor is constituted by the items measuring the attitude towards sustainable tourism (Cronbach's alpha = 0.674).

Results

Profiles of demand-side-defined product clusters

The profiles, described by the items used for clustering, of the six identified demand-side-defined product clusters are listed in Figure S1 (found in the Supplemental data section of the web-based

Table 3. PCA of recreation experience preference (REP) scales.

Indicators (n = 453)	REP Factor 1: Recovery	REP Factor 2: Nature	REP Factor 3: Sports	REP Factor 4: Fun/relax	REP Factor 5: Adventure
			Factor loadings		
...to recover physically	0.789				
...to recover mentally	0.768				
...to escape the daily routine	0.732				
...to learn about nature		0.850			
...to study nature		0.833			
...to be close to nature		0.595			
...to watch the scenery		0.592			
...to increase my endurance/my sporting skills			0.882		
...to stay physically fit			0.869		
...to get well entertained/to just have fun				0.656	
...to get to know new people/to make new contacts				0.632	
...to have something to talk about				0.570	
... to relax and to laze around				0.502	
...to experience an adventure					0.759
...to experience something new					0.752
Cronbach's alpha	0.742	0.747	0.845	0.596	0.718
% of variance explained	24.624	13.419	10.524	8.635	6.758

Principal component analysis extracted five factors with 63.96% of variance explained; Bartlett's test of sphericity is statistically significant at the 0.01 level; KMO value of 0.733.

version of this paper). Its size and complexity prevent its reproduction in the print version of the paper. The black bars in each cluster display the average values of an item of the total sample. The gray bars show the values of the respective segment. All items show significant differences between the clusters (chi square; $p < 0.01$). All segments are described by the variables, which were used for clustering and the two NI scales (F-test; $p < 0.01$) and the concept of "national-park affinity" (chi square; $p < 0.01$). In the text product clusters are described by the percentage of tourists who stated that the activity played an important or even very important role (referred to by (very) important) in their visit. Furthermore, they are characterized by additional background variables that show significant differences between the clusters including the REP factors (the nature factor (F-test; $p < 0.01$), sports factor (F-test; $p < 0.01$), recovery factor (F-test; $p < 0.05$), adventure factor (F-test; $p < 0.05$) and (with limitations) the fun/relax factor (F-test; $p < 0.1$)), socio-demographic variables (education (chi square; $p < 0.01$) and age (F-test; $p < 0.01$)) and travel arrangements ((independent vs. packaged (chi square; $p < 0.01$), several à la carte products (chi square; $p < 0.01$), type of accommodation (chi square; $p < 0.01$) and number of overnights (chi square; $p < 0.01$)).

Furthermore, to give an answer to research question number two, product clusters are compared by the NEP (F-test; $p < 0.05$) and the two sustainability scales, measuring the general attitude towards sustainable tourism (F-test; $p < 0.01$) and the willingness to pay more for a sustainable holiday (F-test; $p < 0.1$). And products are compared by expenses for overnights (F-test; $p < 0.01$), whereas product clusters did not show significant differences concerning expenses for gastronomy.

Cluster 1: "Nature experience (hiker)" (29.8%)
Tourists of this product cluster are interested in activities in the nature experience category such as seeing the wilderness and undisturbed nature (76.6% (very) important) or taking photos of wildlife and landscape (55.7% (very) important). The same applies for hiking (83.7% (very) important). Other activity-participation rates are comparatively low. Nature-orientation is displayed by an NI scale (five motives) slightly above average ($z = 0.09$). Because of the general high interest in nature compared to other activities, this cluster can be positioned in the "nature experience" category of the typology. The "recovery" factor is the only REP factor showing a loading above average ($z = 0.08$). Hence, most of the tourists in this cluster visit the area to be in nature in order to recover physically and mentally.

Concerning socio-demographic variables, travel arrangements and sustainability aspects tourists of the "nature experience (hiker)" product type show rather averaged values.

Cluster 2: "Hedonistic (high activity)" (16.5%)

Tourists in this cluster also have comparatively a high general interest in nature experience activities, but show less interest in learning or studying nature. This leads to an NI scale (based on motives) with the second lowest score ($z = -0.35$) compared to the other clusters while the second NI scale with activities is rather average ($z = -0.07$). Additionally, they have above-average participation rates in activities of the hedonistic category: for 42% of them "making a shopping trip" and for 29% "visiting recreational and leisure facilities" play a (very) important role (compared to 18.6% and 10.1% of the total sample). Concerning the REP scales, the factors "fun/relax" ($z = 0.15$) and "recovery" ($z = 0.19$) show positive scores. The first one has the highest score compared to all other clusters. All in all, this cluster can be positioned in the "hedonistic" category. Concerning socio-demographic variables and travel arrangements tourists of the "hedonistic (high activity)" product type show rather average values. Concerning the sustainability items, it is obvious that all are below average (NEP score: $z = -0.03$; willingness to pay more for a sustainable holiday: $z = -0.23$; attitude towards sustainable tourism: $z = -0.13$).

Cluster 3: "Special nature experience" (ecotourists) (16.3%)

Tourists of this product cluster are the only NP visitors with motivations and activities above average in the nature protection category. However, the participation rate in nature protection activities is still rather low (4.5% compared to 0.8% in the total sample). But to learn about nature is very important for them: almost 70% totally agreed that this motivational factor influenced them to visit the destination (17.7% on average). They take part, above average, in environmental education (41.6% (very) important compared to 11.6% on average). They have the highest probability of taking part in guided hikes organized by the PA (10.7% compared to 4.0% on average) and almost 70% of them visited the main visitor information center of the park, compared to about 40% of the total sample. Tourists of the "special nature experience" product type score highest on the NI scales ($z = 1.24$ (motives); $z = 1.39$ (motives and activities)). They consistently show a high loading on the "nature" ($z = 0.84$) REP factor and this cluster has the highest share of visitors with a high "national-park affinity" with almost 70% compared to an average of about 28% (Job et al., 2016). They are slightly older than average (54.6 compared to the 50.6 years of the total sample). And they show above-average values on all sustainability scales (NEP score: $z = 0.14$; willingness to pay more for a sustainable holiday: $z = 0.15$; attitude towards sustainable tourism: $z = 0.32$). In sum, this cluster predominantly resembles the "nature protection" and "nature experience" category of the typology framework. And as tourists of the product type are interested in learning about nature and are more sustainable than average, they can also be termed "ecotourists".

Cluster 4: "Nature experience (sports)" (14.6%)

Tourists of this product type have a rather high general interest in nature. Seeing the wilderness and undisturbed nature and visiting natural attractions were (very) important for 88.7% and 71.1%, respectively. So, their NI scale scores are comparably high ($z = 0.25$ (motives); $z = 0.29$ (motives and activities)). Parallel to this cluster, tourists are more interested in sporting activities: for about 20%, biking or mountain-biking, and, for almost 40%, mountaineering, plays a (very) important role during their holiday. This pattern is also reflected by the REP scales, with rather high loadings on the "sports" ($z = 0.36$) and "adventure" ($z = 0.27$) scales. Concerning travel arrangement, a rather high percentage of that group stays in holiday flats, overnights in mountain huts and takes part in guided climbing/mountaineer tours. Furthermore, they are comparably young (mean = 46.7 compared to 50.6 in the total sample). Concerning sustainability aspects, they reveal a rather disperse pattern with a NEP score slightly above ($z = 0.14$) and the other two around average ($z = 0.06$; $z = -0.09$). As the cluster is a rather hybrid type, it can be placed in between the "nature experience" and "sports and adventure" category.

Cluster 5: "Hedonistic (low activity)" (14.2%)

Tourists of this product type have relatively low activity preferences. The only activities that are comparably high are hedonistic interest activities such as making a shopping trip (20% (very) important) and "to go for a walk" (50% (very) important). The last named activity played a major role compared to hiking, the only case for this cluster and the "hedonistic (high activity)" cluster. Consequently, tourists of this cluster also have the lowest NI scores ($z = -1.46$ (motives); $z = -1.53$ (motives and activities)) and 92% can be described as visitors with a low "national-park affinity". Concerning the recreation experiences, the "fun/relax" factor is the only factor that is above average ($z = 0.26$). The "nature" factor shows the lowest score ($z = -0.85$). Overall, this cluster can be positioned in the "hedonistic" category. Regarding travel arrangements, tourists of the "hedonistic (low activity)" product category predominantly stay overnight in hotels (46% compared to 25% in the total sample). So, their expenses for accommodation are slightly above average (44.0€ compared to 37.7€ in the total sample). And "hedonistic (low activity)" tourists typically have sustainability scale values below average (NEP score: $z = -0.39$; willingness to pay more for a sustainable holiday: $z = -0.33$; attitude towards sustainable tourism: $z = -0.36$).

Cluster 6: "Mountaineers" (8.7%)

The smallest cluster of all shows a relatively clear activity pattern, dominated by trekking (61.6%) and mountaineering (50.5%) ((very) important). This is also reflected in the REP scales with high loadings on the factors "sports" ($z = 0.50$) and "adventure" ($z = 0.36$). Hence, tourists of this product cluster clearly belong to the "sports and adventure" category. Mountaineers mostly travel independent and predominantly stay overnight in mountain huts (about 50%). Hence, they also show the lowest mean daily expenditure for accommodation (24.5 €). Compared to all other clusters, tourists in this cluster are the youngest (mean = 43.7 years) and the best educated, with 61% having a university degree (compared to 44% in the total sample). Concerning sustainability aspects, typical mountaineers show above-average values on all indicators (NEP score: $z = 0.28$; willingness to pay more for a sustainable holiday: $z = 0.10$; attitude towards sustainable tourism: $z = 0.21$).

The data-driven segmentation, performed to identify product clusters, was followed by a simple common sense segmentation differentiating the products further into "fully standardized" and "customized" on the one side and "independent" product category on the other side. Clusters 1, 2 and 3 have a rather averaged portion of packaged and customized travel arrangements (7.2%, 8.5% and 9.2% compared to 8.1% on average), whereas the "sports and adventure" clusters 4 and 6 have the highest portion of independent travelers (94.2% and 96.8%). Cluster 5, the "hedonistic (low activity)" cluster, has the highest percentage of package tourists (15.4%)). Dependent on relative product sizes and activity and motivational characteristics, the product clusters can be positioned in the typology framework displayed in Figure 1 (in Figure 1, some definitional and terminological changes are applied to the typology, which will be elaborated later). The "fully standardized"/"customized" products can further be regarded form the supply side, which is done in the following section.

Profiles of supply-side-defined products

In Table 4, several "fully standardized"/"customized" products and "à la carte" products are listed, following the logic of the specific view on tourism products. Due to the small number of respondents, the two types of NP excursions had to be merged into one category covering the "nature protection" and the "nature experience" categories. That the excursionists belong to the "nature protection" and "nature experience" categories is revealed by their activity patterns and REP factors similar to the demand-side-defined ecotourism product cluster. About 90% agree that "to learn about nature" was a motive to visit the NP destination and 72.8% claim that visiting natural attractions plays a (very)

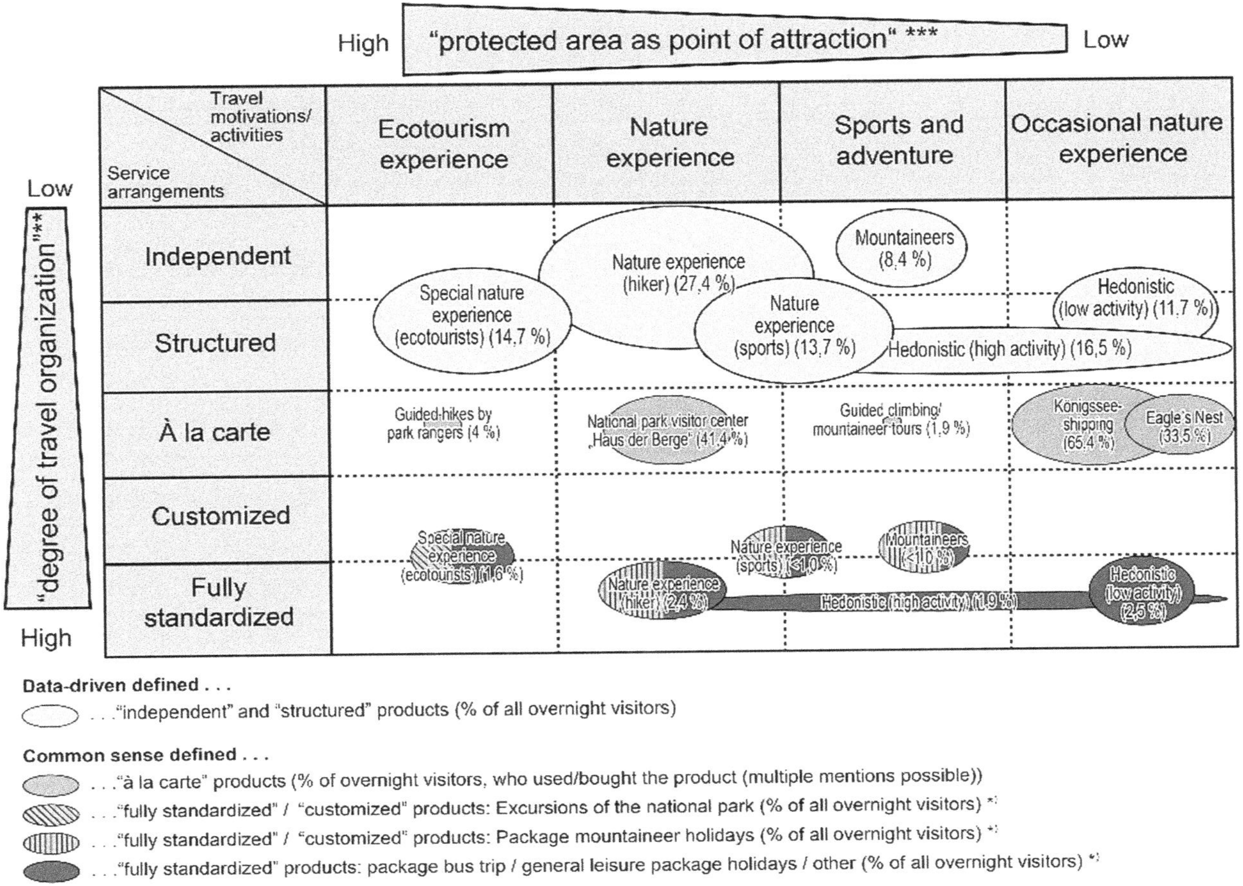

51

A typology diagram (rotated). Column headers (Travel motivations/activities): Ecotourism experience, Nature experience, Nature experience, Sports and adventure, Occasional nature experience.

Row headers (Service arrangements): Independent, Structured, À la carte, Customized, Fully standardized.

Axis labels: "protected area as point of attraction" *** (High to Low); "degree of travel organization" ** (Low to High)

Ellipse labels:
- Special nature experience (ecotourists) (14.7 %)
- Nature experience (hiker) (27.4 %)
- Nature experience (sports) (13.7 %)
- Mountaineers (8.4 %)
- Hedonistic (high activity) (16.5 %)
- Hedonistic (low activity) (11.7 %)
- Königssee-shipping (65.4 %)
- Eagle's Nest (33.5 %)
- Guided hikes by park rangers (4 %)
- National park visitor center "Haus der Berge" (41.4 %)
- Guided climbing/ mountaineer tours (1.9 %)
- Special nature experience (ecotourists) (4.6 %)
- Nature experience (hiker) (24.9 %)
- Nature experience (sports) (5.0 %)
- Mountaineers (< 1.0 %)
- Hedonistic (high activity) (19.3 %)
- Hedonistic (low activity) (25.5 %)

Data-driven defined . . .
. . . "independent" and "structured" products (% of overnight visitors)

Common sense defined . . .
. . . "à la carte" products (% of overnight visitors, who used/bought the product (multiple mentions possible))
. . . "fully standardized" / "customized" products: Excursions of the national park (% of all overnight visitors) *)
. . . "fully standardized" / "customized" products: Package mountaineer holidays (% of all overnight visitors) *)
. . . "fully standardized" products: package bus trip / general leisure package holidays / other (% of all overnight visitors) *)

*) Product segments are derived from the data-driven defined segments on a common sense basis by displaying the proportion of "fully standardized" and "customized" travel arrangements. Hence the sum of all "independent"/"structured" and "customized"/"fully standardized" adds to 100 % of all overnight visitors. The shadings imply, which common sense/ supply-side defined "customized"/"fully standardized" products were used. The sizes of the ellipses are schematic and do not mirror the exact percentages of the segments.
**) The extent to which service arrangements are used increases
***) Relevance of the protected area for the product decreases

Data: overnight visitors of summer season 2014 (n = 1092)
Source: Adapted from Amegger et al. (2010); own calculations. Draft: E. Butzmann, Department of Tourism, MUAS; H. Job, Institute of Geography and Geology, JMU Würzburg, 2016

Figure 1. A typology for "sustainable protected area tourism products" with empirically derived products in Berchtesgaden National Park.

Table 4. Common sense, a priori defined "fully standardized"/"customized" products and "à la carte" products/intermediate outputs.

Motivation/activity-categories of the product-based typolgy for nature-based tourism (Arnegger et al., 2010)	Fully standardized or customized					à la carte/intermediate outputs[a]					
Product category	Nature protection/ experience	Sports and adventure	Hedonistic		Test	Nature protection/ experience		Sports and adventure		Hedonistic	
Product name	Excursions of the national park	Package mountaineer holidays	package bus trip /general leisure package holidays	other package holidays		Guided hikes by park rangers	National park information center "Haus der Berge"	Guided climbing/ mountaineer-tours	Königssee-shipping	Eagle's Nest	Watzmann thermal spring
	z-Score (n = 11)	z-Score (n = 25)	z-Score (n = 53)	z-Score (n = 34)	F-test	z-Score (n = 63)	z-Score (n = 429)	z-Score (n = 26)	z-Score (n = 626)	z-Score (n = 331)	z-Score (n = 181)
Nature interest scales (NI scales)											
NI scale (five motives)	1.10	−0.22	−0.46	0.05	7.49***	0.66***	0.24***	−0.12	0.06**	0.13***	0.07
NI scale (five motives and four activities)	1.28	−0.35	−0.45	0.10	9.15***	0.81***	0.33***	−0.17	0.13***	0.25***	0.13
REP factor loadings	z-score (n = 7)	z-score (n = 22)	z-score (n = 16)	z-score (n = 12)	F-test	z-score (n = 37)	z-score (n = 156)	z-Score (n = 13)	z-score (n = 352)	z-score (n = 114)	z-score (n = 75)
REP Factor1: Recovery	−0.62	−0.62	−0.49	−0.14	0.55	−0.06	0.12	−0.46*	0.06	0.10	0.15
REP Factor2: Nature	1.43	−0.16	−0.35	−0.03	5.32***	0.76***	0.28***	−0.15	0.00	0.06	−0.01
REP Factor3: Sports	−0.50	0.94	−0.43	−0.62	10.29***	−0.38**	−0.07	0.38*	−0.12**	−0.22*	−0.02
REP Factor4: Fun/relax	−0.21	−0.01	0.70	0.01	2.38*	0.17	−0.01	−0.19	0.02	0.04	0.20
REP Factor5: Adventure	−0.03	0.59	−0.30	−0.20	3.17**	−0.05	−0.01	0.28	0.12**	−0.05	−0.15
NEP scale	1.12	−0.11	−0.49	−0.41	5.48***	0.19	0.04	0.05	−0.03	−0.01	0.14
Intended behavior to pay more for a sustainable holiday	1.24	0.57	−0.24	−0.09	5.01***	0.09	0.02	−0.11	−0.02	−0.16	0.09
General attitude towards sustainable tourism	0.91	0.08	−0.21	−0.65	2.57*	0.14	0.11	−0.08	0.00	0.06	0.00

[a]Test statsitic is based on t-tests comparing visitors who used the respective product with non-users.

$*p < 0.1; **p < 0.05; ***p < 0.01.$

important role during their vacation. As a consequence, participants of the NP excursions show the highest NI scores compared to all other products. Looking on environmental attitudes and attitudes towards sustainable tourism, the excursionists also show the highest scores (see Table 4).

The "package mountaineer holidays" tourists show similar activity-participation rates to the demand-side-defined mountaineer's product cluster, with trekking (96.2%) and mountaineering (38.5%) as the most important activities. And they load highest on the "sports" and "adventure" REP factors. The NI scores are comparatively low, placed between the excursionists and the "general leisure package holidays" tourists. Interestingly, concerning the sustainability scales, users of "package mountaineer holidays" score rather high on the willingness to pay more for a sustainable holiday, whereas they score rather average on the other two scales.

"General leisure package holidays" tourists have similar interests as the demand-side-defined "hedonistic" product categories, also loading comparatively high on the "fun/relax" REP factor. Furthermore, they have rather low NI scores and the lowest scores on environmental attitudes and attitudes towards sustainable tourism in general.

Looking at the "fully standardized"/"customized" products defined from the supply side as specific tourism products, the hypothesized pattern of the typology is displayed. They can be matched with the demand-side-defined product clusters and assigned respectively as depicted in Figure 1. The sizes of the product clusters are derived from the representative demand-side assessment of visitors. For example, 9.2% of the ecotourism products can be classified as "fully standardized"/"customized" products. In total, they account for about 1.6% of all overnight visitors. One-third of them took part in NP excursions; two-thirds bought a general leisure holiday package. Another example is the demand-side-defined "fully standardized"/"customized" mountaineers product category with less than 1% of all visitors. More than half of them bought their product from a supplier specialist on "package mountaineer holidays".

"À la carte" products can also be placed in the PTNT framework (see Table 4 and Figure 1). For example, participants in "guided hikes by park rangers" or visitors of the "national park information center" are more interested in nature, expressed by comparatively high loadings on both NI scales, whereas participants in "guided climbing/mountaineer tours" are more focused on sports, shown by high loadings on the sports factor. More general products such as the "Königssee-shipping" or the "Watzmann thermal spring" do not reveal a dominating pattern. Concerning sustainability aspects, no differences can be seen between any of the "à la carte" products.

Discussion

The results show that the methodological frame developed for operationalizing the PTNT framework is able to identify distinct products which can be assigned to the categories of the typology. Furthermore, it justifies Arnegger et al.'s (2010) deductive framework as appropriate for looking at tourism products from two perspectives: the supply side and the demand side. These perspectives are suitable to define and to locate tourism products on the specific and the total level. This combined approach has earlier been implicitly proposed (Eagles, 1992; Weaver, 2005), but has, until the present case study, never been empirically applied in nature-based tourism contexts.

Even though the cluster structure identified in this study resembles the demand side of the typology framework, the results of the case study imply some definitional changes to the categories, which have to be made for formulating a typology for "Sustainable Protected Area Tourism Products" (see Figure 1). The category "nature protection", as defined in the original typology framework, is a very small niche market. Hence, in this case study, this cluster is smaller than, for example, the "enthused" cluster in Weaver's (2013) study. This is not surprising as Weaver (2013) asked about willingness to engage in protection activities, whereas in this study, actual behavior was captured. Therefore, a definitional broadening of the "nature protection" category is proposed. One suitable definitional concept is that of "ecotourists", as tourists of that cluster fulfill all three core ecotourism criteria.

Due to the proposed definitional broadening of the first category, the second category "nature experience" has also to be adapted. This category covers general nature-related experiences (seeing wilderness, undisturbed nature, visiting natural attractions, etc.), which are not learning-oriented or sustainable in the first place. Therefore, this category covers only the first criteria of the three core ecotourism criteria.

Products of the "sports and adventure" category can be clearly separated using the study data and very clearly presented by the demand-side-defined "mountaineers" cluster and the supply-side-defined "package mountaineer holidays" product. The à la carte "guided climbing/mountaineer tours" can be also be matched with this category concerning motivations and activities. However, the second sports cluster (nature experience (sports)) shows a rather hybrid form, also with a rather high interest in nature. Furthermore, the sports REP factor loadings are higher compared to the adventure factor, which is in line with prior research about mountaineering, showing that the role of taking risk or experiencing an adventure is subordinated compared to skill development (Pomfret, 2011). Hence, as implicated by the category's name, it is more a sports than an adventure category.

The "hedonistic" category of the typology can also be captured using the study data, but with some shortcomings, caused by the inadequate definition and operationalization of the category in the original typology framework, where it is more or less defined as a residual or hybrid category containing leisure as well as cultural and general tourists not primarily interested in nature. Hence, in a further development of the typology framework, the "hedonistic" category should be termed, as suggested by Marques et al. (2010), as a general "occasional interest in nature" category (see Figure 1).

All product categories defined in this study can very well be differentiated via the two NI scales, derived from Juric et al. (2002) and Mehmetoglu (2010). And the concept of "national-park affinity" (Job et al., 2009; Mayer et al., 2010) correlates with these results, with the "ecotourists" product cluster showing the highest share of visitors with a high "national park affinity". In general, the NI scales reproduce the sequence of product categories of the dimension "nature as a point of attraction". However, deviations are observable comparing the "sports and adventure" and the "hedonistic" category, dependent on how the NI scale is calculated. Looking on the NI scale measured by motives and activities, the mountaineer products show a lower score, because mountaineers are less engaged in activities such as taking photos of wildlife or taking part in environmental education. This is partly caused by their form of travel, as mountaineers wander in more peripheral areas of the NP (Job, Schamel, & Butzmann, 2016). So looking on the NI scale by motives alone, the mountaineers score higher than the hedonistic clusters. Furthermore, tourists in the "nature experience (sports)" product cluster are a hybrid form, showing rather high NI scales. This implies that, in the context of alpine tourism, activities as mountaineering or trekking, which were assigned in the typology framework to the sports and adventure category, also have a strong relation to nature experience, in line with previous research (Rupf, 2015, p. 229).

Looking at the service arrangement categories, adaptations can be proposed. It is proposed to rename the "individuality" dimension as "degree of travel organization" as the further down a product is placed on the axis, the more service arrangements are bought, or components of the trip are organized, by tour operators or other service providers (see Figure 1). The other proposed changes relate to the distinct categories. The dominating category in the case study is the "independent" category. Packaged and customized travel arrangements play a minor role, with a share of roughly 10%. Especially, the "customized" category is rather small and had to be merged with the "fully standardized" category. For destinations where the "independent" category dominates, a further service arrangement category could be proposed, namely "structured" (see Figure 1). This category builds on the definition of "structured ecotourists" by Weaver and Lawton (2002) and is represented by tourists, who independently plan their trip, but are more dependent on service arrangements and comfort. The independent category could be reserved for tourists who discover the PA destination completely on their own, with a low service orientation. This can be further illustrated by the "ecotourists" product cluster. Those tourists take part in ecotourism activities in the NP (as guided hikes) and at the same time pursue activities that are dependent on the tourism infrastructure of the destination as a whole (e.g. for 22.5% using wellness facilities is (very) important) and overnight in rather comfortable lodgings as hotels (23.7%) or guesthouses (26.6%). Hence, tourists of that product cluster could also be termed "structured ecotourists". "Hard ecotourists", as defined by Weaver and Lawton (2002), are rather a small sub-segment of the "ecotourists" product cluster, overnighting in mountain huts in the NP.

Another aim of this study was to test if the typology is suitable to identify sustainable tourism products that can foster the fulfillment of the double mandate of PAs. Regarding the NEP and sustainability scales, data-driven and common sense-defined products show significant differences. Concerning demand-defined clusters, the "mountaineers" and the "special nature experience (ecotourists)" product clusters have the highest NEP as well as sustainability scale scores. Comparing "customized" and "fully standardized" products, tourists taking part in NP excursions show the highest scores for all scales. This is congruent with previous research, confirming a positive relation between interest in nature and pro-environmental values and beliefs (Luo & Deng, 2007). The high NEP and sustainability scores of the "mountaineers" are in contrast to earlier research, which indicates that more adventurous or active tourists are less concerned about the environment (Dunlap & van Liere, 1978). The lowest scores are attained by products of the "hedonistic" category, which load comparatively high on the "fun/relax" factor. This is in line with previous research showing that classical leisure tourists generally have lower pro-environmental attitudes and affinities for sustainable tourism products (Perkins & Brown, 2012). However, these product clusters are those with comparatively high spending for accommodation. The opposite applies for the "mountaineers" with rather low spending and high affinities for sustainability. This implies a sustainability–profitability tradeoff for park tourism. But, as in the general tourism context (Moeller, Dolnicar, & Leisch, 2011), this can be overcome. From a sustainable destination marketing perspective, the "structured ecotourists" product cluster seems to be the favorable product characterized by comparable high environmental attitudes and a higher affinity for sustainable tourism. Furthermore, tourists using this product type engage in ecotourism activities, using guided hikes by park rangers and at the same time stay overnight outside the PA using the general tourism infrastructure and service arrangements of the destination. Therefore, tourists of this product type are not only environmental conscious, they also spend money in the PA's adjacent communities. This product type shows patterns of "enlightened mass tourism", the form of tourism which combines the positives aspects of "hard ecotourism" and "mass tourism" (Weaver, 2015; Weaver & Lawton, 2002) and can help to fulfill the double mandate of PAs. Destination marketing should try to focus on that product type by specific marketing initiatives and product design (Dolnicar & Leisch, 2008). One possible way could be the promotion of accommodation with ecolabels in combination with ecotourism products. Currently, in the Berchtesgaden PA destination, less than two percent of the hotels or guesthouse accommodation and only two out of 14 mountain huts are eco-certified (own estimations based on BGLT (2016) and DAV (2015)). Tourists using the ecotourism and mountaineers product types could be good markets for further sustainable offers. This is line with research indicating that in Germany in general (FUR, 2014) and in PA destinations in Europe (EUROPARC, 2013, p. 28), a rather high-demand potential for sustainable tourism products exists; however, often both the supply and the marketing of relevant sustainable tourism products is insufficient to fulfill that demand. As in the case study region, where visitation has increased significantly in recent years, the targeting of more sustainable PA products and visitor segments should be preferred to efforts at just increasing PA visitor numbers, which – if continued at the present pace – could cause environmental damage and social problems by crowding effects.

Conclusion

The overall aim of this study was, for the first time, to operationalize the PTNT and to apply it empirically. Berchtesgaden National Park (Germany) served as the case study area. In all six distinct product clusters, the "independent" category and the "fully standardized"/"customized" categories could be extracted from the demand side by activities and motivations, and matched with the typology framework. Products of the "fully standardized"/"customized" category could further be differentiated by the supply side. Demand- and supply-side-defined products were successfully profiled by two scales measuring the dimension "nature as a point of attraction" as well as by further socio-demographic variables, experience preferences and service arrangements.

Furthermore, the derived operationalization was tested to ascertain if it is suitable to identify sustainable tourism products that can foster the fulfillment of the double mandate of PAs. Products of the "ecotourism experiences" category and especially the "structured" ones seem to be favorable segments from a sustainable development perspective showing patterns of "enlightened mass tourism" by combining positive attributes of "hard ecotourism" and "mass tourism". Therefore, focusing on that kind of tourism products could help PAs to fulfill their double mandate.

And, finally, adaptations were discussed to further develop the PTNT to create a typology for "Sustainable Protected Area Tourism Products". Figure 1 summarizes the proposed adaptations and displays the data-driven and common sense-defined products identified in this study. The "nature protection" category should be broadened and be termed "ecotourism experiences". The last category "hedonistic" should be termed "occasional nature experience" so that it covers cultural and not only hedonistic interests. Furthermore, an additional service category namely "structured" is needed.

Management bodies can use the new typology framework for sustainable PA tourism products to develop a comparable segmentation base to enable them to monitor which products of the specific and the total level are offered in PA destinations. Furthermore, products of different categories can be evaluated to check if customers have an affinity for sustainable product design. Monitoring results can be used to support management objectives by marketing of tourism products using ecolabels to attract additional visitors in favorable segments. Greater market knowledge, and the skilled use of that knowledge, could be a new way for PA management bodies to fulfill the double mandate of PAs.

Acknowledgments

The authors like to thank the Berchtesgaden National Park authorities for supporting the visitor survey as well as students of the Würzburg University and the Department of Tourism of the Munich University of Applied Sciences for conducting interviews on-site. The authors also like to thank Susanne Becken for her valuable advice on earlier drafts of this paper, as well as the editor and three anonymous reviewers for their helpful comments.

Disclosure statement

No potential conflict of interest was reported by the authors.

Funding

This project was partly funded by the German Federal Ministry for the Environment, Nature Conservation and Nuclear Safety under the UFO-Plan [number 3512 87 0100].

References

Ajzen, I. (2011). *Attitudes, personality and behavior*. Maidenhead: Open University Press.

Arnberger, A., Eder, R., Allex, B., Sterl, P., & Burns, R. (2012). Relationships between national-park affinity and attitudes towards protected area management of visitors to the Gesaeuse National Park, Austria. *Forest Policy and Economics*, *19*, 48–55.

Arnegger, J., Woltering, M., & Job, H. (2010). Toward a product-based typology for nature-based tourism: A conceptual framework. *Journal of Sustainable Tourism*, *18*(7), 915–928.

Balmford, A., Green, J.M.H., Anderson, M., Beresford, J., Huang, C., Naidoo, R., ... Manica, A. (2015). Walk on the wild side: Estimating the global magnitude of visits to protected areas. *PLoS Biology*, *13*(2), 1–6.

Beaumont, N. (2011). The third criterion of ecotourism: Are ecotourists more concerned about sustainability than other tourists? *Journal of Ecotourism*, *10*(2), 135–148.

Becken, S., & Job, H. (2014). Protected areas in an era of global–local change. *Journal of Sustainable Tourism*, *22*(4), 507–527.

Beh, A., & Bruyere, B.L. (2007). Segmentation by visitor motivation in three Kenyan national reserves. *Tourism Management*, *28*(6), 1464–1471.

Berchtesgadener Land Tourismus GmbH (BGLT) (2016). Gastgeberverzeichnis 2016. Ihre Gastgeber in Berchtesgaden-Königssee, Bad Reichenhall/Bayerisch Gmain und im Rupertiwinkel [Accommodation register. Your accommodation in Berchtesgaden-Königssee, Bad Reichenhall/Bayerisch Gmain and Rupertiwinkel]. Retrieved June 9, 2016, from http://www.berchtesgadener-land.com/emags/Berchtesgadener_Land_Gastgeber_2016/#/218/

Bundesministerium für Umwelt, Naturschutz und Reaktorsicherheit (BMU) (2007). *Nationale Strategie zur biologischen Vielfalt* [National biodiversity strategy]. Berlin: Author. Retrieved June 6, 2016, from http://www.bfn.de/fileadmin/MDB/documents/themen/landwirtschaft/nationale_strategie.pdf

Butzmann, E., & Schamel, J. (2014). Proposed methodological framework for empirical testing the product-based typology for nature-based tourism. In M. Reimann, K. Sepp, E. Pärna, & R. Tuula (Eds.), Proceedings of the 7th international conference on monitoring and management of visitors in recreational and protected areas (MMV), August 20–23, 2014, Tallinn, Estonia (pp. 37–39). Retrieved June 6, 2016, from www.tlu.ee/UserFiles/Konverentsikeskus/MMV7/MMV%20PROCEEDING.pdf

Convention on Biological Diversity (CBD) (2010a). Aichi biodiversity targets. Retrieved June 9, 2016, from https://www.cbd.int/sp/targets/#GoalD

Convention on Biological Diversity (CBD) (2010b). Protected areas – an overview. Retrieved June 6, 2016, from https://www.cbd.int/protected/overview/

Crompton, J.L. (1979). Motivations for pleasure vacation. *Annals of Tourism Research*, *6*(4), 408–424.

Deutscher Alpenverein (DAV) (2015). Mit dem Umweltgütesiegel ausgezeichnete AV-Hütten [Eco-certified AV-huts]. Retrieved June 9, 2016, from http://www.alpenverein.de/chameleon/public/6031584f-0495-c427-897e-35d0cd9170db/Huetten-m-UGS-2014_20511.pdf

Dolnicar, S. (2004). Beyond "commonsense segmentation": A systematics of segmentation approaches in tourism. *Journal of Travel Research*, *42*(3), 244–250.

Dolnicar, S. (2010). Identifying tourists with smaller environmental footprints. *Journal of Sustainable Tourism*, *18*(6), 717–734.

Dolnicar, S., & Leisch, F. (2007). An investigation of tourists' patterns of obligation to protect the environment. *Journal of Travel Research*, *46*(4), 381–391.

Dolnicar, S., & Leisch, F. (2008). Selective marketing for environmentally sustainable tourism. *Tourism Management*, *29*(4), 672–680.

Dudley, N. (Ed.). (2008). *Guidelines for applying protected area management categories*. Gland: IUCN.

Dunlap, R.E., & Van Liere, K.D. (1978). The "new environmental paradigm". *The Journal of Environmental Education*, *9*(4), 10–19.

Dunlap, R.E., Van Liere, K.D., Mertig, A.G., & Jones, R.E. (2000). Measuring endorsement of the new ecological paradigm: A revised NEP scale. *Journal of Social Issues*, *56*(3), 425–442.

Eagles, P.F.J. (1992). The travel motivations of Canadian ecotourists. *Journal of Travel Research*,*31*(2), 3–7.

Eagles, P.F.J. (2014). Research priorities in park tourism. *Journal of Sustainable Tourism*, *22*(4), 528–549.

EUROPARC (2013). Protected area visitors views on sustainable tourism approaches. Retrieved October 22, 2015, from http://www.european-charter.org/charter-projects/steppa-sustainable-tourism

Fairweather, J.R., Maslin, C., & Simmons, D.G. (2005). Environmental values and response to ecolabels among international visitors to New Zealand. *Journal of Sustainable Tourism*, *13*(1), 82–98.

Fennell, D., & Weaver, D.B. (2005). The ecotourium concept and tourism-conservation symbiosis. *Journal of Sustainable Tourism*, *13*(4), 373–390.

Floyd, M.F., Jang, H., & Noe, F.P. (1997). The relationship between environmental concern and acceptability of environmental impacts among visitors to two national park settings. *Journal of Environmental Management*, *51*(4), 391–412.

Fodness, D. (1994). Measuring tourist motivation. *Annals of Tourism Research*, *21*(3), 555–581.

Forschungsgemeinschaft Urlaub und Reisen (FUR) (2012). *Reiseanalyse 2012* [Travel analysis 2012]. Kiel: Author.

Forschungsgemeinschaft Urlaub und Reisen (FUR) (2014). *Abschlussbericht zu dem Forschungsvorhaben: Nachfrage für Nachhaltigen Tourismus im Rahmen der Reiseanalyse* [Final report to the research project: Demand for sustainable tourism as part of the travel analysis]. Kiel: Author.

Hvenegaard, G.T. (2002). Using tourist typologies for ecotourism research. *Journal of Ecotourism, 1*(1), 7–18.

Job, H. (2010). Welche Nationalparke braucht Deutschland? [Which National Parks does Germany need?]. *Raumforschung und Raumordnung, 68*(2), 75–90.

Job, H., Mayer, M., Woltering, M., Müller, M., Harrer, B., & Metzler, D. (2008). *Der Nationalpark Bayerischer Wald als regionaler Wirtschaftsfaktor* [The Bayerischer Wald National Park as a regional economic impact factor]. Grafenau: Bayerischer Wald National Park.

Job, H., Merlin, C., Metzler, D., Schamel, J., & Woltering, M. (2016). *Regionalwirtschaftliche Effekte durch Naturtourismus* [Regional economic impacts of nature-based tourism]. Bonn: Bundesamt für Naturschutz. Retrieved June 15, 2016, from https://www.bfn.de/fileadmin/BfN/service/Dokumente/skripten/Skript_431.pdf

Job, H., Metzler, D., & Vogt, L. (2003). *Inwertsetzung alpiner Nationalparks. Eine regionalwirtschaftliche Analyse des Tourismus im Alpenpark Berchtesgaden* [The valorisation of Alpine National Parks. A regional economic analysis of tourism in Berchtesgaden National Park]. Kallmünz: Lassleben.

Job, H., & Paesler, F. (2013). Links between nature-based tourism, protected areas, poverty alleviation and crises – the example of Wasini Island (Kenya). *Journal of Outdoor Recreation and Tourism, 1*(1/2), 18–28.

Job, H., Schamel, J., & Butzmann, E. (2016). Besuchermanagement in Großschutzgebieten im Zeitalter moderner Informations- und Kommunikationstechnologien [Visitor management in large-scale protected areas – deploying modern information and communication technologies]. *Natur und Landschaft, 91*(1), 32–38.

Job, H., Woltering, M., & Harrer, B. (2009). *Regionalökonomische Effekte des Tourismus in deutschen Nationalparken* [Regional economic impacts of tourism in German national parks]. Bonn: Bundesamt für Naturschutz.

Juric, B., Cornwell, T.B., & Mather, D. (2002). Exploring the usefulness of an ecotourism interest scale. *Journal of Travel Research, 40*(3), 259–269.

Kerstetter, D.L., Hou, J.-S., & Lin, C.-H. (2004). Profiling Taiwanese ecotourists using a behavioral approach. *Tourism Management, 25*(4), 491–498.

Küpfer, I. (2000). *Die regionalwirtschaftliche Bedeutung des Nationalparktourismus. Untersucht am Beispiel des Schweizerischen Nationalparks* [The regional economic importance of national park tourism. The case of Swiss National Park]. Zernez: Forschungskommission des Schweizerischen Nationalparks.

Luo, Y., & Deng, J. (2007). The new environmental paradigm and nature-based tourism motivation. *Journal of Travel Research, 46*(4), 392–402.

Magidson, J., & Vermunt, J.K. (2001). Latent class factor and cluster models, bi-plots and related graphical displays. *Sociological Methodology, 31*, 223–264.

Mair, J. (2011). Exploring air travellers' voluntary carbon-offsetting behaviour. *Journal of Sustainable Tourism, 19*(2), 215–230.

Manfredo, M.J., Driver, B.L., & Tarrant, M.A. (1996). Measuring leisure motivation: A meta-analysis of the recreation experience preference scales. *Journal of Leisure Research, 28*(3), 188–213.

Marques, C., Reis, E., & Menezes, J. (2010). Profiling the segments of visitors to Portuguese PAs. *Journal of Sustainable Tourism, 18*(8), 971–996.

Mayer, M., & Job, H. (2014). The economics of PAs. A European perspective. *The German Journal of Economic Geography, 58*(2/3), 73–97.

Mayer, M., Müller, M., Woltering, M., Arnegger, J., & Job, H. (2010). The economic impact of tourism in six German national parks. *Landscape and Urban Planning, 97*(2), 73–82.

McKercher, B. (2005). Destinations as products? A reflection on Butler's life cycle. *Tourism Recreation Research, 30*(3), 97–102.

Mehmetoglu, M. (2007). Typologising nature-based tourists by activity—theoretical and practical implications. *Tourism Management, 28*(3), 651–660.

Mehmetoglu, M. (2010). Accurately identifying and comparing sustainable tourists, nature-based tourists, and ecotourists on the basis of their environmental concerns. *International Journal of Hospitality & Tourism Administration, 11*(2), 171–199.

Metzler, D., Woltering, M., & Scheder, N. (2016). Naturtourismus in Deutschlands Nationalparks [Nature-based tourism in German National Parks]. *Natur und Landschaft, 91*(1), 8–14.

Middleton, V. (1988). *Marketing in travel and tourism*. Oxford: Heinemann Professional.

Middleton, V. (1989). Tourist product. In Stephen F. Witt & Luiz Moutinho (Eds.), *Tourism marketing and management handbook* (pp. 573–576). New York, NY: Prentice Hall.

Moeller, T., Dolnicar, S., & Leisch, F. (2011). The sustainability–profitability trade-off in tourism: Can it be overcome? *Journal of Sustainable Tourism, 19*(2), 155–169.

Moscardo, G., Morrison, A.M., Pearce, P.L., Lang, C.-T., & O'Leary, J.T. (1996). Understanding vacation destination choice through travel motivation and activities. *Journal of Vacation Marketing, 2*(2), 109–122.

Müller, M., & Job, H. (2009). Managing natural disturbance in PAs: Tourists' attitude towards the bark beetle in a German national park. *Biological Conservation, 142*(2), 375–383.

Palacio, V., & McCool, S.F. (1997). Identifying ecotourists in Belize through benefit segmentation: A preliminary analysis. *Journal of Sustainable Tourism, 5*(3), 234–243.

Perkins, H.E., & Brown, P.R. (2012). Environmental values and the so-called true ecotourist. *Journal of Travel Research, 51* (6), 793–803.

Perkins, H.E., & Grace, D.A. (2009). Ecotourism: Supply of nature or tourist demand? *Journal of Ecotourism, 8*(3), 223–236.

Pomfret, G. (2011). Package mountaineer tourists holidaying in the French Alps: An evaluation of key influences encouraging their participation. *Tourism Management, 32*(3), 501–510.

Rupf, R. (2015). *Planungsinstrumente für Wandern und Mountainbiking in Berggebieten. Unter besonderer Berücksichtigung der Biosfera Val Müstair* [Planning instruments for hiking and mountainbiking in mountain regions. With special consideration of the Biosfera Val Müstair]. Bern: Haupt Verlag.

Sirakaya, E., Sasidharan, V., & Sonmez, S. (1999). Redefining ecotourism: The need for a supply-side view. *Journal of Travel Research, 38*(2), 168–172.

Smith, A.J., Tuffin, M., Taplin, R.H., Moore, S.A., & Tonge, J. (2014). Visitor segmentation for a park system using research and managerial judgement. *Journal of Ecotourism, 13*(2–3), 93–109.

Smith, S.L.J. (1994). The tourism product. *Annals of Tourism Research, 21*(3), 582–595.

Vermunt, J.K., & Magidson, J. (2013). *Technical guide for latent GOLD 5.0. Basic, advanced, and syntax*. Belmont. Retrieved June 2, 2015, from https://www.statisticalinnovations.com/wp-content/uploads/LGtecnical.pdf

Weaver, D.B. (2000). A broad context model of destination development scenarios. *Tourism Management, 21*(3), 217–224.

Weaver, D.B. (2001). Ecotourism as mass tourism contradiction or reality? *Cornell Hotel & Restaurant Administration Quarterly, 42*(2), 104–112.

Weaver, D.B. (2005). Comprehensive and minimalist dimensions of ecotourism. *Annals of Tourism Research, 32*(2), 439–455.

Weaver, D.B. (2013). Protected area visitor willingness to participate in site enhancement activities. *Journal of Travel Research, 52*(3), 377–391.

Weaver, D.B. (2015). Volunteer tourism and beyond: Motivations and barriers to participation in protected area enhancement. *Journal of Sustainable Tourism, 23*(5), 683–705.

Weaver, D.B., & Lawton, L.J. (2002). Overnight ecotourist market segmentation in the Gold Coast Hinterland of Australia. *Journal of Travel Research, 40*, 270–280.

Wight, P.A. (1996). North American ecotourism markets: Motivations, preferences, and destinations. *Journal of Travel Research, 35*(1), 3–10.

Woltering, M. (2012). Tourismus und Regionalentwicklung in deutschen Nationalparken [Tourism and regional development in German National Parks]. Retrieved June 10, 2016, from http://opus.bibliothek.uni-wuerzburg.de/volltexte/2012/7189/.

Zografos, C., & Allcroft, D. (2007). The environmental values of potential ecotourists: A segmentation study. *Journal of Sustainable Tourism, 15*(1), 44–66.

The effects of local context on World Heritage Site management: the Dolomites Natural World Heritage Site, Italy

Maria Della Lucia and Mariangela Franch (iD)

ABSTRACT

This paper investigates the influence of local context on World Heritage Site (WHS) management. Building on the managerial literature about natural WHSs, protected areas and sustainable tourism, it explores the effects that tourism-driven socio-economic conditions have on stakeholder engagement in decision-making and governance both inside and neighboring a WHS. Mixed methods and tools collected case evidence at the lowest administrative level in Italy's Dolomites natural WHS. Results show that UNESCO recognition seems to be creating a deeper divide between areas inside and those outside the WHS than already produced by the diversity of tourism development types. Within the WHS and where tourism-driven socio-economic well-being is already high, there has been reinforced stakeholder indirect participation in decision-making and raised expectations of further development. In contrast, it seems not to have affected the area outside the WHS or marginal destinations, either in terms of participatory decision-making, or of development, or of tourism as a development lever. There is strong belief in the value of WHS status, relatively little opposition to regulation, but a need for reduced reliance on financial incentives, and a better business culture. Preliminary managerial recommendations and suggestions on involving stakeholders more effectively are given.

Introduction

The number of World Heritage Sites (WHSs) has increased dramatically since the World Heritage Convention launched the concept of the WHS in 1972 (UNESCO, 1972) and UNESCO's World Heritage Committee's strategy on a balanced World Heritage List (WHL) was produced in 1994 (Steiner & Frey, 2011). This can be explained by the opportunities for heritage preservation and socio-economic development offered by the international recognition of an area's significance and outstanding universal value for humanity (Rebanks Consulting Ltd, 2009). Tourism is assumed to be an important lever in this regard (Jimura, 2011; Li, Wu, & Cai, 2008; Scuttari & Della Lucia, 2016; Timothy & Nyaupane, 2009).

An integrated cross-sectoral approach, and stakeholder participation are fundamental tenets of effective WHS management (IUCN, 2008), as they are of the management and governance of any protected area (Conradin, Engesser, & Wiesmann, 2014), as fundamental pillars of sustainable development, implemented by all sectors participating in WHS socio-economic development (Conradin et al., 2014). Achieving stakeholder support and involvement in WHS management is difficult in practice (Strickland-Munro & Moore, 2013; Tosun, 2006) and is impacted by a number of place specific

⓫ Supplemental data for this article can be accessed at ⟳ https://doi.org/10.1080/09669582.2017.1316727

factors (Wiesmann, Liechti, & Rist, 2005) – particularly the prevailing socio-economic conditions at the time of the nomination, and the governance system used (Getzner et al., 2014; Leask & Fyall, 2006; Li et al., 2008; Ryan, Chaozhi, & Zeng, 2011).

This paper assesses the influence that local context – defined as socio-economic and institutional conditions – has on the integrated and participatory management of WHSs. In particular, it investigates whether and how tourism-driven socio-economic conditions affect stakeholder engagement in decision-making and the future development of areas, both with and without UNESCO recognition (institutional conditions). This issue has been investigated at the lowest governance level in Italy (the municipality), by examining the Dolomites natural WHS in Italy, a complex trans-regional, serial site consisting of nine separate areas, and with a pronounced tourism-driven socio-economic divide (see /www.dolomitiunesco.it/en/). It is recognized by UNESCO as a global example of best practice in terms of WHS management planning. Following a literature review, of both the management of WHS sites and stakeholder engagement in that management, there is a review of sustainable tourism management possibilities in WHS sites. The methodology and results of two exploratory quantitative analyses are then described and discussed. The first analysis studied the local Dolomite context in terms of tourism-driven socio-economic development inside and outside the WHS; the second analysis used an online questionnaire addressed to mayors to assess the effects that local context has on stakeholder engagement in decision-making and future development. The research topic is approached by linking the actual configuration of natural WHS management to the evolution of management planning methods in protected areas, stakeholder involvement in WHS management and its importance in the sustainable tourism management of WHSs.

The integrated and participatory management of natural WHSs

The expanded definition of protected areas, which now includes six management categories (Borrini-Feyerabend et al., 2013), has strengthened the link between protected areas and the four selection criteria for the inscription of a natural site on the WHL (see http://whc.unesco.org/en/criteria/). Irrespective of category, natural WHSs may be assumed to be protected areas of exceptional value because almost all of the sites fall within the perimeters of protected areas (Boyd & Dallen, 2006), but only a small percentage of protected areas are recognized as being of exceptional and universal value for humanity (IUCN, 2008). From a technical-operative perspective, this qualification gives a UNESCO property[1] an additional layer of conservation, achieved by giving international political support to conservation regulations, thus strengthening the protection of its integrity and universal value. In management terms, it has allowed the extension to UNESCO sites of both evolving concepts of conservation in the scientific debate on protected areas (Becken & Job, 2014; Scuttari & Della Lucia, 2016), and governance types (Borrini-Feyerabend et al., 2013).

The preservation of protected areas is evolving from management models focused on static conservation to models of innovative and dynamic conservation (Becken & Job, 2014; Mose, 2007). The static preservation model – or segregation approach – adopts a single sector and single level (local) management method, which implies focusing on protected areas only, and top-down management approaches and tools (Mose & Weixlbaumer, 2007). In contrast, the model of innovative and dynamic conservation – or integration approach – adopts an integrated and participatory management method (Mose & Weixlbaumer, 2007). Different sectors (sectoral integration), geographic areas, both protected and polluted (ecosystem approach) and administrative levels (multi-level integration) participate in a combination of top-down and bottom-up management approaches and tools (participatory approach). Sustainable development is the goal of this integrated and participatory management method, through the combination of preservation with suitable economic development and employment opportunities and prosperity for local communities.

The four broad governance types applied to protected area categories (Borrini-Feyerabend et al., 2013) mirror this evolving models of conservation. Ranging from governance by *government* and *shared* governance to *private* governance and governance by *indigenous peoples and local communities*, institutional mechanisms and/or (in)formal processes share authority and responsibility for the protected area management amongst several actors, who are entitled to participate in the planning or implementing of initiatives, involving either weak or strong forms of stakeholder engagement. Shared governance – often referred to as co-management, collaborative management, joint management or multi-stakeholder management – is becoming more common in many other fields, too, (Borrini-Feyerabend, Pimbert, Farvar, Kothari, & Renard, 2008) and is particularly suited to trans-boundary conservation areas (Sandwith, Shine, Hamilton, & Sheppard, 2001) such as serial natural WHSs.

The current configuration of WHS management recommendations incorporated the integrated and participatory planning process adopted in well-managed protected areas by integrating the UNESCO Convention (Conradin et al., 2014) with UNESCO's operational guidelines (UNESCO, 2015) explicitly aimed at WHS sustainable development. Holistic planning and stakeholder participation are fundamental tenets of WHS management and sustainable development (UNESCO, 2015). The management plan is mandatory (UNESCO, 2015); it covers the process of WHS listing and the entire period of the site's protection; and is applicable to all the sectors participating in the site development, tourism included (Landorf, 2009), and to the areas adjacent to the WHS.

Stakeholder engagement in WHS management

Like shared governance in protected areas (Borrini-Feyerabend et al., 2013), WHS planning frames a site within the wider process of territorial governance and allows space for the broad based, equitable decision-making needed for effective WHS management and sustainable development (Jamal & Watt, 2011; UNESCO, 2015; Yüksel, Bramwell, & Yüksel, 1999). The effectiveness of the planning process depends on the forms of stakeholder participation involved (Tosun, 2006). Active and direct participation is an ideal type (*spontaneous form*): stakeholders have full managerial responsibility and authority in decision-making. The commonest type is top-down, passive and indirect: the breadth and equality of decision-making processes are shaped by the objectives of powerful interest groups (Aas, Ladkin, & Fletcher, 2005; Trau & Bushell, 2008). Indirect participation usually ranges from a *coercive form*, in which the needs and desires of powerful interest groups are primary, to an *induced form* in which stakeholder participation is restricted to consultation and/or participation for material gain (Tosun, 2006). Efforts to encourage deeper and more active participation are more likely to be made at the local governance level, where the scope and nature of sectoral interdependence and stakeholder interactions are scaled down (Kaltenborn, Thomassen, Wold, Linnell, & Skar, 2013). Capacity building is necessary to encourage dialogue, partnerships and stakeholder engagement and to root sustainable development in a sense of place, culture and identity (Moscardo, 2008).

Failure to engage and empower local stakeholders can exclude many from the potential benefits of WHS recognition and reduce the effectiveness of sustainable development initiatives (Kaltenborn et al., 2013; Simpson, 2001; Tourtellot, 2006). Perceptions of exclusion can generate negative reactions in local stakeholders or antagonistic attitudes towards the development paths envisioned by governments or other lead institutions (Strickland-Munro & Moore, 2013). Under-engagement may also result from stakeholder concern that WHS nomination could limit local stakeholders' freedom to carry out future private initiatives (Magi & Nzama, 2009; Nicholas, Thapa, & Ko, 2009).

There are many different levels at which participation in protected area management can take place. Much depends on local context. This paper looks at the opinions of the mayors of over 300 traditional small communities. This journal issue also has a paper which, in contrast, examines the role and views of representatives of immigrant communities neighboring a peri-urban National Park in Canada (Khazaei, Elliot, & Joppe, 2017).

Sustainable tourism management in WHSs

Tourism is assumed to be the sector that contributes the most to the socio-economic development of WHSs, through the marketing and branding of its universal value (Jimura, 2011; Li et al., 2008; Su &

Lin, 2014). However, the relationship between WHS recognition and tourism is still controversial because the designation involves both positive implications for tourism development and negative impacts on, and threats to, a site's integrity (Buckley, 2004; Hall & Piggin, 2002; Leask & Fyall, 2006). The economic benefits are undoubted, in particular the enhancing of the WHS's symbolic value, image and visibility (Bianchi, 2002; Smith, 2002); increased tourist flows (Su & Lin, 2014); and job creation, infrastructure, business and service development (Leask & Fyall, 2006). However, the extent to which WHSs actually benefit from an increase in international tourism and a heightened visitor profile is not necessarily clear (Cellini, 2011; Shackley, 2006). Quite apart from the issue of growth, tourism places additional environmental and social pressures on valuable sites with extremely fragile natural and cultural ecosystems. WHS status, however, should help to ensure a site's heritage preservation, whether already well looked after, or at risk (Rebanks Consulting Ltd, 2009). Environmental and social impacts are often associated with antagonistic or negative attitudes on the part of local people towards visitors (Conradin et al., 2014).

The nature and intensity of both effects depend both on the ways in which an area's tourism development was being managed at the time of the WHS nomination, and on its governance system (Leask & Fyall, 2006; Li et al., 2008; Ryan et al., 2011). After WHS designation, managing tourism sustainably through participatory planning is crucial to the sustainable development of the whole WHS (Landorf, 2009; Pedersen, 2002), due to the highly fragmented, cross-sectoral nature of tourism, which spans diverse policy domains and has transversal impacts.

The Weaver broad context model of destination development scenarios (2000) (from now on "the Weaver model") and its integration with destination development paths (Weaver, 2012) are managerial tools which can be used to support and manage stakeholder participation in decision-making about tourism development at both the planning and evaluating/monitoring stages. The Weaver model (2000) is a qualitative framework which depicts four inclusive tourism ideal types defined by combining the scale/intensity and sustainability-conducive regulation of the sector. It allows us to study the tourism-driven socio-economic development of an area in terms of the (un)sustainability of different tourism development types (mass, niche) at different stages of a tourism area life cycle (Butler, 1980). Mass tourism destinations are sustainable (SMT) when the impact of large-scale tourism is highly regulated and kept within the carrying capacity threshold; it is unsustainable (UMT) when insufficiently regulated. Niche tourism destinations are unsustainable (circumstantial alternative tourism – CAT) when the growth of economically sustainable small-scale tourism is not regulated or supported; in sustainable niche destinations (deliberate alternative tourism – DAT) increasing small-scale tourism and effective quality regulation foster both economic and environmental balance (Weaver, 2000).

An integration of the model (Weaver, 2012) identifies the norms driving the destination development paths toward the desired outcome, assumed to be sustainable mass tourism – pro-growth norm (*organic path*), sustainability-conducive regulation norm (*incremental path*) or hybrid norm (*induced path*). It is interesting to note that DAT is assumed to be a transitional state – for some protected areas or destinations as a whole – or a part of an integrated SMT strategy of spatial differentiation (Weaver, 2012). These development paths depend on different forms of stakeholder participation (Weaver, 2012) which show strong similarities with the categories introduced by Tosun (2006) – spontaneous, coercive and induced. Similarly to *coercive participation*, in the *organic path* powerful interest groups, led by the pro-growth norm, thwart deep shareholder participation. In the *incremental path*, participation is empowered and then reinforced by the sustainability norm driven by regulation, but readapted to allow for additional growth (combination of *active and induced participation*). In the *induced path*, participation is initially and intentionally displaced by the pro-growth norm, and must, therefore, subsequently be reshaped to allow sustainable development (as must participation into an *induced form*).

By combining Weaver's frameworks (2000, 2012), tourism development types – the socio-economic conditions – may be linked to stakeholder engagement in decision-making (Tosun, 2006) about their future (tourism) development paths envisioned with WHS status (Rebanks Consulting Ltd., 2009).

The Dolomites WHS

The Dolomites natural WHS set in the northeastern Italian Alps is a serial property of nine non-contiguous mountain areas which was designated in June 2009 in recognition of its superlative beauty, its outstanding universally valuable limestone geomorphology and its universal geological value (Gianolla, Panizza, Micheletti, & Viola, 2008) (Figure 1). It was chosen as a case study to investigate the influence of local context on integrated, participatory WHS management because the site is a complex trans-regional area lying almost entirely within protected areas of different types and showing a pronounced tourism-driven socio-economic divide.

The network of institutions which the site straddles comprises three regions, five provinces (Trento, Bolzano, Belluno, Udine and Pordenone), more than 100 municipalities and a number of territorial organizations, tourism-related and not. The provinces have different institutional, legal and administrative systems – Trento and Bolzano have special status under Italian law, which allows them greater autonomy – and linguistic, social and economic differences. Almost all of the sparsely populated area of the site (95% of around 230,000 ha) is protected at the European, or Italian or local level, thus qualifying the Dolomites WHS as a protected area of exceptional value. The site also lies in a mountain region showing different Weaver tourism ideal types (2000). A number of prosperous community-based mass destinations (Murphy, 1985), whose main attractions are natural resources and associated sports (skiing, climbing, hiking and trekking), coexist with marginal and remote settlements where traditional primary industries (forestry, agriculture and animal husbandry) are in decline and tourism is still only in its early stages. Irrespective of tourism, sustainability must be part of the future development of all these areas as it is crucial to the intrinsic socio-economic vitality of the Alpine regions and its traditional landscape maintenance systems (Gios, Goio, Notaro, & Raffaelli, 2006), and a requirement for the Dolomites' WHS listing (UNESCO, 2015).

As the site was designated after the integration of UNESCO's operational guidelines, with the core tenets of sustainable development (UNESCO, 2015), the Dolomites WHS management has implemented an integrated and participatory planning process, recognized by UNESCO as an example of global best practice. The Dolomites Foundation (http://www.dolomitiunesco.info/?lang = en), with a Steering Committee made up of seven representatives of the five provinces, a President and a Scientific Committee of five experts, is the *ad hoc* coordinating body established to ensure the management of the whole WHS according to this approach. The *Operating Networks*, the *Board of Supporters* and *pioneering participatory tools* are its main managerial instruments (UNESCO Dolomites

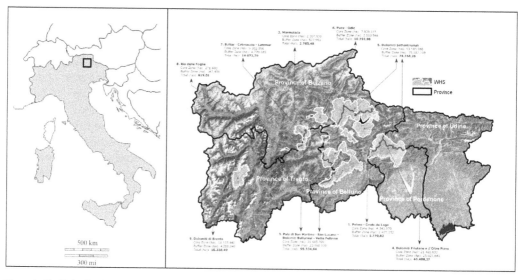

Figure 1. The Dolomites WHS.

Foundation, 2013, 2015). The *Operating Networks* is a matrix organizational structure aimed at harmonizing territorial management policies and actions across sectors (sectoral integration), in and around the site, regardless of provincial administrative boundaries (ecosystem approach). Each of the Foundation's founding members (the five provinces) is responsible – throughout the whole WHS area – for one or more of the following five specific themes: geological heritage; landscape heritage and protected areas; socio-economic development, sustainable tourism and mobility; the promotion of sustainable tourism; and education and scientific research (http://www.dolomitiunesco.info/le-reti-funzionali/?lang=en). The Board of Supporters (currently 128 members) and the organization of public meetings and events are participatory techniques intended to integrate top-down and bottom-up management decision-making levels (multi-level integration) and encourage stakeholder engagement (participatory approach). The Board of Supporters, a body composed mainly of local administrations and associations but also of non-institutional stakeholders and organizations, has an advisory role in the Foundation's activities and programs, and can thus directly influence the harmonization of territorial management policies. The annual DolomitiLab Fest (http://www.dolomitesunescolab fest.it/en/about-us/) and the stakeholder meetings #Dolomites2040 (http://www.dolomitiunesco. info/attivita/dolomiti2040-quali-proposte-per-il-futuro-la-fondazione-lo-chiede-alle-comunita-locali/ ?lang=en) are examples of pioneering participatory tools organized by the Dolomites Foundation to foster stakeholder capacity building and involvement in crucial stages of the WHS management planning process. These meetings were held in May 2015–June 2015 and attended by hundreds of public-private stakeholders, who discussed four themes relevant to the future of the Dolomites (tourism and mobility, socio-economic development, conservation and network building) using participatory techniques (including the World Cafè principles – see http://www.theworldcafe.com/key-concepts-resources/world-cafe-method/).

The results of the #Dolomites2040 meetings (available at the web address given in the previous paragraph) were integrated into the Dolomites WHS General Management Strategy and its Strategy for Sustainable Tourism (http://www.dolomitiunesco.info/wp-content/uploads/2016/09/FD4U-OMS_fi nal_rev20160401_ENG.pdf). In March 2016, seven years after the designation of the WHS, both strategies were approved by the founding members of the Dolomites Foundation and presented to UNESCO. The General Management Strategy is cross-sectoral in nature. The additional sector specific Strategy was prompted by the pre-eminent role of the tourism economy in the development of the Dolomite territories (Elmi & Perlik, 2014). Regional and provincial administrations, and the administrators of Natural Parks and municipalities all share responsibility for the implementation of the two strategies.

Research design and collection of case evidence

The choice of a trans-regional, serial WHS to investigate the effects that local context has on WHS management has necessitated the drawing of clear boundaries within a holistic single case study design (Yin, 2014). Local context is described in terms of the tourism-driven socio-economic conditions of territories, both with and without UNESCO recognition. WHS management entails stakeholder participation in decision-making about WHS planning and development. As decisions can be made, and produce effects, at trans-regional, regional and local levels (i.e. Dolomites Foundation, provinces and municipalities), the research focused on the lowest governance level. The assumption is that where people actually live and work, and sectoral and stakeholder interactions are scaled down – like in the small Dolomites mountain municipalities (ISTAT, 2014) – people have an intimate knowledge of local needs and resources and deeper, more active participation can be expected, aimed at making the most of WHS status (Kaltenborn et al., 2013).

The single case study combines two phases as follows.

(1) *The research design* grounds the empirical analysis, focusing on the theoretical constructs best suited to the analysis of the issue under investigation. The Weaver model (2000) is assumed to

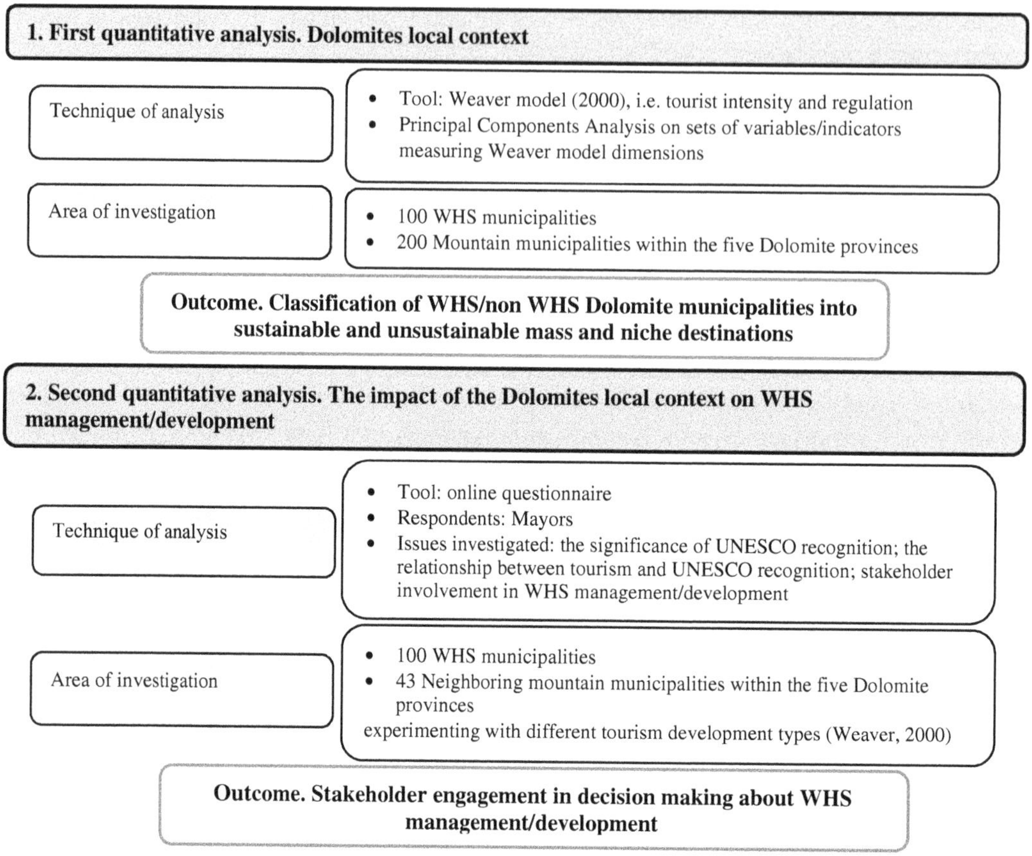

1. First quantitative analysis. Dolomites local context

Technique of analysis
- Tool: Weaver model (2000), i.e. tourist intensity and regulation
- Principal Components Analysis on sets of variables/indicators measuring Weaver model dimensions

Area of investigation
- 100 WHS municipalities
- 200 Mountain municipalities within the five Dolomite provinces

Outcome. Classification of WHS/non WHS Dolomite municipalities into sustainable and unsustainable mass and niche destinations

2. Second quantitative analysis. The impact of the Dolomites local context on WHS management/development

Technique of analysis
- Tool: online questionnaire
- Respondents: Mayors
- Issues investigated: the significance of UNESCO recognition; the relationship between tourism and UNESCO recognition; stakeholder involvement in WHS management/development

Area of investigation
- 100 WHS municipalities
- 43 Neighboring mountain municipalities within the five Dolomite provinces
experimenting with different tourism development types (Weaver, 2000)

Outcome. Stakeholder engagement in decision making about WHS management/development

Figure 2. Empirical analyses and outcomes.

be a particularly effective framework within which to study the local context of municipalities, both within and outside the WHS. The effects of local context on WHS management are studied by levering the connection between the integration of the Weaver model with the different norms driving the destination's development (Weaver, 2012) and stakeholder engagement in decision-making (Tosun, 2006) about the area's future development envisioned with WHS status (Rebanks Consulting Ltd., 2009).

(2) *The collection of case evidence* was carried out by applying this two-step research design in Dolomite municipalities through two exploratory quantitative analyses – the first of secondary data, the other of primary – collected from multiple sources (Yin, 2014). The triangulation of data sources and types provides reliable empirical evidence (Knafl & Breitmayer, 1989). Figure 2 depicts the empirical analysis phases and core outcomes; the time period and the techniques of data collection and analyses and data sources are presented in the following sub-sections.

First quantitative analysis: Dolomites local context

The Weaver model (2000) is quantitatively applied to assess the tourism-driven socio-economic development of municipalities (both within the WHS and not), thus identifying Dolomite tourism development types – sustainable or unsustainable mass or niche destinations. In order to transform it into an operative tool, the set of criteria needed to identify the area of investigation and the explanatory variables measuring tourist intensity and regulation – the Weaver model dimensions – must be defined.

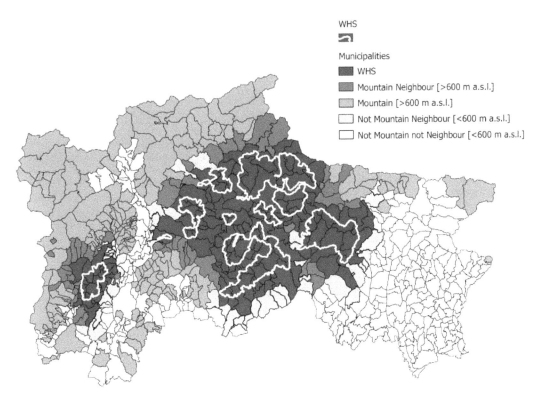

Figure 3. The area of investigation.

The criteria used to define the area of investigation are consistent with the definition of local context given above, and are: being located within the five provinces participating in the Dolomites Foundation (local); being included, or not, in the WHS (institutional condition); among those not included in the WHS, being mountain municipalities (>600 m above sea level), as the economy is more likely to be tourism-driven at higher altitudes (socio-economic condition) or tourism is likely to be a lever with which to tackle their socio-economic marginality. In fact, the physical-geographic criteria (altitude) used to define mountain areas has implications for policies aimed at supporting their development, whether through tourism, or other means (75/268/CEE). Three hundred and six municipalities meet these criteria: all the WHS (100) municipalities, and 206 non-WHS mountain municipalities, 43 of which border on the WHS (Figure 3). 94% of them have less (an average of 1400) than 5000 inhabitants – the Italian National Institute of Statistic (ISTAT) threshold for small municipalities (ISTAT, 2014) and the remaining 6% are medium-sized (average pop. 9500).

The variables used to measure Weaver model dimensions (2000), tourist intensity and regulation in municipalities, were selected by combining criteria of place significance and homogeneity with the availability and quality of the data. These constraints led to the exclusion of many of the variables/indicators often used in studies on sustainable tourism (e.g. McCool & Stankey, 2004). Table 1 provides details of the variables, sources and indicators calculated for each Weaver model dimension. The data were collected from secondary census or sample surveys, from national, regional or provincial statistics offices or from reliable sources on environmental regulation issues. Tourist intensity and second home data refer to 2010/2011, the most up-to-date available interprovincial surveys; the most recent tourist regulation data are from 2012. Although same-day visits are very important in several places within the studied area, they have been left out of the analysis because there is no official data available at the local level. Same-day visit data, taken from sample surveys (https://www.istat.it/it/archivio/178670 for national tourists and https://www.bancaditalia.it/statistiche/tematiche/rapporti-estero/turismo-internazionale/ for international tourists), are not available at the municipal

Table 1. Tourist intensity and regulation. Variables, sources and indicators.

Tourist intensity		
Variables	Sources	Indicators
Tourist flows (arrivals and overnights)	ASTAT – Provincial Institute of Statistics of Bolzano ISPAT – Provincial Institute of Statistics of Trento Regional Institutes of Statistics of Friuli Venezia Giulia and Veneto Regions)	Average length of stay; tourist rate (overnights per inhabitant); tourist density (overnights per km^2); bed occupancy rate
Accommodation capacity (bed places)	ISTAT – National Institute of Statistics. *Accommodation capacity survey 2011*	Accommodation rate (beds per 10,000 inhabitants); accommodation density (beds per km^2)
Tourism businesses (enterprises and establishments)	ISTAT – National Institute of Statistics. *Population Census 2011*	Tourism businesses/total businesses
Tourism employment (employees)	ISTAT – National Institute of Statistics. *Population Census 2011*	Tourism specialization
Regulation		
Variables	Sources	Indicators
Protected areas	ARPA – Regional Environmental Protection Agencies of Friuli Venezia Giulia and Veneto Regions APPA– Environmental Protection Agencies of Trento and Bolzano Provinces	Percentage of protected areas
Environmental certifications	ISPRA – National Institute for Environmental Protection and Research	EMAS and Ecolabel certifications per inhabitant
Differentiated waste collection	ISPRA – National Institute for Environmental Protection and Research	Differentiated waste collection/urban waste
Second homes	ISTAT – National Institute of Statistics. *Population Census 2011*	Second homes/population; second homes/total homes

level and it is difficult to estimate them for over 300 municipalities of the five different provinces. An attempt to provide a regional estimate has been made using tourism-traffic analysis (Scuttari, Della Lucia, & Martini, 2013), but it refers to the South Tyrol region, and is on a wider scale than the UNESCO WHS area. An indication of day visitation is given by a survey on sustainable tourism in the Dolomites carried out in summer 2013 (Elmi, 2014); the average ratio of same-day visitors to tourists detected at selected access points to the Dolomites WHS was around 34%. This ratio results from a sampling plan in which a number of assumptions were made to estimate the site's unknown visitor population.

Municipalities were positioned within the Weaver matrix by using principal component analysis (Rencher, 1998), a particularly suitable statistical technique used for data reduction when variables are highly correlated, as in this case study. The first (F1) and second (F2) principal components, identified by using STATA 13.0 statistics software, are the leading eigenvectors of the eigenvalue decomposition of a correlation matrix of five variables (Table 2). They explain 55.5% of the total data variability: the first principal component explains 33.48%, and the second 21.96%, of the total variance. The appropriateness of the two components is shown by their structure – the contribution that each

Table 2. Principal components. Variables, loadings and correlation.

	Components			
	Loading		Correlation (cos^2)	
Variables	F1	F2	F1	F2
Beds per 10,000 inhabitants	0.64	−0.19	0.69	0.04
Second homes per inhabitant	0.58	−0.21	0.55	0.05
% of protected areas	0.36	0.61	0.22	0.41
Average length of stay	0.29	0.47	0.14	0.25
Overnights per km^2	0.21	−0.57	0.07	0.35

variable gives to the components (loading or scoring coefficient) – and the correlation between the components and the original variables (cos^2). All variables contribute positively to F1: the accommodation rate and second homes per inhabitant have the highest loadings (0.64; 0.58) – the weights by which each standardized original variable should be multiplied to get the component score – and are the variables mostly closely correlated with the first component (cos^2 equals 0.69 and 0.55, respectively). F1 (x-axis) is assumed to be a proxy of the tourism intensity dimension of the Weaver model, as its composition is consistent with mountain municipalities whose (protected) landscape is already, or may become, a tourist attraction, leading to both tourist flows and the development of tourist accommodation establishments and/or their capacity. On the other hand, the variables contribute both positively and negatively to F2, i.e. have both positive and negative loadings and scores. The percentage of protected areas and the average length of stay have the highest values among the variables with positive loadings (0.61; 0.47); overnights per km^2 are among those with negative weights (-0.57). The percentage of protected areas and tourist density are also the variables most closely correlated with the second component (cos^2 equals to 0.41 and 0.35 respectively). F2 (y-axis) is assumed to be a proxy of the regulation dimension; however, its interpretation has to take into account the fact that the position of a particular data point/municipality may result from different situations, which are consistent with the regulation as it has been described. For example, heavily regulated municipalities contain a high proportion of protected areas and attract loyal tourists with long average stays; or have low tourist intensities in terms of overnights per km^2. The matching of the Cartesian plane and the Weaver Model was established by setting the thresholds that discriminate between high and low levels of tourist intensity and regulation in correspondence with the median value of each principal component.

Second qualitative analysis. The impact of local context on WHS management

The effects of local context on stakeholder engagement in decision-making about WHS management and development were investigated through an online questionnaire sent out to the mayors of WHS municipalities and mountain municipalities bordering on the WHS, all of which are experiencing a variety of tourism-driven socio-economic conditions, which were identified through the first quantitative analysis. The area includes almost exactly half of the small municipalities analyzed in the first quantitative analysis: all the 100 WHS municipalities and the 43 neighboring mountain municipalities (Figure 3). Mountain neighbors are assumed to be most interested in, and affected by, the WHS management and development due to their physical-geographical proximity: the history of the Dolomites has shown that proximity matters in local development, as it leads to imitation and/or diffusion (Bätzing, 2005).

Notwithstanding the fact that there is ample evidence that the views of local politicians tend to be shared by only a limited number of their constituents, in this case the mayors' expectations are assumed to be a reasonable proxy for those of the population of which he/she is the elected representative. This is firstly because mayors have been elected by universal suffrage since 2000, instead of being appointed by councilors, as previously happened (Italian Legislative Decree 276/2000). The electoral process – mayoral candidates present a list of people that support him/her and a program that brings together the visions and expectations of the business and civic representatives of the community – enforces the links between mayoral candidates and councilors and between mayoral candidates, councilors and local stakeholders. Once elected, mayors assume positions of local responsibility towards the community and its representatives. The ability of mayors – as members of a "political elite" at the most local of levels – to represent their electorate, has been confirmed by comparative case study analyses on governance, participation and sustainability carried out in European rural areas (Italian, French and English) (Ruzza, Bozzini, Crivellari, and Petrella (2009). Secondly, if mayors are actually in charge of, and responsible for, the implementation, at local governance level, of both the WHS General Strategy and the Strategy for Sustainable Tourism, it is to be hoped, indeed, that they are legitimate spokespersons for the territories they administer/represent. Thirdly, the questionnaire was administered in the early stages of the WHS management, when the role of local administrations in the planning process was mainly advisory; only later, in 2015, were pioneering participatory tools implemented to promote wider stakeholder participation.

The questionnaire included 24 closed-ended questions, either single replies or a two-point agreement/disagreement scale. It was created using the software LimeSurvey, tested with 15 mayors, and then sent out in April/May 2013 as an e-mail link, with an explanation of the purpose of the study. Questions were divided into four parts. Excluding the sections examining the mayoral profile (province, age, gender, education, work sector) (preliminary question group) and the economy of the municipality (diversified, tourism-driven and tourism seasonality) (first question group), question design combined and adapted content from studies which, however, only investigated the following issues singly: the significance of UNESCO recognition for WHS future development (Rebanks Consulting Ltd., 2009) (second question group); the relationship between tourism and UNESCO recognition (Jimura, 2011; Kaltenborn et al., 2013; Strickland-Munro & Moore, 2013) (third question group); and local stakeholder involvement in WHS (tourism) management and development (Tosun, 2006) (fourth question group).

A copy of the questionnaire can be found in the Supplemental Data for the web based version of this paper.

The questionnaire redemption rate was 60% (85 out of 143 mayors): 61 out of 100 were mayors from the WHS and 24 out of 43 were from the municipalities bordering on the WHS. The responding municipalities, both in and around the site, well represent the whole area investigated at this stage (143 municipalities), both in terms of socio-economic conditions (Weaver model's tourism ideal types) and population, i.e. over 95% of replies come from very small municipalities (average pop. less than 1500).

The responses to the questionnaire addressing the core research questions – WHS management (fourth question group) and WHS future development (second and third question groups) – were analyzed separately using descriptive statistics. Exploratory general information on mayoral perceptions was provided by average values; similarities and differences between WHS and non-WHS municipalities and tourism development types were identified by comparing the values evaluating the related mayoral perceptions concerning the issues at hand. Information on WHS management was extracted by analyzing and comparing the values linked to who should make decisions in local development and the role of stakeholders in decision-making (questions D14 and D15). This data was complemented by information on the obstacles impeding (D17), and means fostering (D16), the most effective stakeholder participation. Information on future development was provided by analyzing and comparing the values linked to the opportunities connected to WHS status (D4) and the role of tourism in this regard (D8). The desired model of tourism development (D9) and reservations about UNESCO recognition (question D5) complemented this information.

Main findings

The tourism-driven socio-economic development in and around the WHS – the socio-economic and institutional context – and the effects that this local context have on stakeholder engagement in decision-making about WHS management and on development at the municipal level are the three main outcomes of the research. The first resulted from the first exploratory quantitative analysis and the other two from the questionnaire to mayors.

Tourism-driven socio-economic development in the Dolomites

The municipalities in the Dolomite study area classified into the ideal types of the Weaver model (2000) confirm that tourism-driven socio-economic development has diversified and prosperous mass destinations currently coexist with marginal and remote emerging destinations. One striking finding was the fact that WHS municipalities were revealed to be mainly sustainable and prosperous,

Table 3. Tourism-driven socio-economic development in the Dolomites.

Tourism development	Weaver model ideal types	WHS municipalities	Non-WHS municipalities
Sustainable	DAT	21.0	13.5
	SMT	53.0	24.0
Unsustainable	CAT	4.0	49.5
	UMT	22.0	13.0
Total (%)		100.0	100.0
Total number		100	206

in contrast with non-WHS municipalities, most of which were found to be unsustainable and marginal, and in need of additional support (e.g. from local public institutions, or WHS recognition) to foster local development by promoting place attractiveness, accommodation capacity and supporting services (Table 3).

Sustainability in WHS destinations – mostly mass destinations (53%) and, to a lesser extent, niche destinations (21%) – relies on pre-existing norms covering protected areas whose integrity entitled them to inclusion within the UNESCO property, and will continue to be protected by WHS regulation. Unsustainability, on the other hand, is seen in the mass destinations which already face environmental challenges (22%), i.e. environmentally damaged destinations, which were included in the property to meet the selection criteria for universal geological value (e.g. Marmolada), or those bordering on well-known Dolomite destinations, whose development began during the tourism boom of the 1970s (Bätzing, 2005). This proximity has meant that private accommodation and second homes were (and still are) a preferred form of accommodation, which the analysis considers indicators of low regulation/unsustainability.

The prevailing unsustainability discovered in non-WHS neighboring mountain municipalities is found in the niche destinations (50%) coping with traditional sectors in decline and/or the early stages of tourism development; and in environmentally impacted mass destinations (13%) whose original development was based on nearby demand markets. Sustainability, in turn, depends on a considerable number of sustainable mass (24%) and niche (14%) destinations.

Local context and WHS management

Municipalities envision a multi-level (province and municipality) and multi-stakeholder decision-making process in which both public and private actors are expected to participate. Municipalities largely agreed that the provinces – as founding members of the Dolomite Foundation – should not have the leading role in decision-making (25%) (Table 4); they themselves want to be decision-makers, in consultation with the other stakeholders, i.e. the Destination Management Organizations (DMOs) (either regional, provincial or local) (87%) – which are the institutionally legitimate bodies charged with

Table 4. The impact of local context on WHS management. Percentage values.

Issues investigated	Total	WHS recognition		Tourism development type			
		WHS	Non-WHS	CAT	DAT	SMT	UMT
Leading role in decisions							
The province	24.7	23.0	29.2	45.5	23.1	22.0	20.0
Municipality with province	57.6	57.4	58.3	72.7	61.5	56.1	50.0
Municipality with DMOs	87.1	86.9	87.5	81.8	92.3	87.8	85.0
Municipality with residents	74.1	72.1	79.2	81.8	69.2	75.6	70.0
Stakeholder participation							
Direct participation	30.6	32.8	25.0	26.4	23.1	36.6	20.0
Consultation on decisions	76.5	80.3	66.7	63.6	84.6	75.6	80.0
Being informed about decisions	80.0	85.2	66.7	72.7	92.3	75.6	85.0
Total number	85	85	85	85	85	85	85

tourism management – local residents (74%), and the province (58%). Stakeholders, however are only expected to participate indirectly (31%) in decision-making, and it is considered sufficient that they merely be informed of (80%), and/or consulted on (77%), decisions taken by the municipality.

Both whether or not a municipality was part of the WHS, and its tourism development type, was found to affect these general perceptions (Table 4). The expectation that the province, either alone, or in consultation with municipalities and residents, should have a leading role in decisions is much higher in non-WHS municipalities than in WHS municipalities, and in marginal destinations (CAT) than in mass or niche destinations (DAT). The (advisory) role of DMOs is perceived to be greater where tourism-driven socio-economic well-being is growing (DAT), or is already evident (mass destinations), than in marginal areas. On the other hand, there is more direct participation, consultation and informing of stakeholders within the multi-level and multi-stakeholder decision-making process in WHS municipalities and mass destinations than in neighboring municipalities and other type of destinations. The greater autonomy and power of stakeholders in decision-making that is perceived in these destinations is probably the result of both the general level of tourism-driven socio-economic well-being (particularly in the case of cable car operators and hotel owners), and the opportunities envisioned by UNESCO recognition. Equally, weak socio-economic conditions, together with lack of UNESCO recognition, seem to disempower stakeholders, resulting in their under-participation, and dependence on public institutions.

Many economic and social obstacles need to be overcome in all municipalities in order to empower stakeholders to participate effectively; marginal destinations face the greatest hurdles. Their weak business culture, and lack of skills and public and private resources limits their ability to achieve the critical mass of firms and entrepreneurs needed for meaningful participation in decision-making. In these (and, indeed, other) destinations, public incentives for entrepreneurship have continued to be widely requested, and are considered the main tool to increase local stakeholder involvement, ranked above training and education on key local development issues, or the use of Internet platforms.

Local context and future development in the Dolomites

UNESCO recognition was considered an opportunity by all the municipalities surveyed: as a catalyst for socio-economic development and the improvement of quality of life, (60%), and as a guarantee of heritage conservation (35%) – considered at risk (24%) or still intact (11%). Place marketing and branding under the UNESCO label is undervalued (6%) (Table 5). Tourism is considered the sector most likely to foster the socio-economic development envisioned with WHS status (82% of the surveyed mayors), as part of a diversified economy, however, rather than one which has put all its eggs in the basket of tourism.

Table 5. Local context and future development. Percentage values.

Issues investigated	Total	WHS recognition		Tourism development type			
		WHS	Non-WHS	CAT	DAT	SMT	UMT
1 Possible use of WHS status							
Heritage conservation – intact	10.7	6.6	21.7	27.3	7.7	5.0	15.0
Heritage conservation – at risk	23.8	21.3	30.4	36.4	23.1	22.5	20.0
Place branding	6.0	4.9	8.7	0.0	7.7	10.0	0.0
Socio-economic development	59.5	67.2	39.1	36.4	61.5	62.5	65.0
Total %	100.0	100.0	100.0	100.0	100.0	100.0	100.0
2 Future role of tourism							
No change of present role	17.6	13.1	29.2	18.2	7.7	22.0	15.0
Main economic activity	9.4	11.5	4.2	9.1	0,0	7.3	20.0
Important but ancillary	28.2	26.2	33.3	45.5	46.2	17.1	30.0
As important as other activities	44.7	49.2	33.3	27.3	46.2	53.7	35.0
Total %	100.0	100.0	100.0	100.0	100.0	100.0	100.0
Total number	85	85	85	85	85	85	85

General perceptions in this second set of results are also affected by whether or not a municipality is within the WHS, and by the local type of tourism development (Table 5). In addition to scoring above average, expectations of socio-economic development are much higher in WHS municipalities than in non-WHS municipalities and in mass destinations and sustainable niche destinations (DAT) than in marginal destinations (CAT), where expectations are very low. Neighboring mountain municipalities, where few environmental protection restrictions are in place, and marginal destinations, with a strong sense of place, in contrast, sense greater opportunities in heritage conservation than do WHS municipalities, or those in which tourism-driven socio-economic well-being is already evident.

The role of tourism, envisioned within a diversified economy, confirms these different perceptions. WHS municipalities and mass destinations expect the role of tourism to be greater than do the other municipalities/destinations: non-WHS municipalities and niche destinations (CAT and DAT) believe that tourism's role will not change, or that it will grow, but will continue to be ancillary in the local economy. In the case of the unsustainable mass destinations, the expectations of no change reflect the reality that tourism already plays a prominent role.

It is interesting how the data on the desired models of tourism development, and on the mayors' main reservations about UNESCO recognition (expressed by only 18 mayors), complement these insights. Irrespective of the municipality's WHS status or tourism development type, mayors expected tourism development to be sustainable (98.5%), embracing the valorization of traditions and folklore, rather than just relying on the site's natural beauty. Although mass tourism models seem to have been superseded in the envisioned tourism development, mayors, in WHS municipalities and mass destinations in particular, perceived WHS environmental protection measures to be potentially restrictive of autonomous local decision-making and future planning. This concern suggests that these municipalities – irrespective of whether or not they fall within the perimeters of protected areas – are not fully aware of the fact that most of the WHS measures rely on existing regulations protecting sensitive areas. Mayors also sense a top-down approach to both WHS conservation and valorization measures and disparities between and within Dolomite provinces about the application and the potential benefits of these measures.

Discussion

Tourism development in the Dolomites is revealed to have diversified, both within and outside the WHS; most of the WHS municipalities are sustainable and prosperous, contrasting with the non-WHS municipalities, which are mainly unsustainable and marginal. The exploratory analysis shows that, in this case study, perceptions about WHS management and future development are influenced by local context. Although a multi-level and multi-stakeholder decision-making process was expected by most municipalities, stakeholders' ability to make decisions was perceived as higher in the municipalities within the WHS, and where tourism-driven socio-economic well-being is already evident (mass destinations). In these municipalities, more socio-economic development opportunities are expected, with tourism driving development as part of a diversified economy. In contrast, in marginal destinations and outside the WHS, stakeholders are perceived not to have sufficient scope for participation, and to be dependent on public policies and financial incentives. However, greater heritage conservation opportunities are expected in these municipalities, where the (ancillary) role of tourism is not expected to change to any great extent since these municipalities seem to have little awareness of the extent to which the positive ecological image of WHS status can be used to catalyze local tourism development. However, as (peripheral) regions adjacent to protected areas have demonstrated (Getzner et al., 2014), strategies can be designed to extract value from this positive image, thereby creating (or attracting) new businesses and attracting new visitors, and residents. These results have been interpreted in the light of the connection between the integration of the Weaver model with the different norms driving destination development (Weaver, 2012) and stakeholder engagement in decision-making (Tosun, 2006). In recent years, UNESCO recognition has accelerated the incipient sustainability norm and redirected the growth-driven development (organic path) of mass destinations towards hybrid development (induced path) (Weaver, 2012). Stakeholders who had previously been marginalized by powerful interest groups with pro-growth goals (cable car operators and hotel owners in particular) have been drawn into the process of managing this adaptation, at least to the extent that they are informed and consulted. The territorial administrations of most of the mass destinations have introduced a range of environmental regulatory tools, intended to guarantee long-term economic development by maintaining the integrity of the environment. WHS areas have been included in an integrated SMT strategy of spatial differentiation (Weaver, 2012), as sustainable (niche) destinations, also regulated by WHS measures, within a larger area including mass tourism destinations. The minor reservations expressed by the mayors of some of the mass destinations about UNESCO recognition suggest a (cultural) past path dependence on growth-driven paths, which were both more extensive and less regulated than those now available. The reluctance of powerful interest groups to give up the considerable autonomy they previously enjoyed in planning and driving development is closely connected to this path dependence. These concerns also suggest that public institutions somehow perceive sustainable development to entail limitations or ethical imperatives, rather than long term opportunities to gain and add value to WHS recognition, e.g. through the travel and tourism experience.

Marginal (economically and socially unsustainable) areas, and/or emerging destinations, have now ceased to imitate the mass tourism model which, while it undoubtedly brought certain benefits in the past, had done so at considerable environmental and social cost. Their awareness of environmental and social concerns, along with the UNESCO recognition, is leading them to follow an incremental path of regulation-driven development towards sustainable development. However, conserving the natural mountain landscape, revitalizing primary industries and, potentially, integrating them with tourism, seem to remain challenges for them rather than opportunities. It appears that a combination of demographic, social and economic obstacles is preventing stakeholders from really understanding, and exploiting, the WHS recognition as a catalyst of socio-economic well-being. Capacity building is needed: first, to increase awareness and, second, to empower stakeholders to move towards this development path. Local institutions are perceived as responsible for creating the necessary economic (e.g. incentives) and social conditions for stakeholder empowerment and engagement.

Conclusion

This paper investigates the influence of local context on WHS management and development, a new study area in the literature and practice of this domain. Its value is theoretical, methodological and potentially practical. It connects and tries to cross-fertilize the relevant bodies of managerial literature – focused on protected areas, natural WHS and sustainable tourism – to explore the effects that tourism-driven socio-economic conditions have on stakeholder engagement in decision-making and future (sustainable tourism) development, inside and outside a WHS. Building on selected theoretical constructs from this literature the Weaver model (2000), its integration with destination development paths (Weaver, 2012), and forms of participation in decision-making (Tosun, 2006) – an integrated research design has been developed with mixed methods and tools of analysis used to apply it to the collection of case study evidence. The most innovative is an exploratory framework measuring the (un)sustainability of niche and mass destinations, built by quantitatively applying the Weaver model (2000), and used to study the tourism-driven socio-economic conditions within and outside a WHS.

The results show that in the Dolomites natural WHS – a complex trans-regional, serial site – local context affects stakeholder engagement in decision-making and planning future development. The UNESCO recognition seems to be creating a deeper divide than that already produced by the current diversity of tourism development types in the Dolomites. UNESCO has strengthened direct and induced forms of stakeholder participation in decisions, and has raised expectations of further socio-economic development in the WHS and especially where tourism-driven socio-economic well-being is already high. However, it seems not to have affected the area outside the WHS, or the marginal destinations, either in terms of participatory decision-making, or in terms of the economic development envisioned by stakeholders, or, indeed, of tourism as a lever of development. Pronounced socio-economic divides are the main barriers.

It is important to remember that this evaluation reflects the perceptions of mayors in the very early stages of the Dolomites WHS planning process (2013), which has lasted six years (2009–2016), during which top-down approaches have prevailed, and the role of the local public administrations has only been advisory. Pioneering participatory planning techniques were introduced in the intermediate and final planning phases (2015–2016), recognized by UNESCO as an example of global best practice. The evolution of the process of management planning over time shows the active involvement of a more complex pool of public and private stakeholders at different governance levels, inside and outside the WHS. This evolution allows the municipalities' wish to be involved in the planning process in consultation with other stakeholders (i.e. local DMOs) to be interpreted, not as questioning the entire governance system of the WHS, but as directly supportive of the central coordinating and harmonizing role played by the Provinces in the Dolomites Foundation.

Although exploratory, the lessons learned from the case study allow for some preliminary managerial recommendations and practical suggestions on how to involve stakeholders and communities more effectively. A first set of considerations concerns the territorial and sectoral integration of WHS management. Stronger trans-regional and interregional collaboration – between and within provinces and between and within WHS and non-WHS municipalities – is challenging, but necessary for the effective integrated management of this trans-regional, serial site. The perceived disparities between Dolomites territories which this study has detected have been confirmed by analyses carried out after this study (Scuttari & Della Lucia, 2016). In the opinion of the stakeholders participating in the #Dolomites2040 meetings held in 2015, transcending the administrative boundaries between the Dolomite territories is even more necessary than moving away from the traditional touristic monoculture of mass destinations. These boundaries, in fact, may be the current determinants of the asymmetries in WHS management, and their effects – perceived and actual – on all Operating Networks (e.g. education and scientific research, landscape heritage and protected areas, the promotion of sustainable tourism, mobility) and, in particular, on local development, both tourism-driven and not.

A second set of considerations concerns participatory WHS management. During the management planning process, the Dolomite Foundation took steps towards the co-management of resources by government and local resource users (Borrini-Feyerabend et al., 2013): multi-level and multi-stakeholder participation was introduced along with pioneering participatory approaches. This natural evolution inevitably reflects the considerable time required to deal with the obstacles encountered in designing and implementing effective channels for sharing power and responsibility among several stakeholders, especially in trans-regional and serial WHSs. Operational difficulties in adopting broad direct stakeholder participation will inevitably remain due to the complexity of the site; this fact partially supports the theoretical insight that, even in developed countries, participation is partly induced (Tosun, 2006). The main obstacles to stakeholder participation that the empirical analysis has revealed – the inadequacy or even non-existence of knowledge, competences, resources and socio-economic networks in local communities – suggest that tools for capacity and networking building may be a crucial lever for increasing stakeholder awareness of the opportunities linked to WHS status and getting them more, and more widely, involved, whether directly or indirectly. Different players can play a role in this process: local public institutions – to overcome their incentive logic; European Union projects fostering rural development and social inclusion, e.g. Leader projects (European Commission, 2006); local business networks and/or associations. Examples of participatory engagement tools include education and training, the establishment of local committees, discussion forums, stakeholder dialogue and the involvement of young people (e.g. schools) in projects connected to the future; visitor education and information is a complementary tool to foster both tourist engagement and to create selling propositions which embed the site's ecological value. The Provinces of Bolzano and Trento – which have taken a lead (because of their greater autonomy) in the development of integrated, participative processes in which tourism is an agent in rural community development – may play a leading role. Greater territorial collaboration would also foster the exploitation of synergies in these domains, allowing best practices to be shared, the creation of joint standards, and possibly cooperative marketing and operational innovation.

Limitations and future research

Further research on both local context and its effects is needed to overcome the limits of this exploratory study. With respect to local context, the range of data on tourism intensity and regulation (Weaver model) used to assess tourism-driven socio-economic conditions at the inter-provincial municipal level must be updated and the set of explanatory variables of regulation must be widened. The development of qualitative analysis to understand the specific determinants affecting the classification of the Dolomite municipalities into the different tourism development types is also necessary. Regarding the effects of local context, a more complex pool of public-private stakeholders at different governance levels (e.g. members of the Board of Supporters, representatives of different economic sectors, etc.) needs to be included when evaluating shareholder involvement perceptions of WHS decision-making and future development. A greater variety of tools for, and approaches to, information collection should be used (e.g. in depth interviews, focus groups involving experts/representative stakeholders, etc.). Longitudinal analyses should be made to monitor both the evolution of local contexts and the effects that WHS management will have on future development, socio-economic conditions and stakeholder engagement. More information on local-specific and stakeholder-specific issues and their collection over time will allow us to experiment with factorial analysis techniques (multiple correspondence analysis for example) to detect and represent potential data structures.

Note

1. The word property is often used by UNESCO to describe the land designated as a WHS. It does not imply UNESCO's ownership of that land (which normally remains in its pre-designation ownership).

Acknowledgements

The authors would like to acknowledge methodological support received from Professor Pier Luigi Novi Inverardi (University of Trento), Dr Anna Scuttari (Eurac Research), Dr Pietro Marzani (University of Trento) and Professor Giuliana Passamani (University of Trento). Helpful comments and assistance from the journal editor, the guest editors and the anonymous reviewers in improving this paper are also gratefully acknowledged.

Disclosure statement

No potential conflict of interest was reported by the authors.

ORCID

Mariangela Franch (iD) http://orcid.org/0000-0002-2312-2823

References

Aas, C., Ladkin, A., & Fletcher, J. (2005). Stakeholder collaboration and heritage management. *Annals of Tourism Research, 32*(1), 28–48.

Bätzing, W. (2005). *Le Alpi. Una regione unica al centro dell'Europa* [The Alps. A single region in the center of Europe]. Torino: Bollati Boringhieri.

Becken, S., & Job, H. (2014). Protected Areas in an era of global–local change. *Journal of Sustainable Tourism, 22*(4), 507–527.

Bianchi, R.V. (2002). The contested landscape of world heritage on a tourist Island: The case of Garajonay national park, La Gomera. *International Journal of Heritage Studies, 8*(2), 81–97.

Borrini-Feyerabend, G., Dudley, N., Jaeger, T., Lassen, B., Pathak Broome, N., Phillips, A., & Sandwith, T. (2013). *Governance of protected areas: From understanding to action.* Best Practice Protected Area Guidelines Series No. 20. Gland: IUCN-Protected Areas Programme.

Borrini-Feyerabend, G., Pimbert, M., Farvar, M. T., Kothari, A., & Renard, Y. (2008). *Sharing power: A global guide to collaborative management of natural resource.* London: Earthscan.

Boyd, S.W., & Dallen, J.T. (2006). Marketing issues and World Heritage Sites. In A. Leask & A. Fyall (Eds.), *Managing World Heritage Sites* (pp. 55–68). London: Butterworth-Heinemann.

Buckley, R. (2004). The effects of World Heritage Listing on tourism to Australian National Parks. *Journal of Sustainable Tourism, 12*(1), 70–84.

Butler, R.W. (1980). The concept of a tourist area cycle of evolution: Implications for management of resources. *Canadian Geographer, 24*(1), 4–12.

Cellini, R. (2011). Is UNESCO recognition effective in fostering tourism? A comment on Yang, Lin and Han. *Tourism Management, 32*, 452–454.

Conradin, K., Engesser, M., & Wiesmann, U. (2014). Four decades of World Natural Heritage – How changing protected area values influence the UNESCO label. *Die Erde, 146*(1), 34–46.

Elmi, M. (2014). *Turismo Sostenibile nelle Dolomiti, approfondimento dell'analisi. Questionario rivolto ai turisti nella stagione estiva 2013* [Sustainable tourism in the Dolomites. Summer season 2013]. Bolzano: EURAC.

Elmi, M., & Perlik, M. (2014). From tourism to multilocal residence?. *Journal of Alpine Research, 102*(3), 2–14.

European Commission. (2006). *The leader approach. A basic guide.* Luxembourg: Office for Official Publications of the European Communities.

Getzner, M., Lange Vik, M., Brendehaug, E. & Lane, B. (2014). Governance and Management Strategies in National Parks: Implications for Sustainable Regional Development. *International Journal of Sustainable Society, 6*(1/2), 82–101.

Gianolla, P., Panizza, M., Micheletti, P., & Viola, F., (2008). Nomination of the Dolomites for inscription on the World Natural Heritage List UNESCO. Retrieved March 1, 2017, from http://www.provincia.belluno.it/nqcontent.cfm?a_id=4029

Gios, G., Goio, I., Notaro, S., & Raffaelli, R. (2006). The value of natural resources for tourism: A case study of the Italian Alps. *International Journal of Tourism Research, 8*, 77–85.

Hall, C.M., & Piggin, R. (2002). Tourism business knowledge of World Heritage Sites: A New Zealand case study. *International Journal of Tourism Research, 4*(5), 401–411.

ISTAT. (2014). *Il territorio* [The Italian national territory]. Roma: Author. Retrieved March 1, 2017, from http://www.istat.it/it/files/2014/11/C01.pdf

IUCN-Protected Areas Programme. (2008). *Management planning for Natural World Heritage properties.* Gland: Author.

Jamal, T., & Watt, M. (2011). Climate change pedagogy and performative action: To-wards community based destination governance. *Journal of Sustainable Tourism, 19*(4/5), 571–588.

Jimura, T. (2011). The impact of World Heritage site designation on local communities. A case study of Ogimachi, Shirakawa–mura, Japan. *Tourism Management, 32*(2), 288–296.

Kaltenborn, B.P., Thomassen, J., Wold, L.C., Linnell, J.D.C., & Skar, B. (2013). World Heritage status as a foundation for building local futures? A case study from Vega in Central Norway. *Journal of Sustainable Tourism, 21*(1), 99–116.

Khazaei, A., Elliot, S., & Joppe, M. (2017). Fringe stakeholder engagement in protected area tourism planning: Inviting immigrants to the sustainability conversation. *Journal of Sustainable Tourism.* doi:10.1080/09669582.2017.1314485

Knafl, K., & Breitmayer, B.J. (1989). Triangulation in qualitative research: Issues of conceptual clarity and purpose. In J. Morse (Ed.), *Qualitative nursing research: A contemporary dialogue* (pp. 193–203). Rockville, MD: Aspen.

Landorf, C. (2009). Managing for sustainable tourism: A review of six cultural World Heritage Sites. *Journal of Sustainable Tourism, 17*(1), 53–70.

Leask, A., & Fyall, A. (2006). *Managing world heritage sites*. Oxford: Butterworth-Heinemann.

Li, M., Wu, B., & Cai, L. (2008). Tourism development of World Heritage Sites in China: A geographic perspective. *Tourism Management, 29*(2), 308–319.

Magi, L., & Nzama, T.A. (2009). Tourism strategies and local community responses around the World Heritage Sites in Kwazulu-Natal. *South African Geographical Journal, 91*(2), 94–102.

McCool, S.F., & Stankey, G.H. (2004). Indicators of sustainability: Challenges and opportunities at the interface of science and policy. *Environmental Management, 33*(3), 294–305.

Moscardo, G. (2008). *Building community capacity for tourism development*. Wallingford: CABI.

Mose, I. (2007). *Protected areas and regional development in Europe. Towards a new model for the 21st century*. Hampshire: Ashgate.

Mose, I., & Weixlbaumer, N. (2007). A new paradigm for protected areas in Europe? In I. Mose (Ed.), *Protected areas and regional development in Europe. Towards a new model for the 21st century* (pp. 3–19). Hampshire: Ashgate.

Murphy, P. (1985). *Tourism: A community approach*. London: Methuen.

Nicholas, L.N., Thapa, B., & Ko, Y.J. (2009). Residents' perspectives of a World Heritage Site. The Pitons management area, St. Lucia. *Annals of Tourism Research, 36*, 390–412.

Pedersen, A. (2002). *Managing tourism at World Heritage Sites*. Paris: World Heritage Centre.

Rebanks Consulting Ltd. (2009). *World Heritage status. Is there opportunity for economic gain? (Research and analysis of the socio-economic impact potential of UNESCO World Heritage Site Status)*. Newcastle upon Tyne: Author and Trends Business Research.

Rencher, A., (1998). *Multivariate statistical inference and applications*. New York, NY: Wiley.

Ruzza, C., Bozzini, E.. , Crivellari, P., & Petrella, A. (2009). Europa e Territorio: Governance rurale, partecipazione e sostenibilità [Europe and Territory: Governance of rural areas, participation and sustainability]. Cosenza (I): Rubettino.

Ryan, C., Chaozhi, Z., & Zeng, D. (2011). The impacts of tourism at a UNESCO Heritage Site in China – A need for a meta-narrative? The case of the Kaiping Dialolou. *Journal of Sustainable Tourism, 19*(6), 747–766.

Sandwith, T., Shine, C., Hamilton, L., & Sheppard, D. (2001). Transboundary protected areas for peace and cooperation. Retrieved March 17, 2017, from http://journal-iostudies.org/sites/journal-iostudies.org/files/JIOSfinal_4_0.pdf

Scuttari, A., & Della Lucia, M. (2016). Principi di management delle aree protette e patrimonio naturale. L'efficacia della gestione integrata e partecipata delle Dolomiti UNESCO. [Management of protected areas and natural WHSs. The effectiveness of Dolomites WHS management]. In H. Pechlaner, M. Valeri, & M. Gon (Eds.), *Innovazione, Sostenibilità e Competitività. [Innovation, Sustainability and Competitiveness]. Teoria ed esperienze per la destinazione e l'azienda* (pp. 113–124). Torino (I): Giappichelli.

Scuttari, A., Della Lucia, M., & Martini, U. (2013). Integrated planning for sustainable tourism and mobility. A tourism traffic analysis in Italy's South Tyrol region. *Journal of Sustainable Tourism, 21*(4), 614–637.

Shackley, M. (2006). Visitor management at World Heritage Sites. In A. Leask & A. Fyall (Eds.), *Managing World Heritage Sites* (pp. 83–94). London: Butterworth-Heinemann.

Simpson, K. (2001). Strategic planning and community involvement as contributors to sustainable tourism development. *Current Issues in Tourism, 4*(1), 3–41.

Smith, M. (2002). A critical evaluation of the global accolade: The significance of world heritage site status for maritime Greenwich. *International Journal of Heritage Studies, 8*(2), 137–151.

Steiner, L., & Frey, B. (2011). Correcting the imbalance of the World Heritage List: Did the UNESCO strategy work? Retrieved March 1, 2017, from http://journal-iostudies.org/sites/journal-iostudies.org/files/JIOSfinal_4_0.pdf

Strickland-Munro, J., & Moore, S. (2013). Indigenous involvement and benefits from tourism in protected areas: A study of Purnululu National Park and Warmun Community, Australia. *Journal of Sustainable Tourism, 21*(1), 26–41.

Su, Y.-W., & Lin, H.-L. (2014). Analysis of international tourist arrivals worldwide: The role of world heritage sites. *Tourism Management, 40*, 46–58.

Timothy, D.J., & Nyaupane, G.P. (2009). *Cultural heritage and tourism in the developing world: A regional perspective*. London, Routledge.

Tosun, C. (2006). Expected nature of community participation in tourism development. *Tourism Management, 27*(3), 493–504.

Tourtellot, J. (2006). World heritage destinations rated. *The National Geographic Traveler, 23*(8), 12–124.

Trau, A., & Bushell, R. (2008). Tourism and indigenous people. In S.F. McCool & R.N. Moisey (Eds.), *Tourism, recreation and sustainability: Linking culture and the environment* (pp. 260–282). Wallingford: CABI Publishing.

UNESCO Dolomites Foundation. (2013). Management progress report 2013. Retrieved March 17, 2017, from http://www.dolomitiunesco.info/wp-content/uploads/2015/05/FD4U_management-progress-report-2013.02.15_EN.pdf

UNESCO Dolomites Foundation. (2015). *#Dolomites 2040*. Retrieved March 1, 2017, from http://www.dolomitiunesco.info/wp-content/uploads/2015/11/FD4U-D2040_PP_report-finale1.pdf

UNESCO. (1972). Convention concerning the protection of the world cultural and natural heritage. Retrieved March 1, 2017, from http://whc.unesco.org/en/conventiontext/

UNESCO (2015). Operational guidelines for the implementation of the World Heritage Convention. Retrieved March 1, 2017, from http://whc.unesco.org/en/guidelines/

Weaver, D.B. (2000). A broad context model of destination development scenarios. *Tourism Management, 21,* 217–224.

Weaver, D.B. (2012). Organic, incremental and induced paths to sustainable mass tourism convergence. *Tourism Management, 33*(5), 1030–1037.

Wiesmann, U., Liechti, K., & Rist, S. (2005). Between conservation and development. *Mountain Research and Development, 25*(2), 128–138.

Yin, R.K. (2014). *Case study research: Design and methods* (5th ed.). Thousand Oaks, CA: Sage Publications.

Yüksel, F., Bramwell, B., & Yüksel, A. (1999). Stakeholder interviews and tourism planning at Pamukkale, Turkey. *Tourism Management, 20*(3), 351–360.

Estimating the value of the World Heritage Site designation: a case study from Sagarmatha (Mount Everest) National Park, Nepal

Nabin Baral, Sapna Kaul, Joel T. Heinen and Som B. Ale

ABSTRACT

This paper estimates the economic value of World Heritage Site (WHS) designation for the Sagarmatha (Mount Everest) National Park, Nepal. In 2012, entrance fees were $30 per international visitor; lower fees apply to South Asian visitors, and no fees to domestic visitors. We surveyed 522 international visitors to the Park in 2011 to elicit their willingness to pay (WTP) for access, using the contingent valuation method. Logistic regression results show that bid amounts, gender, age, educational attainment, use of a guide, length of stay in the park, information about park substitutes, and knowledge about the park's WHS designation predicted visitors' WTP decisions. The median WTP amount was US$90.93 per trip; 63.8% of visitors were willing to pay more than the existing entry fee. The revenue maximizing entry fee was $80 per trip. Knowledge about the park's WHS designation prior to their trip contributed $16.39 to the median WTP: better marketing of the site's WHS status could add up to US$ 566, 619 to the site's annual income. Given that many protected areas now suffer falling public sector financial help, accurate knowledge of WTP is increasingly key to supporting sustainable management in WHS sites, and in justifying tourism to them.

Introduction

The World Heritage Convention of 1972 is instrumental in protecting natural and cultural areas that are invaluable to humankind by including them on the World Heritage Site (WHS) list (UNESCO, 2014). To date, 165 countries are parties to the Convention, and 1052 sites fall under WHS designation worldwide. Of these, most (814) are cultural sites, while 203 are natural sites. The remainders (13) are mixed sites containing both cultural and natural amenities. Most natural sites are additionally protected within their countries as conventional protected areas, and the vast majority are national parks. The WHC is important in recognizing sites that are of unique international importance and its headquarters in the United Nations Education, Scientific and Cultural Organization (UNESCO) in Paris maintains a trust fund to aid developing countries in maintaining their WHSs (whc.unesco.org/en/list). Countries are motivated to inscribe natural and cultural sites on the World Heritage List because listed sites often receive prestige, increased tourism, international cooperation, and financial assistance for conservation/preservation from the World Heritage Fund. As such, the World Heritage List

Supplemental data for this article can be accessed at https://doi.org/10.1080/09669582.2017.1310866.

itself can become a "brand" for marketing sites and attracting visitors (Heinen, 1995a; Ryan & Silvanto, 2009; Shackley, 1998) and WHS designation has substantially contributed toward promoting tourism to listed sites (Frey & Steiner, 2011). An empirical analysis shows that the number of WHSs within a country is positively correlated with international tourist arrivals (Su & Lin, 2014).

The development of tourism in WHSs could be a boon or bane for several reasons. An increase in visitor numbers due to the WHS designation can positively influence employment creation and revenue generation (Heinen, 1995a; Nyaupane, Lew, & Tatsugawa, 2014) and, because the development of tourism requires good infrastructure and civil order, WHSs can promote development, build support for conservation and advance social harmony (Jha, 2005). On the other hand, tourism development can negatively affect conservation and protection of properties included on the World Heritage List (Batisse, 1992; Okech, 2010). For example, a 130-fold increase in tourist arrivals at the Jiuzhaigou WHS in China over a 30-year period has caused pollution and poses threats to biodiversity (Gu, Du, Tang, Qiao, Bossard, & Deng, 2013). This leads to two important questions: what is the economic value of the WHS designation to visitors and how can an understanding of this value improve the management of listed sites?

WHSs in the developing world frequently suffer from a lack of effective site management (Landorf, 2009) and, as a consequence, maintaining their unique characteristics might be difficult if not impossible (Alberts & Hazen, 2010). Although tourism contributes substantially toward revenue generation for managing protected areas, including WHSs (of WH natural sites), the vast majority are national parks and all are under some form of national protection (Heinen, 1995a), lack of adequate financial resources is one of the major reasons for ineffective management of protected areas in developing countries (Heinen, 2012). It is now well-accepted that protected area managers must devise alternative mechanisms to pay for the maintenance and protection of areas, including the concessions and other infrastructure used by tourists, given the global trend toward decreasing governmental budgets in the conservation sector (Whitelaw, King, & Tolkach, 2014). In a major review of research priorities regarding tourism in protected areas, Eagles (2014) discussed 10 separate issues of importance. Two of those, protected area financing and pricing policy, address budget issues directly. The other issues he discussed, such as assessing economic impacts, promoting public support and building professional competence in dealing with tourism, all require expanded budgets for their successful implementation.

While there is a growing literature on methods to capture rents in support of protected areas from indirect users in the form of carbon sequestration and other ecosystem services (Whitelaw et al., 2014), one of the easiest ways to increase protected area budgets is to increase direct user fees for those who visit (Heinen, 2010). Numerous studies worldwide have shown that there is a general acceptance of higher user fees for many protected areas in many contexts (e.g. Casey, Brown, & Schuhmann, 2010; Lee, Lee, Kim, & Mjelde, 2010), but such acceptance typically varies by any number of factors including the type of protected area, the average length of visit, the gender and age of respondents (see Kline, Cardenas, Duffy, & Swanson, 2012) and less tangible issues (for international tourists) such as the historical, economic, and political context of the society involved (Buckley, 2003).

For these reasons, instituting appropriate entry fees for sites that attract a large number of visitors is an effective way to generate adequate financial resources to mitigate some negative impacts of tourism. Increasing any positive impacts for the site and visitors alike, while reducing negative impacts of tourism are critical management goals in WHSs (Borges, Carbone, Bushell, & Jaeger, 2011). One way of achieving these goals is to identify and examine the willingness to pay (WTP) of visitors to visit any protected area. Additionally, those with WHS designation offer extra opportunities to study WTP in the context of the international designation. In all cases, information generated can be used to determine the extent of extra revenue generation with potentially increased entry fees in the future (e.g. Casey et al., 2010).

This study examines international visitors' WTP to visit Sagarmatha (Mount Everest) National Park, Nepal, which was inscribed on the World Heritage List in 1979 for its extraordinary scenery and geological features. Domestic visitors, or visitors from other Southeast Asian countries, are not considered here because they are relatively few in number and park entry fees are much lower for them than for people from other parts of the world. Nepal has had a long history of adopting national polices and innovative strategies to foster conservation and manage its national parks and reserves and their surrounding buffer zones and community forests (Heinen & Rayamajhi, 2001; Timilsina & Heinen, 2008) and it is party to a number of international conservation agreements. But, as a least-developed country, Nepal is dependent on foreign aid for up to 80% of its development budget and has a history of facing strong barriers to implementing a number of national and international conservation policies, in part due to political unrest (Baral & Heinen, 2006), but mostly due to weak agencies and limited finances in the conservation sector. This has, for example, hindered the implementation of the Convention on International Trade in Endangered Species (CITES) within the country (Dongol & Heinen, 2012; Heinen, Yonzon, & Leisure, 1995) and its own national non-timber forest products policy (Shrestha-Acharya & Heinen, 2006).

There is increasing awareness worldwide about the importance of properly evaluating natural resources for national economies and considering ways to increase revenues to better manage natural areas and species (Barbier, 2014; Heinen, 1995b). Here we survey international visitors to Mount Everest National Park to elicit their WTP for increased entry fees and we identify visitor characteristics that explain variations in WTP for increased entry fees. We then estimate the mean and median WTP amounts and the use value that could be generated by knowledge of WHS designation using the contingent valuation method because of its design simplicity and suitability to evaluate policy-relevant questions (Hanemann, 1994). Finally, we propose a revenue-maximizing entry fee and discuss policy implications for the sustainable management of World Heritage Natural Sites in general. Specifically, the goals of this study are to survey international tourists to Mount Everest National Park to assess: (1) knowledge about the park's World Heritage Status; (2) added value that visitors place on World Heritage status; (3) WTP more for entry/user fees overall, and because of World Heritage status; and (4) any demographic trends (age, income, gender, etc.) that relate to WTP.

The study area

Sagarmatha (Mount Everest) National Park (E 86° 30'53" to E 86° 99'08" and N 27° 46'19" to N 27° 6'45") was officially established in 1976 to protect unique Himalayan ecosystems within an area of 1148 km^2 (Figure 1), particularly from the negative impacts of tourism that flourished after the first

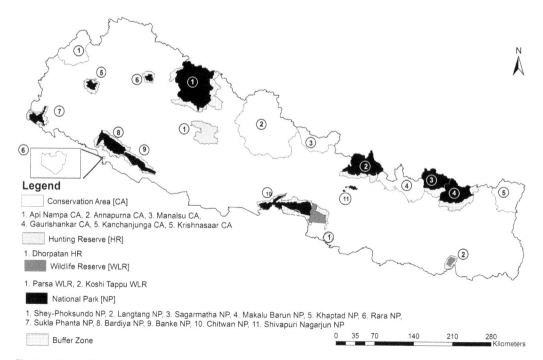

Figure 1. Sagarmatha National Park and its buffer zone (in black represented by 3) along with networks of protected areas in Nepal.

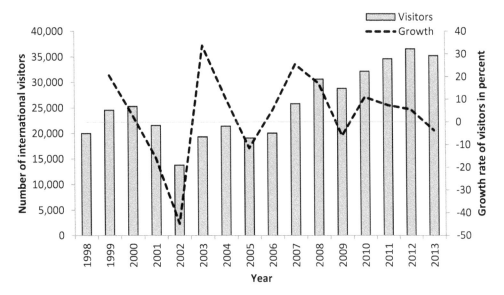

Figure 2. Number of international visitors and their growth rate in Sagarmatha National Park (SNP), Nepal since 1998. The sharp decline in visitor numbers in 2002 can be attributed to the events such as the Royal Palace massacre in Nepal and the 9/11 terrorist attack in the United States. Data source: SNP Tourist Information Center, January 2014.

ascent of Mount Everest in 1953. The park is enclosed by high mountain ranges and elevations vary from 2845 to 8848 m atop Mount Everest, the world's highest peak, which is flanked by 25 or more peaks over 6000 m. An additional area of 275 km^2 was designated as a buffer zone in 2002. The park's rugged terrain and the unique culture of the Sherpa people attract visitors from all over the world. Important wildlife species include Himalayan tahr (*Hemitragus jemlahicus*), serow (*Capricornis thar*), musk deer (*Moschus leucogaster*), goral (*Naemorhedus goral*), red panda (*Ailurus fulgens*) and Asiatic black bear (*Ursus thibetanus*); snow leopards (*Uncia uncia*) recolonized the region in the early 2000s from neighboring reserves after a 40-year absence (Ale, Yonzon, & Thapa, 2007).

The park is part of Nepal, India and China's Sacred Himalaya Landscape, a trans-boundary conservation area (World Wildlife Fund, 2013), and it is bordered by Makalu National Park to the east, Gauri Shanker Conservation Area to the west, and Qomolongma Nature Reserve in the Tibetan Autonomous Region of the People's Republic of China to the north, the largest single protected area in Asia. The park is a popular tourist destination and attracted more than 35,000 international visitors in 2013 (Figure 2), although there was volatility in visitor numbers in some years in the early 2000's due to political unrest caused by the Maoist insurgency that ended in late 2006 (Baral & Heinen, 2006). The primary sources of livelihoods include potato cultivation, livestock herding, trade and tourism. Of them, tourism is a major driver of the local economy. The provision of guides, porters, lodges, and other trekking and mountaineering services are major sources of employment for local people. About 3500 Sherpas reside in five main settlements and other small hamlets. The park's alpine/subalpine grasslands and scrublands (28%) and forests (3%) provide much needed grazing land for livestock, firewood and leaf-litter, and timber. No strict grazing rules were imposed but the firewood collection, timber harvesting, and use and collection of non-timber products, as in any national park, have been strictly banned or regulated.

Methods

Sampling and data collection

Based on our prior experience and literature review, a questionnaire was drafted that was reviewed by 12 experts in the social science field for content, clarity, and organization. The review panel had expertise in sustainable tourism, protected areas management, non-market valuation and question-naire design. After incorporating the experts' comments and suggestions, the revised instrument was tested with a convenience sample of 30 tourists in Kathmandu. In light of respondents' feedback, some minor changes were made to the instrument for clarity. The final four-page questionnaire con-tained 42 questions and was divided into five sections starting with trip characteristics, perceptions of the park, contingent valuation of entry fees, and background information. Both closed-ended and open-ended questions were posed. The questionnaire was written in English and variables were mea-sured on dichotomous, ordinal and continuous scales. The questionnaire is available upon request and the contingent valuation portion is provided in Supplemental Data as Table S1 in the web-based version of this paper.

The questionnaire was self-administered and took about 30 minutes to complete. We collected primary data through visitor surveys in November 2011. It was not feasible to get an accurate sam-pling frame of international visitors a priori, which precluded the use of easier probability sampling techniques such as simple random sampling. We surveyed visitors to the park in two phases, imple-menting cluster sampling and systematic random sampling strategies in sequence. Our sampling frame consisted of 99 hotels in total within the major trekking routes in the park. We performed clus-ter sampling of visitors by randomly selecting 3–5 hotels each day through lottery (in total 20 hotels were surveyed at least once). Because visitors are often distributed indiscriminately among the hotels, each hotel cluster can be seen as heterogeneous. A research assistant was trained to carry out field surveys, and collect accurate and complete data following standard survey procedures. He approached all visitors staying in the sampled hotels, briefed them about the research purpose and requested their participation. After soliciting verbal consent, the research assistant handed question-naires to visitors to fill out on their own and remained in the hotel to answer questions and provide clarification while questionnaires were being completed.

After completing 174 observations, this sampling strategy became increasingly difficult to imple-ment and inefficient. We thus instructed the research assistant to implement a systematic random sampling of visitors at tourist checkpoints within the park, resulting in 348 additional observations. No statistically significant differences were found in important variables between the samples col-lected through cluster sampling and systematic random sampling.

The self-administration strategy for questionnaire survey not only reduced any potential biases associated with the interviewer but also provided greater anonymity for information people may consider sensitive. Of the 614 visitors who were requested to fill out the questionnaire, 522 actually returned the questionnaires and 51 declined to participate. Declining visitors reported that they were tired, not interested in taking the questionnaire, or lacked the experience as they had just began the trek. The overall response rate was 85%. Eighty-nine questionnaires had missing data on one or another key variable, but there was no systematic pattern in missing data points. We used only complete questionnaires, so the effective sample size for our analyses was 433. Data imputation techniques were not used to replace missing values as they do not add information and this sample size was considered adequate for our purposes.

Theoretical and statistical models

The contingent valuation method was used to estimate the WTP of park visitors for increased entry fees. This method has been successfully implemented to assess entry fees for gaining access to pro-tected areas worldwide (e.g. Baral & Dhungana, 2014; Baral, Stern, & Bhattarai, 2008; Bhat, 2003; Mmopelwa, Kgathi, & Molefhe, 2007; Wang & Jia, 2012; Xuewang, Jie, Ruizhi, Shi'en, & Min, 2011). WTP is defined as the amount that an individual will pay for environmental quality improvement by remaining on the same indifference curve, that is

$$V(y - WTP, q^1, p, X; \alpha, \varepsilon) = V(y, q^0, p, X; \alpha, \varepsilon), \tag{1}$$

where V is the indirect utility function, y is the individual income, q corresponds to environmental quality that increases from q^0 to q^1, p denotes the vector of prices, X is a matrix containing respondent characteristics, α denotes preference parameters, and ε is the error component.

The WTP question in our questionnaire asked visitors if they were willing (or not) to pay stated bid amounts as new entry fees to provide more funds to cater to visitors' needs, conserve biodiversity and address management problems. As such, the contingent valuation question was framed using the single-bounded closed ended format (Supplemental Data Table S1).

A visitor's response to a closed-ended contingent valuation question is such that

$$R = \begin{cases} \text{"Yes" if } WTP > Bid \\ \text{"No" if } WTP \leq Bid \end{cases}, \tag{2}$$

where R denotes the response and Bid is the stated entry fee posed in the WTP question. We randomly allocated 10 bid amounts ($20, $30, $40, $50, $60, $70, $80, $90, $100, and $120) in equal proportion to surveyed visitors. Because the entry fee at the time of the survey was $13 and we targeted visitors who were already inside the park, we chose the above bid amounts (or potential future entry fees) higher than the entry fee based on the pilot survey, prior experience of conducting similar studies (Baral & Dhungana, 2014; Baral et al., 2008) and literature review. In the final samples, the frequency distribution of the 10 bid amounts ranged between 6.6% and 12.4%.

The probability of "yes" and "no" responses to the contingent valuation question can be modeled as

$$Pr(R = \text{"Yes"}) = 1 - F(\alpha + \gamma Bid + X'\beta), \tag{3}$$

$$Pr(R = \text{"No"}) = 1 - Pr(R = \text{"Yes"}) = F(\alpha + \gamma Bid + X'\beta), \tag{4}$$

where $F(\cdot)$ is the cumulative density function, γ is the coefficient for the Bid variable, and β is the vector of coefficients for other independent variables. In our analysis, the variables in matrix X belonged to the following broader categories: visitor characteristics, park utilization, perceptions regarding the park. Visitor characteristics included the variables such as age, gender, educational attainment, and employment status of the visitors (see Supplemental Data Table S2). Park utilization included variables such as whether visitors hired a guide while in the park, the number of days spent in the park and the visitors' overall satisfaction with the trip. Variables such as visitors' knowledge about the park's WHS designation and substitution for the park – and their perceptions regarding the authenticity, unimpaired condition and outstanding universal value of the park – constituted the category "perceptions regarding the park". We also included a variable asking whether the visitors were members of any environmental organizations as a proxy for their attitudes towards the environment. All the above mentioned variables were included in the model because previous studies have shown statistically significant relationships of the WTP to such variables (Baral & Dhungana, 2014; Baral et al., 2008; Chen & Jim, 2012; Lee & Han, 2002; Moran, 1994; Richardson, Rosen, Gunther, & Schwartz, 2014; Shultz, Pinazzo, & Cifuentes, 1998; White, Bennett, & Hayes, 2001). We assume that V is linear in income, therefore "y" drops out of Equation (1) (Hanemann, Loomis, & Kanninen, 1991).

Since the response variable was measured on a dichotomous scale, a logistic regression was used to model the data and estimate the mean/median WTP. We used the logistic cumulative density function: $F(\cdot) = 1/\left[1 + e^{\alpha + \gamma Bid + X'\beta}\right]$ in Equations (3) and (4). With that, the log likelihood function for utility maximization for the two sets of responses (yes and no) is

$$\ln L(Bid, X; \alpha, \gamma, \beta) = \sum_i R^y \times \ln \pi^y + R^n \times \ln(1 - \pi^y), \tag{5}$$

where R^y is "1" if individual "i" says "yes" to the stated Bid amount, otherwise "0." Similarly, R^n takes the value "1" if the individual does not agree to pay the bid and "0" otherwise. π^y is the $Pr(R = \text{"Yes"})$ from Equation (3). Equation (5) is solved using a maximum likelihood estimator and the mean/median WTP is $(\alpha + \bar{x}'\beta)/\gamma$, where $\bar{x}$ is the vector containing the means of independent variables.

All analyses were conducted in Stata 13.0 (StataCorp LP, College Station, TX). The confidence intervals for mean/median WTP were estimated using bootstrap simulations. The "wtpcikr" command in Stata was used to estimate confidence intervals using Krinsky and Robb's method (Lyssenko & Martinez-Espineira, 2012). This method uses parametric bootstrapping to estimate the empirical distribution of a measure, which is non-linear in parameters. At first, the coefficients of the WTP model and their variance-covariance matrix are estimated. Then a new vector of coefficients is estimated by randomly drawing from the standard normal distribution given the estimated coefficients and variance-covariance matrix. Finally, the new vector of coefficients is used to re-estimate the mean/median WTP. Using these procedures, we ran 5000 simulations to obtain the distribution of mean/median WTP that was used to compute 95% confidence intervals of the corresponding WTP summary measure.

Results

Visitors' characteristics

A majority of international visitors (81.9%) stated that the primary purpose of their trip to Nepal was to visit Mount Everest National Park. On this trip, 7.7% of visitors were traveling by themselves, 12.5% with family, 31.1% with friends, 7.5% with family and friends, 39.2% with a tour group, and 2.1% responded "others". On average, the travel group size was 6.73 ± 5.33. International visitors spent 13.02 ± 5.32 days, on average, inside the park. Their reported average expenditure was US $57.01 ± 160.44 per day, but the median expenditure was $30 per day, which was more reasonable than the mean. About 15% of visitors had completed high school, 17.9% had associate's degrees, 30.9% had undergraduate degrees, 30.7% had master's degrees, and 4.7% had doctoral degrees (see Discussion).

About two thirds of the visitors (65.8%) were employed full time, 6.0% were employed part time, 10.4% were temporarily unemployed, 8.1% were retired, 3.9% were students, and 3.2% were homemakers. Few visitors (2.5%) chose the "other" category to report their employment status. If the respondents were currently in the labor force (i.e. employed full or part time) then their economic status was considered active. Respondents were from 34 countries throughout the world. The top five countries in terms of visitor numbers were the United Kingdom (23.2%), Australia (13.4%), the United States (9.4%), France and Germany (6.7%), and Canada (6.5%), which also match the distribution of actual visitors to the park. See Supplemental Data Table S2 for more information.

Knowledge of and experience with WHS

Many international visitors (47.3%) reported that they learned about Mount Everest National Park from family and friends and 44.3% mentioned the web/internet, 31.5% guidebooks, 16.4% travel agencies, 13.9% magazines, and 2.1% other sources. More than three fourths of visitors (76.2%) were aware of the World Heritage designation but fewer (57.3%) knew that Mount Everest National Park was a WHS prior to their trip. Furthermore, 65.1% stated that the WHS designation did not influence their decision to visit, 22.9% stated that it influenced their decision to a limited or moderate extent, and only 12.0% stated the WHS designation influenced their decision to a large or very large extent. A large majority (81.4%) reported that they did not see the WHS logo inside the park.

Almost all visitors (95.8%) mentioned that they would recommend their family and friends to visit Mount Everest National Park. In general, visitors agreed that the park had outstanding universal value, they reported the park experience to be authentic and they considered the park to be whole and intact (Supplemental Data Table S2).

Predictors of willingness to pay

The logistic regression model appears to be valid and reliable because the statistically significant value of hat ($z = 7.28$, $p < 0.01$) and the insignificant value of hat squared ($z = 1.1$, $p = 0.266$) indicate that the major statistical assumptions were met and the model was correctly specified ($\chi^2 = 122.15$,

Table 1. Logistic regression of willingness to pay on trip characteristics and socio-demographic variables. Bold typeface indicates statistical significance.

Willingness to pay	Coefficient	Std. Error	z	p	Effect size
Bid amount	**−0.034**	**0.005**	**−7.41**	**0.001**	**−3.4%**
Gender	−0.485	0.255	−1.90	0.057	−38.4%
Education	0.198	0.112	1.77	0.077	21.2%
Age	−0.001	0.011	−0.05	0.957	−0.1%
Economic status	0.036	0.289	0.13	0.900	3.7%
Environmental membership	−0.473	0.321	−1.47	0.141	−37.7%
Use of a guide	**1.182**	**0.304**	**3.88**	**0.001**	**226.1%**
Visitor days	**0.060**	**0.026**	**2.33**	**0.020**	**6.2%**
Park substitute	**−0.580**	**0.296**	**−1.96**	**0.050**	**−44.0%**
Knowledge of SNP's WHS designation	**0.559**	**0.254**	**2.20**	**0.028**	**74.9%**
Satisfaction with the trip	0.104	0.108	0.96	0.336	11.0%
Unimpaired condition of the park	0.083	0.072	1.16	0.248	8.6%
Outstanding universal value of the park	0.011	0.051	0.22	0.827	1.1%
Authenticity of the park	−0.004	0.133	−0.03	0.976	−0.4%
Constant	−0.504	1.325	−0.38	0.704	
Model fit statistics	Likelihood ratio χ^2 = 122.15, df = 14, p < 0.001, n = 417, Log likelihood = −203.03, pseudo R^2 = 0.23, correctly classified = 76.7%				

df = 14, p < 0.001). In an auxiliary regression, taking predicted values as an explanatory variable and the actual outcome as a response variable, hat is a coefficient of the predicted values and hat squared is a coefficient of the quadratic term. Furthermore, the model correctly specified 76.7% of all cases. Of the 14 explanatory variables included in the logistic regression model, seven significantly predicted the people who were more likely to agree to pay higher entry fees (Table 1). Both visitor characteristic variables (gender and education) were statistically significant at the 10% significance level. Males were less likely to agree to pay higher entry fees than were females and the chance of saying "yes" to the WTP question decreases by 38.4% if the respondent was male. Similar gender-specific results have been found in other WTP studies (e.g. Kline et al., 2012).

Higher levels of formal education positively influenced WTP. Moving from one lower level to the next higher level of education, the likelihood of a 'yes' response to the WTP question increased by 21.2%. Visitors who hired a guide on their trip to the park were about three times more likely to agree to pay higher entry fees compared to those who did not hire a guide. Those visitors who stayed longer in the park were also more likely to agree to pay higher entry fees than those who stayed for shorter duration. If the visitors thought that a close substitute existed for the park for a similar experience, the odds of their WTP higher entry fees decreases by 44.0%. Prior knowledge of the designation of the park as a WHS also had a positive influence on WTP. For visitors who had heard of Mount Everest's WHS designation, the chance of saying yes to the WTP question increased by 74.9% compared to those who had not. All other variables in our model were insignificant in explaining variation in WTP responses.

As expected, the variable "Bid amount" had a significantly negative effect on the WTP responses. For each one dollar increase in entry fees, the odds of saying yes to the WTP question decreased by 3.4%. On plotting the observed probability of "yes" responses against the bid amounts, there was a downward sloping demand curve (Figure 3). This finding, along with the narrow gap between the observed and predicted probabilities, provided some evidence for the theoretical validity of the model.

In response to a follow-up question about why visitors were willing (or not) to pay higher entry fees, 81.9% of respondents provided written responses (Table 2). The three most important reasons for WTP higher entry fees were that they would like to pay for the protection of the park, the proposed entry fee was affordable and they thought the park to be so unique (in terms of scenery, mountains and trekking opportunities, etc.) as to justify higher entry fees. A few visitors were willing to pay higher entry fees on the condition that the money be used properly. The top three reasons for visitors' unwillingness to pay higher entry fees were that they could not afford the proposed entry fee, they were willing to pay lower amounts than the proposed entry fee and they distrust government officials for the proper use of money.

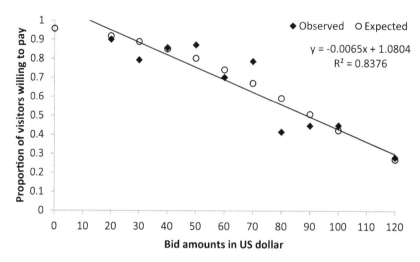

Figure 3. A demand function derived from the observed frequency of "yes" responses (y) to various bid amounts (x) presented as potential future entry fees.

Value of WHS designation

At the time of the survey, 63.8% of visitors were willing to pay higher entry fees than the (then) park entry fee of US $13 per visit. The mean and median WTP was $91.45, and the 95% confidence intervals for the mean ranged from $83.56 to $102.82. More than two thirds of respondents (68.4%) thought that the new entry fee would be acceptable to visitors. To examine the association between visitors' income levels and their WTP decisions, we also specified another WTP model by including the income variable. As expected, income had a positive and significant impact on WTP (Figure 4). The fitted regression line between median WTP and different income categories showed that the median WTP amount would increase by about $6 while moving from any one lower level of income bracket to the next higher level.

We also evaluated the impact of the WHS designation on WTP. There was a statistically significant difference between visitors who were aware of the WHS designation prior to arriving in the park and those who were not regarding the WTP. Thus, one way to assign the value for the WHS designation was to predict the median WTP amounts separately for each type of visitor. The difference between the two median WTP amounts could be assigned as the value of WHS designation. The median WTP amount for visitors who were aware of the park's WHS designation was $97.73 compared to $81.34 who were unaware. Thus, the median difference due to the WHS designation was $16.39. Given 34,571 international visitors in 2011, the total economic value that could be generated by the WHS designation was US $566,619. Hence, the WHS designation alone accounted for 21.3% of the median WTP amount in this case.

Table 2. Summary of visitors' stated important reasons for their WTP decisions regarding increased park entry fees.

Emerging categories of the stated reasons	Frequency
Individuals who responded "Yes" to the WTP question, $n = 281$	
Pay for preserving, protecting or conserving the park	23.5%
Proposed entry fee is affordable	13.9%
Pay for the uniqueness of the park	10.7%
Pay only if the money is used properly	3.6%
Help the region financially or otherwise	1.1%
Individuals who responded "No" to the WTP question, $n = 145$	
Proposed entry fee is not affordable	44.1%
Willing to pay a lower amount than the proposed entry fee	8.3%
Distrust of government officials for the proper use of money	5.5%
Entry fee hike reduces tourism because only wealthy people can visit the park	3.4%

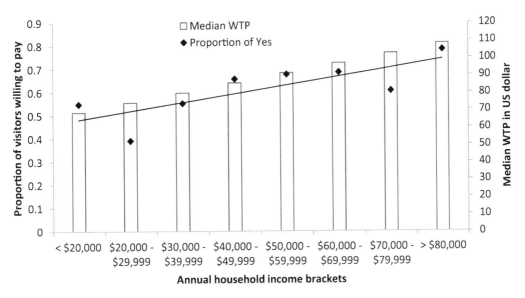

Figure 4. Relationship between visitors' income levels, their WTP decisions and median WTP amounts.

Finally, based on our WTP results, we found that there is an opportunity to raise park entry fees. We conducted simulations with four potential future entry fees between $40 and $91 to project revenues and identify the revenue-maximizing fee. We projected the expected number of visitors and expected revenue with these entry fees and computed the revenue surplus in comparison to the status quo scenario (that is, the entry fee at the time of the survey) for seven years into the future (Table 3). If the park entry fees were increased to $40, on average, total revenue would increase by 12.5% compared to the status quo over the next seven years. Similarly, if entry fees were increased to $60, $80 and $91, the average increase in the revenue would be 39.4%, 45.7% and 43.2%, respectively, compared to the status quo over seven years (Figure 5). The estimated revenue maximizing entry fee was $80.

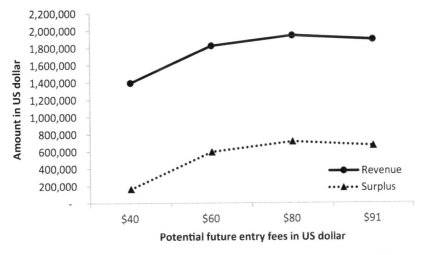

Figure 5. Average amount of revenue generated per year over the next seven years (2014 to 2020) and surplus revenue compared to the current entry fee under the scenarios of four potential future entry fees.

Table 3. The expected number of visitors, the amount of revenue generated in the status quo scenario of present entry fee, and the expected revenues and surpluses compared to the status quo scenario under the four potential future entry fees calculated based on the parameters estimated by the WTP model.

Year	Predicted visitors	Status quo revenue	Fee increased to US $40			Fee increased to US $60			Fee increased to US $80			Fee increased to US $91		
			Visitors	Revenue	Surplus	Visitors	Revenue	Surplus	Visitors	Revenue	Surplus	Visitors	Revenue	Surplus
2014	36,539	1,096,159	31,065	1,242,606	146,447	27,082	1,624,946	528,787	21,635	1,730,762	634,603	18,558	1,688,776	592,617
2015	37,927	1,137,813	32,246	1,289,825	152,012	28,112	1,686,694	548,881	22,457	1,796,531	658,718	19,263	1,752,949	615,136
2016	39,368	1,181,050	33,471	1,338,838	157,788	29,180	1,750,789	569,739	23,310	1,864,799	683,749	19,995	1,819,561	638,511
2017	40,864	1,225,930	34,743	1,389,714	163,784	30,289	1,817,319	591,389	24,196	1,935,662	709,732	20,755	1,888,705	662,775
2018	42,417	1,272,515	36,063	1,442,523	170,008	31,440	1,886,377	613,861	25,115	2,009,217	736,702	21,544	1,960,475	687,960
2019	44,029	1,320,871	37,433	1,497,339	176,468	32,634	1,958,059	637,188	26,070	2,085,567	764,696	22,362	2,034,973	714,102
2020	45,702	1,371,064	38,856	1,554,238	183,174	33,874	2,032,465	661,401	27,060	2,164,819	793,755	23,212	2,112,302	741,238

Predicted visitors: The average growth rate of visitors from 1998 to 2013 was 3.8% per year. Based on the assumption that there would be average growth in the future, the estimates of visitors were made for the next seven years.

Status quo revenue: The current revised entry fee is about US $30 (it varies slightly due to exchange rate fluctuations). The multiplication of predicted number of visitors and the current entry fee yielded the status quo revenue.

Visitors: If the entry fee is increased, the econometric model predicts the decrease in the number of visitors. The predicted number of visitors in the status quo scenario is adjusted by the appropriate decline rate for each proposed entry fees.

Revenue: This is the sum of visitors (the expected number of visitors to park in a particular scenario) times the proposed entry fee.

Surplus: This is the additional revenue expected if the entry fee is increased by this particular amount.

Discussion

Our results indicate that, on average, international visitors to Mount Everest National Park were willing to pay much more (over $90) in entry fees to cater to visitors' needs, conserve biodiversity and address management problems, a general result in concordance with other such studies (e.g. Casey et al., 2010; Lee et al., 2010). In 2012, the Government of Nepal increased the park entry fee to approximately US $30 for international visitors. Our results indicate that international visitors were willing to pay about three times higher, on average, than the revised entry fee. Variation in WTP was explained by socio-demographic and trip characteristics such as gender, educational attainment, use of a guide, length of stay within the park, perception of substitution for the park, and knowledge about the park's WHS designation, general results that agree with other related studies (e.g. Buckley, 2003; Kline et al., 2012). In our study, visitors who knew about park's WHS designation prior to their visit were willing to pay about $16 more on average than those who did not. WHSs thus can be seen as unique cases, which in turn can lead to ascribing greater values for designated parks.

As such, visitors who attach a higher value to a specific site are likely to agree to pay higher entry fees. A potential future entry fee of US $80, slightly lower than the median WTP ($91) found here, is the revenue maximizing entry fee based on our model. The results of this study support previous findings that many attractive protected areas have not fully captured the economic values that visitors place on them via entry fees (Baral & Dhungana, 2014; Baral et al., 2008; Casey et al., 2010; Hadker, Sharma, David, & Muraleedharan, 1997; Lee & Han, 2002; Reid-Grant & Bhat, 2009; White & Lovett, 1999). Highlighting the economic value of WH Sites can motivate policy makers and park management to revise entry fees and generate more revenue. This argument is also supported by the finding that the WHS designation positively influences visitors' willingness to visit sites elsewhere (Poria, Reichel, & Cohen, 2011). However, increasing entry fees to public parks can lead to social inequity by preventing access to people who could not afford them. Such inequity is not an issue in this case because there is no entry fee for Nepalese citizens and people from neighboring South Asian countries pay much lower fees (about $3.00 at the time of our study). Compared to the total travel cost to Nepal, even the increased park entry fee is small for visitors from developed countries. Also other studies have indicated a higher WTP for general park admissions by tourists to Nepal (e.g. Baral & Dhungana, 2014; Heinen, 1990; Heinen & Thapa, 1988).

In our study, visitors were willing to pay more toward the entry fees for several reasons. Most visitors reported that they would pay a higher amount in order to preserve, protect and conserve the park, and for the uniqueness of the park. This is consistent as quality improvement is one of the prime motives for WTP. Moreover, many visitors reported that the proposed entry fee was affordable. However, some visitors suggested that they would pay only if the money was used properly implying that the park authorities need to demonstrate greater transparency and accountability in determining an entry fees and spending the revenue. Furthermore, a few visitors who were not willing to pay higher fee reported distrust with the government officials in using the collected money properly. This finding again reinforces the idea of demonstrating greater accountability in revenue and cost management by the park authorities.

Based on the regression results, the management authority can evaluate four variables to gain support of visitors if they decide to increase entry fees. First, the median WTP is the amount that at least 50% of visitors are willing to pay, so we can assume that increasing the fee at a rate lower than the median WTP would receive support from a majority (more than 50%) of visitors. Second, the management authority can encourage visitors to hire a guide and introduce programs to increase the visitors' length of stay within the park. Over 70% of visitors did not hire a guide and more return on investment could be gained by increasing the length of visit. Third, the management authority could focus on marketing the World Heritage designation to international visitors because more than 40% of visitors to Mount. Everest did not know that the park was a WHS before the survey was conducted. It is likely that many visitors might not know about the WHS designation even after their arrival because only 18% of visitors in our sample reported seeing the WHS logo inside the park. Thus, promoting and marketing the park's WHS designation could help to gain visitors' support for increased entry fees. Knowledge of a substitute site is shown to affect park valuation negatively (Willis, 2009). This is potentially because substitute sites can be seen as options for switching in case of rising costs of the primary site. Several high Himalayan sites in Nepal, such as Annapurna Conservation Area and Langtang National Park, are already well-visited and well-known, while others (e.g. Manaslu, Gaurishankar, Api Nampa and Kanchanjunga Conservation Areas) are not (Lama & Job, 2014). However, these are also harder to reach, more challenging logistically and lacking in visitor facilities compared to better-known sites.

Logar (2010) reviewed a number of policy incentives that have been proposed in the literature, which should be evaluated for their effectiveness in managing tourism in case of Mount Everest National Park. An eco-tax levied on tourists is specifically earmarked for improving environment quality and may be beneficial for a low-resource country such as Nepal. Park quotas on the other hand limit the number of visitors that can be admitted to a site during a time period. Quotas may help manage tourism by reducing overcrowding and burden on park management staff, and also provide exclusivity to tourists thereby enriching their experiences. Yet, quotas work on the principle of excluding some visitors and that may be seen as unfair and restraining tourists' rights and equality. These alternative mechanisms need to be evaluated in the light of their feasibility in implementation, the elasticity of demand of park visitors, and how these mechanisms could result in reduced numbers of visitors as well as future revenues.

Although the WTP method is widely used, there are some limitations in this and related studies that should be taken into consideration when interpreting results (e.g. Carson & Mitchell, 1993; Hausman, 2012). First, the contingent valuation method has some potential biases and other sources or error. For example, people are asked if they are willing to pay more knowing that they will not be asked to pay more on this trip if they say yes, which could artificially inflate estimated WTP values. Also, much of the world was still suffering from the Great Recession in 2011, when this study was conducted, and unemployment was high in a number of Western countries, especially for young adults. This may have reduced the numbers of lower earning (and younger) trekkers to Nepal, which may have inflated our estimate. Notice, for example, that about 35% of the respondents in our sample reportedly had advanced graduate degrees (master's or doctorate), which is a much higher average educational attainment than has been reported in other foreign tourism studies in Nepal (e.g. Baral & Dhungana, 2014; Heinen & Thapa, 1988). People at the high end of educational attainment are also at the high end of income distribution in developed economies, which may have further inflated our estimated WTP.

Although we did our best to minimize some biases through a rigorous research design and extensive sampling, the nature of the study itself, and when it was completed, raises these questions. It is typical in contingent valuation literature to evaluate marginal effects of knowledge, attitude, awareness, and socio-demographic variables on WTP estimates (e.g. Buckley, 2003). We used visitors' prior knowledge of the WHS status of the park to evaluate the average marginal impact on WTP, and the economic value of the WHS designation was estimated here through statistical control that is not as powerful as conducting actual experiments. Future studies could use quasi-experimental approaches by administering at least two sets of questionnaires: one in which the World Heritage designation is not highlighted and another in which it is highlighted, to establish the economic value of the WHS designation more directly. Nonetheless, the findings provide insights into potential values of the WHS designation and we would assert that, even with the issues raised above, the park entry fee was well below the amount that could be charged to improve management, a common finding from related studies elsewhere (Asafu-Adjaye & Tapsuwan, 2008; Casey et al., 2010; Shultz et al., 1998). Given that high entry fees only to apply to international tourists who must spend much more than Nepalese or citizens of other South Asia nations in travel to Nepal, the higher suggested rate should not deter tourism or reduce equity in any substantial way.

Conclusions

The findings of this study are important for several reasons. Evaluating WTP associated with the WHS designation is beneficial for examining the overall value of the park to visitors to assess whether entry fees could be raised from existing levels. The attraction that leads to excess tourism can be turned into financial resources to implement more effective management policies to protect the park from potentially negative impacts of tourism.

The study shows that WHS designation has an economic value in and of itself. In order to market WHSs as a brand, it is important to clearly communicate the value of WHS designation to visitors. By doing so, visitors can feel good about making a greater return on their investment to visit the site (Steckenreuter & Wolf, 2013), which in turn can increase the likelihood of agreeing to pay higher entry fees.

Acknowledgments

The authors would like to thank visitors who participated in the survey for their generous help and time. The Snow Leopard Conservancy, US, provided the research grant to conduct the fieldwork. Sirgid Smith, Lukas Rinnhofer, Georgina Cullman, Malcolm McCallum, Kathy Frame, Aeshita Mukherjee, Alexandra Bosbeer, Yaniv Poria, and Yash Veer Bhatnagar provided invaluable comments on the draft survey instrument. Ranju Baral and Prakash Poudel helped with the figures.

Disclosure statement

No potential conflict of interest was reported by the authors.

Funding

The Snow Leopard Conservancy, http://snowleopardconservancy.org/.

References

Alberts, H.C., & Hazen, H.D. (2010). Maintaining authenticity and integrity at cultural world heritage sites. *Geographical Review, 100*(1), 56–73.
Ale, S.B., Yonzon, P., & Thapa, K. (2007). Recovery of snow leopard *uncia uncia* in Sagarmatha (Mount Everest) National Park, Nepal. *Oryx, 41*, 89–92.

Asafu-Adjaye, J., & Tapsuwan, S. (2008). A contingency valuation of scuba diving benefits: Case study in Mu Ko Similan Marine National Park, Thailand. *Tourism Management, 29*(6), 1122–1130.

Baral, N., & Dhungana, A. (2014). Diversifying finance mechanisms for protected areas capitalizing on untapped revenues. *Forest Policy and Economics, 41*, 60–67.

Baral, N., & Heinen, J.T. (2006). The Maoist people's war and conservation in Nepal. *Politics and the Life Sciences, 24*(1–2), 1–11.

Baral, N., Stern, M.J., & Bhattarai, R. (2008). Contingent valuation of ecotourism in Annapurna conservation area, Nepal: Implications for sustainable park finance and local development. *Ecological Economics, 66*(2–3), 218–227.

Barbier, E.B. (2014). Account for depreciation of natural capital. *Nature, 515*, 32–33.

Batisse, M. (1992). The struggle to save our world heritage. *Environment, 34*(10), 12–32.

Bhat, M.G. (2003). Application of non-market valuation to the Florida keys marine reserve management. *Journal of Environmental Management, 67*(4), 315–325.

Borges, M.A., Carbone, G., Bushell, R., & Jaeger, T. (2011). *Sustainable tourism and natural world heritage – Priorities for action*. Gland: IUCN.

Buckley, R. (2003). Pay to play in parks: An Australian policy perspective on visitor fees in public protected areas. *Journal of Sustainable Tourism, 11*(1), 56–73.

Carson, R.T., & Mitchell, R.C. (1993). The issue of scope in contingent valuation studies. *American Journal of Agricultural Economics, 75*, 1263–1267.

Casey, J.F., Brown, C., & Schuhmann, P. (2010). Are tourists willing to pay additional fees to protect corals in Mexico? *Journal of Sustainable Tourism, 18*(4), 557–573.

Chen, W.Y., & Jim, C.Y. (2012). Contingent valuation of ecotourism development in country parks in the urban shadow. *International Journal of Sustainable Development & World Ecology, 19*(1), 44–53.

Dongol, Y., & Heinen, J.T. (2012). Pitfalls of CITES implementation in Nepal: A policy gap analysis. *Environmental Management, 50*(2), 181–192.

Eagles, P.F.J. (2014). Research priorities in park tourism. *Journal of Sustainable Tourism, 22*(4), 528–549.

Frey, B.S., & Steiner, L. (2011). World heritage list: Does it make sense? *International Journal of Cultural Policy, 17*(5), 555–573.

Gu, Y., Du, J., Tang, Y., Qiao, X., Bossard, C., & Deng, G. (2013). Challenges for sustainable tourism at the Jiuzhaigou world natural heritage site in western China. *Natural Resources Forum, 37*(2), S103–S112.

Hadker, N., Sharma, S., David, A., & Muraleedharan, T.R. (1997). Willingness to pay for Borivli National Park: Evidence from a contingent valuation. *Ecological Economics, 21*(2), 105–122.

Hanemann, W.M. (1994). Valuing the environment through contingent valuation. *Journal of Economic Perspectives, 8*(4), 19–43.

Hanemann, M., Loomis, J., & Kanninen, B. (1991). Statistical efficiency of double-bounded dichotomous choice contingent valuation. *American Journal of Agricultural Economy, 73*(4), 1255–1263.

Hausman, J. (2012). Contingent valuation: From dubious to hopeless. *Journal of Economic Perspectives, 26*(4), 43–56.

Heinen, J.T. (1990). The design and implementation of a training program for tour guides in Royal Chitwan National Park, Nepal. Bangkok: FAO. *Tiger Paper, 17*(2), 11–15.

Heinen, J.T. (1995a). International conservation agreements. In W.A., Nierenberg (Ed.), *Encyclopedia of environmental biology (volume 1)* (pp. 375–384). San Diego, CA: Academic Press.

Heinen, J.T. (1995b). Applications of human behavioral ecology to wildlife conservation and utilization programmes in developing countries. *Oryx, 29*(3), 178–186.

Heinen, J.T. (2010). The importance of a social science research agenda in protected areas management. *Botanical Review, 76*, 140–164.

Heinen, J.T. (2012). International trends in protected areas policy and management. Retrieved 16 March 2017 from http://www.intechopen.com/books/protected-area-management/international-trends-in-protected-areas-policy-and-management

Heinen, J.T., & Rayamajhi, S. (2001). On the use of goal-oriented project planning in Nepalese protected area management. *Environmental Practice, 3*(4), 227–236.

Heinen, J.T., & Thapa, B.B. (1988). A feasibility study of a proposed trekking trail in Chitwan National Park. Kathmandu: Tribhuvan University. *Journal of the Forestry Institute, 10*, 19–28.

Heinen, J.T., Yonzon, P.B., & Leisure, B. (1995). Fighting the illegal fur trade in Kathmandu, Nepal. *Conservation Biology, 9*(2), 245–247.

Jha, S. (2005). Can natural world heritage sites promote development and social harmony? *Biodiversity and Conservation, 14*(4), 981–991.

Kline, C., Cardenas, D., Duffy, L., & Swanson, J.R. (2012). Funding sustainable paddle trail development: Paddler perspectives, willingness to pay and management implications. *Journal of Sustainable Tourism, 20*(2), 235–256.

Lama, A.K., & Job, H. (2014). Protected areas and the road development: Sustainable development discourses in the Annapurna Conservation Area, Nepal. *Erdkunde, 68* (4), 229–250.

Landorf, C. (2009). Managing for sustainable tourism: a review of six cultural World Heritage Sites. *Journal of Sustainable Tourism, 17*(1), 53–70.

Lee, C.K., & Han, S.Y. (2002). Estimating the use and preservation values of national parks' tourism resources using a contingent valuation method. *Tourism Management, 23*(5), 531–540.

Lee, C.K., Lee, J.H., Kim, T.K., & Mjelde, J.W. (2010). Preferences and willingness to pay for bird watching tours and interpretive services using a choice experiment. *Journal of Sustainable Tourism, 18*(5), 695–708.

Logar, I. (2010). Sustainable tourism management in Crikvenica, Croatia: An assessment of policy instruments. *Tourism Management, 31*(1), 125–135.

Lyssenko, N., & Martinez-Espineira, R. (2012). Been there done that: Disentangling option value effects from user heterogeneity when valuing natural resources with a use component. *Environmental Management, 50*(5), 819–836.

Mmopelwa, G., Kgathi, D.L., & Molefhe, L. (2007). Tourists' perceptions and their willingness to pay for park fees: A case study of self-drive tourists and clients for mobile tour operators in Moremi Game Reserve, Botswana. *Tourism Management, 28*(4), 1044–1056.

Moran, D. (1994). Contingent valuation and biodiversity: Measuring the user surplus of Kenyan protected areas. *Biodiversity and Conservation, 3*, 663–684.

Nyaupane, G.P., Lew, A.A., & Tatsugawa, K. (2014). Perceptions of trekking tourism and social and environmental change in Nepal's Himalayas. *Tourism Geographies, 16*(3), 415–437.

Okech, R.N. (2010). Socio-cultural impacts of tourism on World Heritage sites: Communities' perspective of Lamu (Kenya) and Zanzibar Islands. *Asia Pacific Journal of Tourism Research, 15*(3), 339–351.

Poria, Y., Reichel, A., & Cohen, R. (2011). World Heritage Site – Is it an effective brand name? A case study of a religious heritage site. *Journal of Travel Research, 50*, 482–495.

Reid-Grant, K., & Bhat, M.G. (2009). Financing marine protected areas in Jamaica: An exploratory study. *Marine Policy, 33*(1), 128–136.

Richardson, L., Rosen, T., Gunther, K., & Schwartz, C. (2014). The economics of roadside bear viewing. *Journal of Environmental Management, 140*, 102–110.

Ryan, J., & Silvanto, S. (2009). The World Heritage list: The making and management of a brand. *Place Branding and Public Diplomacy, 5*, 290–300.

Shackley, M. (Ed.). (1998). *Visitor management: Case studies from World Heritage Sites*. Oxford: Butterworth-Heinemann.

Shrestha-Acharya, R., & Heinen, J.T. (2006). Emerging policy issues on non-timber forest products in Nepal. *Himalaya, 26*(1–2), 51–54.

Shultz, S., Pinazzo, J., & Cifuentes, M. (1998). Opportunities and limitations of contingent valuation surveys to determine national park entrance fees: Evidence from Costa Rica. *Environment and Development Economics, 3*(1), 131–149.

Steckenreuter, A., & Wolf, I.D. (2013). How to use persuasive communication to encourage visitors to pay park user fees. *Tourism Management, 37*, 58–70.

Su, Y.W., & Lin, H.L. (2014). Analysis of international tourist arrivals worldwide: The role of world heritage sites. *Tourism Management, 40*, 46–58.

Timilsina, N., & Heinen, J.T. (2008). Forest structure under different management regimes in the western lowlands of Nepal: A comparative analysis. *Journal of Sustainable Forestry, 26*(2), 112–131.

UNESCO. (2014). Operational *guidelines* for the *implementation* of the World Heritage Convention. Paris: World Heritage Center. Retrieved 24 November 2014 from http://whc.unesco.org/en/guidelines/

Wang, P.W., & Jia, J.B. (2012). Tourists' willingness to pay for biodiversity conservation and environment protection, Dalai Lake protected area: Implications for entrance fee and sustainable management. *Ocean & Coastal Management, 62*, 24–33.

White, P.C., Bennett, A.C., & Hayes, E.J.V. (2001). The use of willingness-to-pay approaches in mammal conservation. *Mammal Review, 31*(2), 151–167.

White, P.C.L., & Lovett, J.C. (1999). Public preferences and willingness to pay for nature conservation in the North York Moors National Park, UK. *Journal of Environmental Management, 55*, 1–13.

Whitelaw, P.A., King, B.E.M., & Tolkach, D. (2014). Protected areas, conservation and tourism – Financing the sustainable dream. *Journal of Sustainable Tourism, 22*(4), 584–603.

Willis, K.G. (2009). Assessing visitor preferences in the management of archaeological and heritage attractions: A case study of Hadrian's Roman Wall. *International Journal of Tourism Research, 11*(5), 487–505.

World Wildlife Fund. (2013). The Sacred Himalaya Landscape. Retrieved 16 March 2017 from http://assets.worldwildlife.org/publications/326/files/original/The_Sacred_Himalayan_Landscape.pdf?1345732409

Xuewang, D., Jie, Z., Ruizhi, Z., Shi'en, Z., & Min, L. (2011). Measuring recreational value of world heritage sites based on contingent valuation method: A case study of Jiuzhaigou. *Chinese Geographical Science, 21*(1), 119–128.

Private protected areas, ecotourism development and impacts on local people's well-being: a review from case studies in Southern Chile

Christopher Serenari, M. Nils Peterson, Tim Wallace and Paulina Stowhas

ABSTRACT

Private protected areas (PPAs) are expanding rapidly in less-industrialized nations. This paper explores cases in Los Ríos, Chile, to understand how local people living in and near three PPAs viewed impacts of tourism development on human well-being and local governance asking: (1) Why and how do governing PPA actors engage local people in conservation and ecotourism? (2) How do local people perceive the impacts of PPAs? (3) How do perceived impacts differ between PPA ownership types and contexts? We used an Opportunities, Security and Empowerment research framework derived from local definitions of well-being. Results suggest that governing PPA actors (PPA administrations and Chilean government officials) viewed local people as threats to forest conservation goals, embraced exclusion from reserve governance, but encouraged self-governance among local people through educational campaigns promoting environmental stewardship and ecotourism entrepreneurship. PPA administrations avoided emerging participatory democracy approaches to ensure local resistance did not threaten their authority. Despite asymmetrical power relations, PPA—community partnerships were viewed locally as both improving and damaging well-being. Our findings suggest that the social impacts and consequences of PPAs facilitating ecotourism development should be subjected to the same level of scrutiny that has been given to public protected areas.

Introduction

There is a growing interest in the private conservation movement among governing conservation actors (including governments, non-governmental organizations [NGOs] and international develop-ment agencies). Like public protected areas, private protected areas (PPAs) have flourished in the last two decades (Igoe & Brockington, 2007), though official global tallies remain elusive. Recent discus-sions about PPAs reason that they fill national biodiversity conservation coverage needs, bolster resource management, enhance citizen participation, promote bottom-up management (Stolton et al., 2014) and are potentially lucrative (Holmes, 2013b). PPAs may be owned by individuals, cooperatives, NGOs or corporations, and diverse ownership types have been linking their reserves with ecotourism initiatives.

Whether it is defined as an investment opportunity, tourism experience, land-use practice or conservation tool, ecotourism is attractive for those interested in private conservation. Ecotourism has become an important part of a global agenda pursuing "weightless capitalism" (Gee, 2000), assigning an economic value to natural resources to further conservation goals (Honey, 2008) and reduce social inequity and overconsumption of natural resources often associated with a free market system (Fletcher, 2011). At national and regional levels, PPA—ecotourism initiatives have garnered the attention of policy-makers interested in promoting development through PPAs and reducing anthropogenic threats to natural resources (Serenari et al., 2015). These initiatives have also inspired reserve creation (Langholz & Krug, 2004) and economic transformation at the local level (Taylor, 2010).

Ecotourism impacts human well-being because it targets biological conservation and community development simultaneously (Ballantyne & Packer, 2013; Jamal, Borges, & Stronza, 2006). Ecotourism's impacts are commonly evaluated in a Panglossian[1] manner by the suite of benefits ecotourism might provide; one of those being sustaining the well-being of local people (Singh, Slotkin, & Vamosi, 2007). Although the term well-being is often referred to in works exploring the linkages between protected areas and ecotourism, there have been few attempts to operationalize the term. Assessing ecotourism universally, Ballantyne and Packer (2013) claimed that if ecotourism is defined as benefits for local communities rather than ecotourism as activity, impacts to well-being can be expressed as protection or respect of local cultures, heightened visitor awareness, local participation and ownership of business ventures, enhanced local pride, and sense of empowerment. Gateway communities, in particular, tend to experience such benefits through increased employment and income, infrastructure development, and cultural renewal (Bennett, Lemelin, Koster, & Budke, 2012). These scholars also noted ecotourism's failures to improve well-being. Our research supported the generic categories of well-being (opportunities, security and empowerment) discernible in the World Bank's (2001) approach.

Though limited, research on PPAs suggests mixed impacts on local people's well-being. Alderman (1991) found that PPAs in Latin America and sub-Saharan Africa resulted in considerable employment increases and opportunities for local recreation, while Langholz (1996) cited the establishment of youth academic scholarships in Costa Rica. Barany and colleagues (2001) identified improved democratic decision-making in Nicaragua, while improved local economies in South Africa and linkages to social empowerment in Columbia were cited by Sims-Castley, Kerley, Geach, and Langholz (2005) and Langholz and Lassoie (2001), respectively. Scholars also uncovered social costs, such as sociopolitical rifts in Africa and elsewhere (Langholz & Lassoie, 2001), loss of cultural identity among local ranch hands in Chile (Jones, 2012), elevated land prices, contests over land-use rights and disturbances to employment conditions (Holmes, 2013b).

Works exploring PPA impacts on well-being tend to comprise macro or PPA owner perspectives and, accordingly, investigations of local perspectives on well-being impacts on those living near, and on rare occasion within, PPA borders are needed. There is relatively little literature exploring the social impacts of PPA ecotourism development on well-being, or how these impacts differ among different PPA ownership types. This study offers, therefore, a first look at PPA impacts on well-being through the eyes of impacted local publics and helps draw attention to and situate the private conservation movement into practice and scholarship. With an emic approach, we privileged rural communities' perceived needs and aspirations (Infield, 1988) and employed a case study in Los Ríos, Chile, to answer three key questions: (1) Why and how do governing PPA actors engage local people in PPA conservation and ecotourism development? (2) How do local people perceive impacts of PPA-ecotourism development on their well-being? (3) How do perceived impacts differ between PPA ownership types and contexts?

The study extends the discussion of future research into protected areas, encouraged by Eagles (2014) in his much cited recent paper on research priorities in park tourism, by exploring issues of park governance, management, evaluation and finance that are typically researched in public protected areas into the neo-liberal world of private protected areas.

Overview of the study area

Since the 1970s, economic expansion and corporate timber have been closely associated in Chile. Decree-Law 701 (i.e. 1974 Forest Law) presented timber companies with incentives to purchase large tracts of forest for timber production, including the biodiverse Valdivian forest in Los Ríos in southern Chile (Klubock, 2006). Alex and Clapp (1998) and others chronicled forceful social disturbances during this era as rural people were displaced or saw the forest in which they depended for resources become degraded. Communities turned to employment with the corporate timber regime (Klubock, 2006). The timber era was an extension of *latifundia,* a period characterized by livelihoods dependent on a single industry, worker exploitation and temporary or seasonal employment on large private estates during the early to mid-twentieth century (Faúndez, 1988). In Los Ríos, timber laborer and *campesino* strikes, evictions, political revolt and violence occurred during this era (Klubock, 2014).

As the timber industry deteriorated in the latter half of the twentieth century, many rural people were left landless and with few income options (Armesto, Smith-Ramírez, & Rozzi, 2001). A relatively recent State-led sustainable development path attempted to address the issues of forest degradation, poverty and emigration in Chile's rural areas. Defunct timber plantations were designated protected areas by various wealthy private actors. Chile has at least seven large PPAs undertaking ecotourism and community engagement which cover over 40,000 km^2/15,444 mi^2 of terrain and impact thousands of natural-resource-dependent people via private development-based conservation. PPA administrations encourage nearby communities to participate in micro-tourism endeavors, wage employment, and provide goods and services to the PPA itself in order to address local material poverty and meet biodiversity conservation goals (Bishop & Pagiola, 2012) or protect or modernize disappearing cultural traditions (e.g. http://fundacion.huilohuilo.com). Yet, some larger Chilean PPAs met resistance over inequitable benefits, livelihood disruption (E. Corcuera, personal communication, 2013), elitism, rising land costs, foreign ownership (Holmes, 2013a) and land rights (Meza, 2009; Wakild, 2009).

Los Ríos is 840 km south of Santiago. It has an area of 18,429 km^2 with an estimated population of 356,396. The region comprises a northern portion of the Valdivian Rainforest Ecoregion, a rare, ancient and threatened temperate rainforest. The forest's condition began declining in the sixteenth century due to Spanish–Indigenous conflict, colonization and subsistence livelihoods, and intensified later with commercial exploitation (Alex & Clapp, 1998; Wilson, Newton, Echeverría, Weston, & Burgman, 2005). The forest is considered a global biodiversity hotspot, harbors one of the highest rates of endemic species in the world, and was targeted for protection by the Worldwide Fund for Nature (WWF) and the World Bank (Dinerstein et al., 1995; Núñez, Nahuelhual, & Oyarzún, 2006).

Methods

Defining human well-being

To assess well-being, we asked the question *Que es para usted una buena calidad de vida* (For you, what is a good quality of life)? Contextualizing well-being in this way is an approach that has been used in other studies (e.g. Fattore, Mason, & Watson, 2007) and helped us understand PPA impacts from the local people's perspective. Thematic analysis of informant responses yielded three dimensions of well-being (Opportunities [the ability to do something desired], Security [free from want and fear], Empowerment [ability to impact outcomes, increase capacity, or obtain power and authority]), similar to those employed by the World Bank (2001). We used a forward and backward translation process to design English and Spanish versions of the interview guide to ensure clarity and validity (Marín & Marín, 1991).

PPA and participant selection

We used a case study design to investigate PPAs intentionally engaging local people in ecotourism development and assess the perceived impacts to well-being under three different PPA private governance types in Los Ríos, Chile, as defined by the International Union for Conservation of Nature (IUCN): NGO (Valdivian Coastal Reserve [RCV]), Corporate (Oncol Park [OP]) and Individual (Huilo Huilo Reserve [HH]). We employed selective (key informant) sampling to obtain "tentative theoretical jumping off points from which to begin theory development" (Thompson, 1999, p. 816), followed by snowball sampling. Sampling was not intended to be representative, rather we targeted people who were most knowledgeable about and experienced with PPAs and who maximized the range of data elicited by providing diverse perspectives (Beebe, 2001; Lincoln & Guba, 1985) within the budding Chilean PPA network (e.g. government officials involved in PPA matters [4], PPA advisors [3], PPA administrators and staff [8; none were owners or executive team members], current or former community leaders [19], and local residents engaged in tourism development [e.g. tour guides (2), tourism enterprise owners (14)], and current or former PPA employees and contractors speaking as citizens [10]).

Data Collection and Analysis

A team of two Chilean researchers fluent in English completed most of the 85 semi-structured interviews between May and August of 2013; a third researcher completed interviews with English speakers. Key informant sampling started with predominately male community leaders which created a similar bias in the overall informant pool (67% [OP: 71%; RCV: 64%; HH: 74%]). The sample achieved at least basic primary education (81% [OP: 82%; RCV: 72%; HH: 96%]), over 45 years of age (55% [OP: 71%; RCV: 64%; HH: 44%]), and half self-reported as Chilean (50% [OP: 14%; RCV: 44%; HH: 91%] while more than one-third identified as Indigenous 37% [OP: 64%; RCV: 56%; HH: 4%].

Continual movement between text and themes with QSR International's NVivo qualitative data analysis software (Version 10, 2012) allowed us to find critical thematic moments and relationships to build a "thematic map" and encourage reflection on our own involvement in the study (Guba & Lincoln, 1985; Petty, Thomson, & Stew, 2012). We used a variety of strategies to ensure accuracy of our findings, including triangulation, member checking and employing external auditors (Creswell, 2013). The sources of results from, and statements by, informants are indicated in the text by the codes given in Table 1.

Results

State motivations for pursuing PPAs and community interactions

Our interviews with Chilean government officials revealed that legislators were aware of PPAs; they were viewed as a way for the country to address its biodiversity representation gaps and offer new development opportunities to rural communities (summarized sentiments of MIN-001 and 002). Respondents noted that Article 35 in the National Environmental Framework Law (1994) recognizes PPAs, but the government does not regulate them or provide tax-based incentives to establish them. Certain PPAs, mainly large ones, attempt to unite global and national sustainable development and biodiversity conservation goals, primarily through ecotourism development. An example was provided by the regional government of Los Ríos, which crafted the Regional Tourism Policy to facilitate sustainable development through public and PPA tourism development. This strategy was reiterated by SIRAP (Sistema Regional de Áreas Protegidas) who claimed on their website that "local development and conservation go hand in hand in the region of Los Ríos".

The regional policy, financially supported by the regional government's Fund for Regional Development (FRD), promoted social participation and community development because of the conventional assumption that local people living near a protected area will degrade it:

> PPAs realized they couldn't be like a bubble. People (would) keep stealing wood or introducing their animals, introducing dogs that kill the fauna...therefore a work strategy was implemented in order to work with the people...to create productive sustainable activities like ecotourism and gastronomic services around the protected area. That way the protected areas turn into a way of development for the communities and (the local people) are not a threat; and at the same time protected area biodiversity...is better protected. (MIN-001)

Table 1. Sources of results from informants.

Informant ID	Qualifying attribute
MIN-001	SIRAP tourism development official
MIN-002	Protected areas official with the Ministry of the Environment
Oncol-001	Lead reserve administrator
Oncol-002	Community relations specialist
TNC-001	Forest engineer
TNC-002	RCV coordinator
TNC-003	Park ranger
TNC-004	Community relations specialist
HH-001	Lead reserve administrator
HH-002	Community relations specialist
NEL-001	Former teacher; tourism business owner
NEL-003	Former community president
NEL-004/M1	Former timber worker; guide business owner
NEL-005	Owner of lodging and food sales business
NEL-006	President of water committee
NEL-007	Former HH employee; relatives work at HH
NEL-008	Tour guide; spouse is a teacher
NEL-010	HH hotel maid
NEL-011	Tourism business owner
NEL-013	Religious figurehead
PF-001-W1 and W2	Citizens with family who were evicted or mistreated
PF-002	School director
PF-003	Former timber employee; community association president
PF-005	Former HH employee
PF-006	Municipal delegate
PF-007	Artisan; Former HH employee
PF-008	Observed peculiar timber operations at HH
CH-002	Community elder
CH-003	Cabaña and restaurant owner
CH-004	Community tourism officer
CH-006	Artisan and chef
CH-008	RCV guide
CH-009	Cabaña owner
Huape-002	President
Huape-003	Former corporate logger; spouse owns restaurant
Huape-007	Artisan; Cabaña owner
Huape-008	Fisherman's union leader
Huape-009	TNC logger
Huiro-001	Former corporate logger; TNC ranger (preferred resident voice)
Cad-002	Stakeholder in livestock remediation project
Bon-001	President
Bon-004	Fisherman/Livestock owner
Pellin-001	Former Arauco logger; life-long resident
Pellin-002	President
Minas-001	President; Oncol ranger
Pilol-002	Collaborated with Oncol on project proposal; Spouse is Oncol employee
NGO-001	Executive for Chilean PPA network

Increased collaboration with other organizations and interests, including the Association of Conservation Initiatives on Private and Indigenous Lands (ASI Conserva Chile), Chilean universities (e.g. Universidad Austral de Chile, Universidad de Chile), various government agencies (e.g. Ministerio del Trabajo y Previsión Social; Corporación Nacional Forestal) and global conservation and development actors (e.g. United Nations Development Program [UNDP]) have influenced PPA conservation-development efforts and internal considerations about the "public politics" of PPA—community interactions (Oncol-002). The Nature Conservancy (TNC)[2] had a more complete documentation of stakeholder involvement than Oncol Park and Huilo Huilo. External partners provided support and funds to help local communities focus on ecotourism to offset the reserve's creation and transition to "more sustainable and resilient livelihoods through a range of income generating opportunities"

(TNC, 2015, p.9). Our research notes that stakeholders have included the WWF, the towns of Corral, La Unión and Valdivia, the World Bank's Global Environmental Facility-Regional System of Protected Areas (GEF-SIRAP) program, and Alerce Coastal National Park (Delgado, 2005; TNC 2015). The UNDP partnered with TNC to establish a US$300,000 fund to forge development pathways for neighboring communities.

Synopsis of PPA context and motivations for engaging local people

A family-owned and funded enterprise, *Huilo Huilo* is 165 km east of the city of Valdivia and located on the northern part of the Patagonia Ecoregion in the Andes Mountains. It is a globally recognized ecotourism destination, resides inside a UNESCO Biosphere Reserve, and harbors the 64-ha Huemul (*Hippocamelus bisulcus*)[3] Conservation Centre (Vidal, Smith-Fleuck, Flueck, & Arias, 2011). Chilean billionaire, Victor Petermann, his partner Fernando Boher and ex-wife Ivonne Reifschneider began the Huilo Huilo project in 1999 after Petermann acquired two *fundos* (estates) covering 104,000 ha. His purchase of the Pilmaiquén *fundo*, a worker-owned estate that became the reserve, resulted in uneven levels of financial capital in the valley as shareholders cashed out of their stakes (Henderson, 2013). Huilo Huilo's operations are based on the three pillars of sustainable development: environmental commitment, social commitment and sustainable economic activity. The Huilo Huilo Foundation (HHF), the project's philanthropic arm, promotes biodiversity conservation; provides support and training for altered livelihoods comprising nature tourism; pursues community activities and projects; and develops a societal ecological conscience. Huilo Huilo officials reported to us that in 2004 the reserve hosted 5068 visitors, over 33,000 in 2011, and between 50,000 and 60,000 in 2013.

The most impacted communities, Neltume (pop. 3495) and Puerto Fuy (pop. 391), were timber communes with local people working for low wages and lacking job security (Klubock, 2004). This area has a history of social unrest (timber worker strikes over unacceptable social conditions, political violence) chronicled by Klubock (2014) and others. A top Huilo Huilo official explained that community interactions take place under the philosophy that "…you have to work with the local communities, for better or for worse. They are there. They are your workers and supporters. They benefit from the park directly or indirectly. You must include them" (HH-001).

TNC and WWF partnered to create RCV, part of a low-altitude coastal mountain range bisecting Valdivia province, in 2003. The Chilean government deemed RCV an underrepresented ecosystem in the National System of Protected Wild Areas of the State and also a priority area for biodiversity conservation by the National Strategy Biodiversity (Delgado, 2005). In an apparent act of good will and support for rural tourism, TNC donated approximately 9300 ha to the Chilean Ministry of National Assets in 2012 to form Alerce Coastal National Park, reducing RCV to 50,251 ha. RCV tourism is growing. TNC officially reported an increase in tourism to RCV from 300 in 2006 to 1794 in 2014 (TNC, 2015).

Huiro (pop. 109) and Chaihuín (pop. 240), followed by Cadillal (pop. 40) and Huape (pop. 233) (Delgado, 2005) engage with RCV the most. These communities employed livelihood diversification strategies, but historically relied on fishing, small-scale agriculture or animal husbandry. TNC identified traditional local livelihood strategies as threats to RCV (Delgado, 2005). TNC has a community ranger who liaises between TNC and local people to enact TNC's Social Participation and Community Development Program (*Programa de Participacion Social y Desarrollo Comunitario*) that helps communities secure FRD and other funds and operates under the assumption provided by a top RCV Official:

> … if the people feel that they are part of a project that uses an important land of which they are neighbors, if you have good relationships and understanding, knowing what the area is for and the benefits they might obtain from this area, they become good neighbors and allies of the project…we create a good relationship with them by being good neighbors. We keep our doors open to them, we listen to them. (TNC-002)

TNC's RCV project was viewed as the "most advanced" model by the regional government as it ponders how to "incorporate the communities into development and strengthen protected areas"

(MIN-001). TNC's "political coordination" with the state (TNC-002) makes it inherently a political endeavor. However, a reserve official affirmed TNC prefers an apolitical stance with community affairs stating, "We don't get into political, religion, or ethnics matters" (TNC-001).

Oncol Park is located 29 km northwest of Valdivia and 5 km from the Pacific Ocean in the coastal mountains. It protects 754 ha of the Valdivian forest on the western flank of Cerro Oncol. In 1985, Forestal Valdivia, a subsidiary of Celco-Arauco, bought Fundo San Ramón and then declared part of the land protected to fulfill international forest certification standards, making Oncol Park one of Chile's first large PPAs. Tourism visitation increased from less than 200 in 1990 to approximately 1200 in 1990–1991 (Corcuera, Sepúlveda, & Geiss, 2002), approached 2400 in 2003, and reached 14,000 in 2007.

We examined Oncol Park's impact on Bonifacio (pop. 167), Los Pellines (pop. 244), Las Minas (pop. 37; gateway community) and Pilolcura (pop. 39). The working population of these communities was primarily Indigenous Mapuche and self-employed (farming, commercial forestry and fishing) (Ponce, 2007). In 2011, the Oncol Park administration hired an anthropologist to engage local communities and help local people apply for FRD funds. The administration embraced a PPA–community strategy inspired by the RCV project called the "Territorial Involvement-Community Management Program". The administration was in the midst of "winning the trust of the communities" at the time of our study. A top Oncol Park official explained the PPA–community philosophy: "It is important that the communities that live in the same biological corridor of the park, in the matrix, are also maintained. They are also impacting the park" (Oncol-001).

Impacts on human well-being

Economic opportunities

Informants stated they initially expected PPAs to create jobs and identified a timber employment culture as responsible for this expectation. A tourism business owner near Huilo Huilo explained: "It's just culture; there's a generation used to receiving everything. (Timber employers) used to give you electricity, water, and your salary. So, having to make your own money is more difficult" (NEL-004/M1). An expectation of jobs was also due in part to a belief that entities capable of buying huge tracts of land should create jobs. Near Oncol Park, a Bonifacio community leader asserted:

> I would have thought the millionaires, the owners, would have come, because they have to be rich, and millionaires, to have (a large PPA), and they would have come to ask if we needed jobs, and we could have said, "Yes, our communities could benefit from Oncol Park," but, no. (Bon-001)

Near RCV, the expectation of job creation reverberated from Huiro to Huape. A former corporate logger from Huape commented: "…we expected that (TNC) would protect the forest and give job opportunities to the community" (Huape-009). For some, this expectation was due to TNC's size and wealth. A community elder in Chaihuín expected a greater influx of money from TNC:

> We know that (TNC) is part of the twelve countries that are the richest, so we thought they would help people with some projects, give them money to build cabins, to fix the cabins they already had, to focus things towards tourism. But, it didn't happen, nothing happened. (CH-002)

To a former logger from Pilolcura, it was imperative that Oncol Park created jobs stating, "Big enterprises have to give jobs to the people" (Pellin-001). Wage employment was an important Huilo Huilo strategy. An official added: "We give 90% of the jobs in the area" and "informally, 95% of the staff is local" (HH-001). Oncol Park administration used local employment to counter forest disturbances previously caused by traditional livelihoods:

> We have always given priority to hiring people from (communities) around the park…We think…that if a family used to raise animals within the park or chopped firewood for Valdivia, and today we employ them, this means a positive impact on the forest because we create a direct economic income for the head of the family…the (protected area) is not being affected…anymore. (Oncol-001)

Local people also praised PPA administrations for providing new options for income where there would otherwise be few. Near RCV, informants credited TNC for helping diversify livelihood strategies through support for ecotourism microenterprises (e.g., craft making, bee keeping) and sustainable living projects (e.g., teaching novel natural resource harvesting techniques). A former corporate logger in Huape told us, "(TNC) insists…on projects…that enable you to live not only from the sea, or from the hills. There are a million things that one can do, starting from projects you can develop yourself" (Huape-003). A Chaihuín guide confirmed,

> Here you don't see poverty, well, at least critical poverty. Here the main income is from the mussels, then tourism in the summer. This community is doing better economically than other places. We have many different ways to make a living, with the mussels, the wood, tourism, fishing, et cetera. (CH-008)

A Huilo Huilo official explained how tourism's arrival helped loggers develop new employable skillsets: "In the past, all the people that lived here worked in timber exploitation operations. They didn't know anything else. There weren't any carpenters, stonemasons…these projects helped them discover their talents" (HH-002).

Local people wanting to engage in entrepreneurial tourism ventures believed they needed more knowledge and financial capital to prosper, however. A cabaña owner near RCV said: "You have to save the money…first" (Huape-007). To obtain seed money, a Huape leader observed that some community members sold their lands remarking, "…to have more money and make their cabins…they sell two pieces of land and they build a cabin…" (Huape-002). Informants felt they needed more intellectual and financial capital from the Huilo Huilo administration saying, "They could prepare people, train them so they can improve their services and…local tourism" (PF-002) and "It's important that people have…money to invest, having more resources so that they can work with tourism" (NEL-011). A Neltume entrepreneur asked, "…if they can't get the means to prepare, to work with tourism, what's it good for?" (NEL-011).

Informants perceived stratification of new economic opportunities as a reason that local people had a difficult time amassing capital. The aforesaid Neltume entrepreneur also questioned why particular people received opportunities from Huilo Huilo to develop their tourism businesses, but not others: "Why give that many opportunities to those that already have them?…so few chances for the smaller ones" (NEL-011). Informants near Oncol Park stated that communities lacking an ocean beach or cabins have fewer opportunities to attract tourists. Near RCV, informants contended certain communities and businesses were favored by RCV administration. A Huiro *quincho* (barbequing structure) owner claimed RCV selectively helped local people, and a tourism officer stated, "I think it's very directed, (RCV officials) have a group of favorite names to call" (CH-004). An RCV official offered a different perspective saying local people who adapted to changes brought by PPAs are "the ones that get economic benefits" (TNC-003). A second RCV official confirmed disengaging from Huape because "strategically it hasn't been given priority stating: it's a little bit farther away, and also the people aren't that easy to work with" (TNC-004).

Renewed human–nature relations

Local people perceived native forest protection under PPAs diminished utilitarian values toward nature. A Chaihuín guide stated that his community now felt "more respect towards nature" (CH-008). The protection and restoration of native forest also meant conceiving nature differently and increasing awareness of "conservation problems" (HH-002). Informants near Huilo Huilo and RCV communities declared sentiments such as, "People…now…favor conservation, and have much more information at their disposal than they did only a few years ago…People have changed the way they conceive their surroundings, the way they see nature" (NEL-007). A long-term resident of RCV community Chaihuín, a rental cabin owner, reflected:

> (Historically), people who lived here didn't care about nature. We used to get a stick and then we made fire with it. Sometimes, we didn't use wood properly since the remains were not burned. Nowadays, we regret that

because we didn't take advantage of the benefits we had. In those times wood was abundant. Now it is not. We have to buy it and it is expensive. (CH-009)

Both Huilo Huilo and RCV officials attributed these changes to a "soft hand" (TNC-003) and patient approach where they "didn't threaten them with calling the police or anything, it was a very slow process of talking to them and convincing them" (HH-002). An RCV official believed "for…touristic development to be really successful, we have to change the ways the people relate to the environment" (TNC-002).

Enriched youth

PPAs provided novel opportunities for historically underprivileged youth. Youth at all three sites participated in PPA sponsored field trips focusing on forest ecology, biology, conservation, recycling, composting and ecotourism. Neltume and Puerto Fuy residents also reported the HHF provided lessons in language and music, and gave educational scholarships.

Some adults claimed youth "take much more advantage of the environment" (NEL-005) than older generations did and, in some cases, now can avoid emigration to the city. For instance, young adults near Oncol Park and Huilo Huilo accepted employment and internships in outdoor recreation and tourism. A former Huilo Huilo employee explained that Huilo Huilo influenced youth to value the forest in a sustainable manner:

> Especially for the younger kids, it has meant a lot. They now have courses in school where they learn what a huemul is…what a reserve is, what the meaning of nature is. (This) translates into a long-term advantage for them because when they grow up they will know the value of their land and the opportunities tourism has to offer. (NEL-007)

Community improvements

Informants linked PPAs to community improvement projects in all study communities. Informants discussed collaborations with Oncol Park's administration to fix a school roof in Los Pellines and make road improvements, while RCV communities gained trash removal, a paved road and bus service. Informants mentioned Petermann (the land owner) and TNC gave water rights to nearby communities, ensuring clean and reliable water. Informants believed Huilo Huilo helped turn the gateway community Neltume from a logging shantytown into an attractive mountain town. An executive within a Chilean PPA network echoed others on this matter stating, "I was in Neltume 10 years ago, so I can tell you there was nothing there, just shacks" (NGO-001). Puerto Fuy and Neltume were furnished with wood sculptures, playgrounds, residences for the aged, cultural centers and community activities. An official added:

> We built the fire station of Puerto Fuy, we work with the fire station of Neltume, we built the radio station, we give financial help to a local folk band, give direct advice on how to tend to different parks in the town. We also built a bike road, and many other things. (HH-001)

Security

Financial security

PPAs addressed rural economic decline at all three sites by driving a shift from timber-based livelihoods to tourism. Huilo Huilo communities were banking on a bright economic future due to the reserve's year-round tourism. A former employee described how the transition already boosted financial security for many local people: "A few years ago, our only work source was the timber industry, and nowadays there's this whole tourism industry going on, which offers much more job opportunities for everyone, not only for the men" (NEL-007). Informants also noted the seasonality of these positions telling us, "…few people work the whole year, most of them just work for the season" (PF-007) and "We have even three times the amount of people during summer" (HH-001). Informants consistently indicated Huilo Huilo is the main employer noting, "I switched to tourism because there was no other option, it was our future" (NEL-004/M1) and "…we can look at it as a monopoly. Here, who doesn't work for Mr. Petermann is either independent or has to work…outside of Neltume or Puerto Fuy" (PF-003; "monopoly" was echoed by NEL-003 and PF-002). The future was uncertain as local people held their collective breath over a rumor that Petermann was considering abandoning his Huilo Huilo venture. A Puerto Fuy leader commented on the precariousness of so many people relying on one person, who owns most of the land and provides most of the jobs: "I think…if (families) do not have the ability to expand (i.e., buy land from Petermann), or if…forestry and (tourism) construction stops, people are going to be forced to sell (their plots)" (PF-003). The consequence is emigration to urban areas where most Chilean jobs are located.

For those residing near RCV, improved social organization helped boost financial security. One informant commented on how observing others' success led to action: "We knew that Chaihuín was doing well with the cooperative…so we inquired…the reserve said, 'Yes…we can form a cooperative here. Let's form a group of…10-12 people and let's talk about it'" (Huape-003). Those near RCV recognized TNC for helping them get organized, pool and distribute resources, give neighboring residents a better chance to keep rather than sell their lands, and win project funds with larger sums of money versus pursuing individual endeavors. An RCV official explained: "…funds were opened mainly for organizations, however, this latest fund for neighboring communities also opened for (individuals); the difference is that the organizations can access higher amounts than (individuals)" and "organizations…above a certain income level are excluded from participating" (TNC-004).

Where local people were not organized they became disenchanted with the competitive funding scheme. Those near Oncol Park reported losing interest because FRD grant proposals failed and PPAs received public relations support even when communities did not reap benefits. A lifelong Bonifacio elder expressed his frustration:

> … they call all the communities to the contest, so with that (Oncol Park/Arauco) say that they are working with all the communities…they make you apply for a project and they never say that they have enough money for (all of the) projects…or this is going to be the amount. And the people already lost interest…In fact, they have come offering projects and nobody (applied)…because they are going to lose…it's the biggest lie because (Oncol Park/Arauco are) deceiving the poor man; they deceive, they deceive, hope, hope, hope, they give and then, when everything is set, they tell you, "Your project didn't come out because it lacked this."…Never. I mean, almost never anybody wins, at least no one here. (Bon-004)

However, an Oncol Park official insisted Oncol Park was winning nearby residents over one funded project at a time:

> I helped (Pilolcura) to create a project, they conceived the idea, but I drafted it and uploaded it…and we won a grant for the project. And so they keep involving me…and that has led to a very close relationship with the people of Pilolcura, because they see you working in a good way, independent of a corporation, or some organism from the government…people see this as an opportunity and they realize, that in this logic of local economic development, or sustainable development, it's not fiction or a declaration of good intentions. (Oncol-002)

Some community members considered security offered by PPAs owned by moguls or corporations as a bribe. For example, Oncol Park communities were fighting the construction of a pulp mill effluent duct proposed by Arauco, a company responsible for the largest environmental disaster in Chile's history, occurring in Los Ríos (see Sepúlveda & Villarroel, 2012). In several of our interviews Oncol Park was linked to this project and the park administration's community support concept was viewed as a "political maneuver" (Pellines-005) and rejected by the indigenous faction "because of a political vision" (Oncol-002). A woman who had worked with Oncol staff elaborated on this tension:

> … we are people that won't sell for a couple of pesos…first we need to know why are they offering us funds because it could be that they are offering this money so they can approve the pipe. That is what we most fear with this relationship between us and the park. (Pilol-002)

An Oncol Park official explained residents in these communities "didn't understand that private wildlife protected areas could be targets of public funds" (Oncol-002). Yet, we also note similar sentiment among Huilo Huilo informants who thought engaging the reserve was akin to being "bought" or a publicity stunt (PF-003). One Neltume leader said: "I haven't given them a chance to buy me" (NEL-006), while a religious figurehead remarked: "I wouldn't ask them for their help, it would make me feel corrupted" (NEL-013).

Access to forest resources

Informants stated PPA creation restricted their freedom to extract resources from the forest, particularly where PPA lands were previously unsanctioned commons. Near RCV, a Chaihuín elder reflected: "Back then, if somebody needed wood to make a house, they just went there and got it" (CH-002). Although TNC donated wood to schools, medical centers and other social institutions, they did not sell timber to local residents and they went so far as posting a public letter threatening penal retaliation for stealing wood. Now, local people must purchase wood from elsewhere. A Huape leader expressed his dismay with TNC:

> … they should give more work so that when there's bad weather (they) offer a solution to not using native wood…but nothing, they just prohibited (wood extraction). They didn't give anything back, something for us to survive off of while the weather, that can get really rough, hits us. (Huape-008)

Near Huilo Huilo, "the heating issue" (PF-006) prompted the Puerto Fuy neighborhood delegate to wonder, "Why can't I get firewood from (the reserve) if I don't have money to buy it? Why can't we take a stick from there if the stick is just lying there?" (PF-006). Residents near Oncol Park had parcels of family-owned land and did not express much dependency on timber residing on park lands. Informants near the park also disclosed they occasionally received wood from park officials for personal construction projects.

Informants near Huilo Huilo recounted that game hunting, fishing the Río Fuy and gathering wild edibles were no longer viable livelihood strategies. Oncol Park local people said they formerly harvested palmitas (Gleichenia quadripartida) and sold them. TNC allowed collection of ferns, fruits and mushrooms within its borders with their permission and guidance.

Changed access to grazing areas across all three sites created perceived threats to human security and resulted in conflict between communities and PPA administration. Informants said, the lands where RCV is located used to be "full of animals" but were "close(d)…with fences" (CH-003) to keep livestock out. Informants said Huiro and Cadillal families were most impacted because they lost their winter grazing lands. A woman working with TNC on the Cadillal project declared: "the reserve cornered us, and isolated us" (Cad-002). A union leader suspected farmers will "have to sell (cattle) to survive" (Huape-008). Some residents responded by decreasing herd size. Ten kilometers down the road, a Huape leader said his community was eliminating cows:

> Everybody who has sheep has (enough) land. The problem is the (cows). Well, mostly it's the people of Huiro who have (cows). Here, there are people (with cows), but they are mostly eliminating all the (cows), because they have already been fined. (Huape-002)

An official responded with a different outlook on the conflict:

> … it's a behavioral change that they have to assume…it's…hard for them, especially the elders. They find it hard to understand why they have to change their ways…because they have always lived here and have always had their cattle free, so (it's like), "Why do these gringos arrive now and tell us to take our cattle out, and want us to change our way of life?" (TNC-004)

In 2014, a cooperative agreement was signed between cattle producers and TNC to help families adopt cattle breeding practices that do not impact RCV and preserve the husbandry heritage and income of impacted Cadillal and Chaihuín families. Residents near Huilo Huilo and Oncol Park experienced change in forest access for livestock when PPAs began, but were reportedly less impacted than RCV communities because livelihoods were less dependent on the forest.

Eviction with few alternatives

Four informants detailed that privatization impacted a number of families near PPAs by removing them from their homesteads. A long-time Puerto Fuy resident stated that the business arm of the Huilo Huilo project, run by Petermann, is "the most serious conflict (because) he is shrinking the communities; he is taking over everything" (PF-008) and "the owner of all Neltume" (NEL-003). Two Puerto Fuy residents explained that after Petermann purchased the *fundos* their family members living on the property for generations, but without deeds, "were kicked out or their houses were torn down… They preferred to leave peacefully" (PF-001/W2). "They weren't given a thing, not even (an option) to build somewhere else, nothing" (PF-001/W1). Informants stated 37 families near Huilo Huilo were landless and priced out of the local housing market, with some moving into a social housing development called Villa El Bosque in Neltume. Respondents also expressed concern about the difficulty of establishing a tourism business because there was no place to build one in a shrinking town.

Empowerment

Stewardship

PPAs provided education and programs help local people adapt to the changes PPAs brought to local communities. According to a HHF official, "The mission of the Foundation is to contribute to the conservation and preservation of the temperate rainforest and its resources, while at the same time educating communities on the true value of their surroundings" (HH-002). A TNC official provided a similar logic: "…the idea was always to conserve…with the communities in mind…conserve with local development, environmental education in schools, work with the communities, because you have to provide new possibilities so they can develop in a sustainable way" (TNC-001). Informants revealed environmental education courses offered by PPAs included lessons in forest ecology and biology and sustainable behavior, and highlighted human-caused degradation of the Valdivian Forest.

Local people believed administrations focused environmental education on youth. A Neltume tour operator, married to a teacher, explained the main objective was to diffuse environmental education throughout the household rather than engage adults directly:

> The Foundation teaches the young boys how to take care of nature, to appreciate it, and how to raise awareness inside their own families, because it's one thing to start that process in a young mind and another completely different to do it in an adult. Through those courses, the Foundation seeks to influence entire households through the younger kids, who have shown to be more permeable to that kind of knowledge. (NEL-008)

A TNC official articulated a similar strategy:

> …the environmental education program has its focus on affecting children because what these kids learn is taken to their homes…effectively, these kids become transmitters of information and also become critics of the negative actions of their parents…this (knowledge)…is radiated to the complete community. (TNC-002)

According to local people, these strategies were effective at the household and community levels. A resident with close ties to RCV responded "yes" when asked if the TNC's program changed his child's perception of nature and conservation saying, **"Kids pick that up much quicker than we do, on the subjects of taking care of nature, the trash, having more consciousness on what it means to protect the environment"** (Huiro-001). A Neltume tourism entrepreneur described how youth react to people, especially parents, who perform behaviors they see as harmful to the environment responding, "…Now if a parent throws a paper to the ground, they're all over them" (NEL-011).

Through these educational efforts adults also came to believe they were threats to the forest. A former logger who participated in PPA environmental education confessed he unknowingly harmed nature and then changed his ways telling us:

> I used to go to the hills with a machete…just cutting branches, any branch, but it turns out that I didn't realize I was hurting a tree that afterwards would give us air. But now the reserve has educated us so much, that instead of hacking the tree we take care of it. We take care of nature now. (Huape-003)

A Chaihuín artisan commented TNC's efforts helped him recognize the impact of historical community practices, "because of ignorance, we planted eucalyptus here, and now vegetables don't grow like they used to" (CH-006). An Oncol Park official found their environmental education programming, usually taking place inside the park, was not as successful with local adults who (1) do not enter Oncol Park or (2) "feel that Oncol Park is for more wealthy people and that they do not have a place (there)" (Oncol-002).

Capacity-building programs offered by all PPAs concentrated on skill building within a market-based economy. Informants noted PPAs offered courses in entrepreneurship and business finance, leadership, sustainable harvesting techniques, apiculture, guiding, as well as culinary and artisan interests. Informants believed these empowerment opportunities enhanced self-pride, self-worth and self-confidence. As tourism projects progressed and garnered visits from regional and national authorities, informants noted resident transformations. A former Huape logger explained how TNC's projects and programs empowered him and his fellow community members: "... it makes you grow as a person...now we have more empowerment. We are not running away anymore...Now...you (desire) to be something more" (Huape-003).

Female empowerment

Informants stated that women gained knowledge and skills from PPAs that allowed them to work for the first time, changing local culture. According to informants, women chose to step out of the shadow of their male partners. This was true of RCV women, less visible among Oncol Park women, and especially true near Huilo Huilo, perhaps due to the HHF's devotion to women's empowerment. A Puerto Fuy woman spoke about the cultural change women were "motivated" to make declaring, "Now we women aren't so submissive or anything. We go out to work and ready (to work)" (PF-001). Women comprised most of the Huilo Huilo hotel staff and can make much more money than their male counterparts. A Huilo Huilo official noted: "...there are even months where women make three times more money than their husbands" (HH-002). Women in Huilo Huilo and RCV communities ran tourism businesses and organizations or earned income on the side making handmade goods. A former timber worker now working in tourism believed Huilo Huilo women to be "better entrepreneurs" than men because women were not entrenched in the previous timber culture that stifled the entrepreneurial spirit (NEL-004/M1).

Women's empowerment ruptured customary household dynamics in some cases, however. A Neltume woman explained she accepted employment as a maid for Huilo Huilo because her husband's wages working in reserve construction for Petermann were not enough to support their family. She linked her subsequent divorce to her decision to keep working for the reserve, telling us, "I think we (women) all have the same problem...men can't stand it, that there's no schedule...they're too macho...and want women to be around for them" (NEL-010). A woman working in tourism noted that the working mother concept impacted children because mothers, who usually stayed at home, were working, "...children are the ones who benefit the least because their mothers are now busier and don't have as much time for them" (NEL-001).

Community disempowerment

Informants across all three cases felt disempowered by exclusion from decision-making. An RCV official stated local people "don't participate in the decisions that we make" (TNC-003). Similarly, a Puerto Fuy resident stated: "I've never seen (Huilo Huilo administration) in a meeting with the community, or asking them, 'We have this project, what do you think?'" (PF-005). Others in the Huilo Huilo area echoed sentiments that they have no power when it comes to the Huilo Huilo administration's actions stating, "they have the power; they get along with the police, so there's nothing we can do" (PF-008) and "If (Petermann) doesn't like something he growls and if he feels like sending them to hell he does" (NEL-010). Despite asymmetrical power relations, resistance to unilateral decisions occurred.

Informants detailed how communities near RCV and Oncol Park wrote denunciation and rejection letters and cut the road in response to PPA actions that threatened local people's well-being.

Informants across all sites also acknowledged community tensions surrounding PPA—community relations, which undermined empowerment efforts. Social relations in Neltume and Puerto Fuy were summarized with two dichotomies. A Huilo Huilo official offered one stating that those working in tourism and conservation rather than timber were "traitors" (HH-002). Perhaps this label was moored to the sale of the worker-owned *fundo* to Petermann in 1988, ending the dream of a commune:

> We were going to build a community near El Salto (waterfall) for those who had shares. We dreamed about doing that, everyone having land where we could grow our crops, etc. But, we never thought so many of our neighbors would sell their shares to foreigners or companies. (NEL-006)

A second theme pits dissenters against those who were thankful for a job. A reserve maid and others revealed that dissenters were commonly fired; for example, those fired for arguing, "(Petermann) isn't paying enough…will realize what they lost and wonder, 'How can I even get food to eat tomorrow?'" (NEL-010). Informants also noted divisions within communities near Oncol Park, with three informants stating Arauco employed a "divide and conquer" strategy to get their way (Minas-001, echoed by Pellin-002 and Bon-004), but divisions were noted less frequently near RCV.

Discussion

Our results suggest that Los Ríos PPA administrations and government officials, i.e. governing actors, pursued PPA—community relations because they believed local people degraded the forests and had insufficient capacity to craft sustainable livelihoods on their own. Centralized environmental governance has historically blamed local people for degrading resources within parks, imposing an ideological separation of culture and nature that rendered local livelihoods a threat to ecological integrity (Adams & Hutton, 2007; Sarkki, Rantala, & Karjalainen, 2015). Our findings differ from the scholarship on public protected areas because local people were not portrayed as the primary source of governance-related conflict, suggesting governing actor beliefs about local people were depoliticized. However, to justify development projects and ideological divides, PPAs, as conservation-development institutions, must be depoliticized (Büscher, 2010). Our findings suggest five factors that may explain how PPA administrations navigated the process of depoliticization. First, PPAs were not viewed by the State as entities to control. Instead, PPAs were tools by which the State could help address species and habitat protection and human development needs. This first factor established our second — PPA owners were able to make virtually all decisions about their land and to what degree they interacted with local people, including closing the unsanctioned commons and directing ecotourism development involving private land. Chile's socially legitimized neoliberal regime, our third factor, includes strong property rights and liberalized markets (Holmes, 2015), making the second factor possible. Fourth, PPA administrations moderated their customarily antagonistic beliefs by characterizing PPA—local relations with apolitical nomenclature (neighbors, supporters, workers). It is debatable whether this step was necessary given our final factor — PPAs were viewed as inevitable by non-PPA local informants, and when people are confronted by things perceived as inevitable, they are inclined to legitimize them to avoid discomfort (Moore, 1978).

While the broader conservation community has shifted toward participatory conservation-development projects since the 1980s (Eagles, 2004; Sarkki, Rantala, & Karjalainen, 2015), all PPAs in our study were carving out a unique position within contemporary protected areas management by excluding local people from governance at the PPA scale, but encouraging them to participate in development projects to inspire self-governance. PPA administrations shirked emerging participatory approaches being adopted in public protected areas, such as devolution of resource rights or community-based conservation (Bixler, Dell'Angelo, Mfune, & Roba, 2015), preferring projects that preserved their decision-making authority. Under the model identified in our study, PPAs carry a unique risk of creating a false sense of financial security and disempowering local people in ways that are less evident in similar efforts promoted by public protected areas. For instance, the PPA owner can, at their sole discretion, withdraw community support after local people become dependent on a PPA. Little has been written about the social concerns surrounding protected areas outside of land acquisition patterns and management (Stolton et al., 2014). Our findings suggest that the social impacts and consequences produced by PPAs facilitating ecotourism development should be subjected to the same level of scrutiny by protected area authorities that have been given to public protected areas.

Although we found PPAs maintained near-exclusive control over reserve-scale governance, they did work successfully to encourage self-governance among local people through educational and livelihood diversification strategies aimed at producing "environmental subjects". Agrawal suggests an environmental subject is a person who has come to show concern for and act in ways beneficial to the environment (2005). This approach of encouraging local people to accept dependence on nature and can benefit from ecotourism is obviously paternalistic (e.g. Cooke & Kothari, 2001), but studies suggest that when people living near protected areas perceive benefits associated with ecotourism, especially socioeconomic ones (Bonet-Garcia et al., 2015; Bottrill et al., 2014), they may turn from dissidents to advocates for protected areas (Hayes, Peterson, Heinen-Kay, & Langerhans, 2015; Liu, Ouyang, & Miao, 2010). Interestingly, informants reported positive gains in empowerment and security, and did not express concerns about the paternalistic nature of PPA efforts to produce environmental subjects and ecotourism advocates. This context may help explain why ecotourism has shown a potential to diminish the probability of conflict between local people and park management (Buckley 2009; Duffy, 2008) despite challenging existing livelihood strategies. Taking these findings in aggregate, it is conceivable that PPA owners will continue to embrace ecotourism development schemes (Langholz & Lassoie, 2001; Stolton et al., 2014), paternalistic as they may be, because the approach can be used to exert influence over local subjectivities and behavior for the benefit of conservation.

Our results highlight at least two ways PPAs may more effectively build ecotourism-based community partnerships in communities receptive to PPAs. First, our informants noted concerns about inadequate levels of startup capital. Therefore, PPA actors will need to work with community leaders to make more readily available the various capitals needed to take advantage of the new ecotourism economy in these areas. Focusing on collective arrangements rather than individual efforts could be a starting point, particularly in indigenous communities where community identity, solidarity, and action were prioritized. Finally, governing PPA actors can build support by reducing perceptions of social stratification in new opportunities and benefits. Our results do not clearly delineate between benefits being provided in biased ways or perceptions of that problem, but more equitable distribution of efforts and more transparent communication about how benefits are allocated would address both potential explanations.

Conclusion

PPA ecotourism ventures can improve the well-being of local inhabitants as well as degrade it. If PPAs reach their projected lucrative state in the future, PPA ecotourism development projects could be a boon to local people, particularly those residing in gateway communities, possessing ample tourism startup capital, and eager to engage in entrepreneurial endeavors. However, this model does not serve everyone, and in some cases may contribute to negative social outcomes.

Our study of PPA ecotourism development illustrates at least three take-home points with implications for scholars and practitioners who contemplate sustainable alternatives to historically unjust conservation-development schemes. First, when hegemonic conservation-development principles are viewed as a source of hope and means to economic and social mobility, and benefits are observable and deemed essential to survival, local people may be willing to reproduce broader conservation and development discourses across space and time and discount or overlook drawbacks (Silva & Motzer, 2015). Second, if PPA actors sincerely endeavor to enhance well-being from ecotourism, they need to pursue equitable capacity-building efforts (Bennett, Lemelin, Koster, & Budke, 2012) that align with local definitions of well-being. Should they fail to give serious consideration to local defini-tions of well-being, dissenters may brandish various "weapons" (Scott, 1985) to rewrite dominant conservation and development discourses to suit their vision of well-being or simply ignore PPAs. Finally, our study demonstrates that despite attempts by PPA administrations, PPA ecotourism could never be apolitical, and a belief that related participatory projects are purely an exercise in altruism or corporate responsibly would be naive. PPAs have strong and undertheorized ties to political econ-omy and environmental histories that shape the design and execution of private development-based conservation. The processes and web of actors that give rise to PPAs and their social impacts must be examined within the political economy of conservation and place-based governance and perceptions.

This paper's introduction noted Eagles' (2014) 10 research priorities for protected areas in the future. While many of the 10 research priorities are common to both public sector and private/NGO sector protected areas, this paper suggests that at least one additional research area is required: the governance, political economy and socio-economic evaluation of PPAs.

Notes

1. Panglossian – optimistic, even naively optimistic
2. The Nature Conservancy is a globally active non-governmental organization based in the United States. Their mis-sion is to "conserve the lands and waters on which all life depends" and use "non-confrontational, pragmatic, mar-ket-based solutions" to achieve their conservation goals. See www.nature.org for further information.
3. The south Andean deer.

Acknowledgments

We thank those informants who participated in this study, as well as our dedicated research team members. We also rec-ognize the following funding sources for their contribution to this research: Tourism Cares; Laarman International Gift Fund; American Alpine Club; Food and Agricultural Sciences National Needs Fellowship Grants Program.

Disclosure statement

No potential conflict of interest was reported by the authors.

Funding

We also like to recognize the following funding sources for their contribution to this research: Tourism Cares; Laarman International Gift Fund; American Alpine Club; Food and Agricultural Sciences National Needs Fellowship Grants Program.

References

Adams, W.M., & Hutton, J. (2007). People, parks and poverty: Political ecology and biodiversity conservation. *Conservation & Society, 5*, 147–183.

Agrawal, A. (2005). *Environmentality: Technologies of government and the making of subjects.* Durham, NC: Duke University Press.

Alderman, C.L. 1991. Privately owned lands: Their role in nature tourism, education, and conservation. In J.A. Kusler (Ed.), *Ecotourism and resource conservation* (pp. 289–323). Madison, WI: Omnipress.

Alex, R., & Clapp, R.A. (1998). Waiting for the forest law: Resource-led development and environmental politics in Chile. *Latin American Research Review, 33*(2), 3–36.

Armesto, J.J., Smith–Ramirez, C., & Rozzi, R. (2001). Conservation strategies for biodiversity and indigenous people in Chilean forest ecosystems. *Journal of the Royal Society of New Zealand, 31*(4), 865–877.

Ballantyne, R., & Packer, J. (2013). *International handbook on ecotourism.* Cheltenham: Edward Elgar Publishing.

Barany, M.E., Hammett, A.L., Shillington, L.J., & Murphy, B.R. (2001). The role of private wildlife reserves in Nicaragua's emerging ecotourism industry. *Journal of Sustainable Tourism, 9*(2), 95–110.

Beebe, J. (2001). *Rapid assessment process: An introduction.* Walnut Creek, CA: AltaMira Press.

Bennett, N., Lemelin, R.H., Koster, R., & Budke, I. (2012). A capital assets framework for appraising and building capacity for tourism development in aboriginal protected area gateway communities. *Tourism Management, 33*(4), 752–766.

Bishop, J., & Pagiola, S. (2012). *Selling forest environmental services: Market-based mechanisms for conservation and development.* London: Earthscan.

Bixler, R.P., Dell'Angelo, J., Mfune, O., & Roba, H. (2015). The political ecology of participatory conservation: Institutions and discourse. *Journal of Political Ecology, 22*, 165.

Bonet-García, F.J., Pérez-Luque, A.J., Moreno-Llorca, R.A., Pérez-Pérez, R., Puerta-Piñero, C., & Rodríguez, R.J.Z. (2015). Protected areas as elicitors of human well-being in a developed region: A new synthetic (socioeconomic) approach. *Biological Conservation, 187*, 221–229.

Bottrill, M., Cheng, S., Garside, R., Wongbusarakum, S., Roe, D., Holland, M., … Turner, W.R. (2014). What are the impacts of nature conservation interventions on human well-being: A systematic map protocol. *Environmental Evidence, 3*(16), 1–11.

Buckley, R. (2009). Evaluating the net effects of ecotourism on the environment: A framework, first assessment and future research. *Journal of Sustainable Tourism, 17*(6), 643–672.

Büscher, B. (2010). Anti-politics as political strategy: Neoliberalism and transfrontier conservation in southern Africa. *Development and Change, 41*(1), 29–51.

Cooke, B., & Kothari, U. (Eds.). (2001). *Participation: The new tyranny?* London: Zed Books.

Corcuera, E., Sepúlveda, C., & Geisse, G. (2002). Conserving land privately: Spontaneous markets for land conservation in Chile. In S. Pagiola & J. Bishop (Eds.), *Selling forest environmental services: Market-based mechanisms for conservation and development* (pp. 127–150). New York, NY: Earthscan.

Creswell, J.W. (2013). *Research design: Qualitative, quantitative, and mixed methods approaches.* Thousand Oaks, CA: Sage.

Delgado, C. (2005). *Plan de conservación: Reserva Costera Valdiviana.* Santiago: The Nature Conservancy.

Dinerstein, E., Olson, D.J., Graham, D., Webster, A.L., Primm, S.A., Bookbinder, M., & Ledec, G. (1995). *A conservation assessment of the terrestrial eco-regions of Latin America and the Caribbean.* Washington, DC: World Bank.

Duffy, R. (2008). Neoliberalising nature: Global networks and ecotourism development in Madagascar. *Journal of Sustainable Tourism, 16*(3), 327–344.

Eagles, P.F.J. (2004). Tourism at the fifth world parks congress, Durban, South Africa, 8–17 September 2003. *Journal of Sustainable Tourism, 12*, 169–173.

Eagles, P.F.J. (2014). Research priorities in park tourism. *Journal of Sustainable Tourism, 22*(4), 528–549.

Fattore, T., Mason, J., & Watson, E. (2007). Children's conceptualisation(s) of their well-being. *Social Indicators Research, 80*(1), 5–29.

Faúndez, Julio. (1988). *Marxism and democracy in Chile: From 1932 to the fall of Allende.* New Haven, CT: Yale University Press.

Fletcher, R. (2011). Sustaining tourism, sustaining capitalism? The tourism industry's role in global capitalist expansion. *Tourism Geographies, 13*(3), 443–461.

Gee, J.P. (2000). The new capitalism: What's new? Proceedings from Working Knowledge: Productive Learning at Work. Sydney: University of Technology.

Guba, E.G., Lincoln, Y.S. (1985). *Naturalistic inquiry.* Beverly Hills, CA: SAGE.

Hayes, M.C., Peterson, M.N., Heinen-Kay, J.L., & Langerhans, R.B. (2015). Tourism-related drivers of support for protection of fisheries resources on Andros Island, the Bahamas. *Ocean & Coastal Management, 106*, 118–123.

Henderson, R.I. (2013). *The dismembered family: Youth, memory, and modernity in rural southern Chile* (Doctoral dissertation). Montréal: Papyrus: Université de Montréal Institutional Repository.

Holmes, G. (2013a). *Private protected areas and land grabbing in southern Chile* (Discussion paper). Leeds: University of Leeds.

Holmes, G. (2013b). *What role do private protected areas have in conserving global biodiversity?* (Discussion paper). Leeds: University of Leeds.

Holmes, G. (2015). Markets, nature, neoliberalism, and conservation through private protected areas in southern Chile. *Environment and Planning A, 47*(4), 850—866.

Honey, M. (2008). *Ecotourism and sustainable development: Who owns paradise?* (2nd ed.).Washington, DC: Island Press.

Igoe, J., & Brockington, D. (2007). Neoliberal conservation: A brief introduction. *Conservation and Society, 5*(4), 432—449.

Infield, M. (1988). Attitudes of a rural community towards conservation and a local conservation area in Natal, South Africa. *Biological Conservation, 45*, 21—46.

Jamal, T., Borges, M., & Stronza, A. (2006). The institutionalisation of ecotourism: Certification, cultural equity and praxis. *Journal of Ecotourism, 5*(3), 145—175.

Jones, C. (2012). Ecophilanthropy, neoliberal conservation, and the transformation of Chilean Patagonia's Chacabuco Valley. *Oceania, 82*(3), 250—263.

Klubock, T.M. (2004). Class, community, and neoliberalism in Chile: Copper workers and the labor movement during the military dictatorship and the restoration of democracy. In P. Winn (Ed.), *Victims of the Chilean miracle: workers and neoliberalism in the Pinochet era, 1973—2002* (pp. 209—260). Durham, NC: Duke University Press.

Klubock, T.M. (2006). The politics of forests and forestry on Chile's southern frontier, 1880s—1940s. *Hispanic American Historical Review, 86*(3), 535—570.

Klubock, T.M. (2014). *La frontera: Forests and ecological conflict in Chile's frontier territory.* Durham, NC: Duke University Press.

Langholz, J. (1996). Economics, objectives, and success of private nature reserves in subs-Saharan Africa and Latin America. *Conservation Biology, 10*(1), 271—280.

Langholz, J.A., & Krug, W. (2004). New forms of biodiversity governance: Non-state actors and the private protected area action plan. *Journal of International Wildlife Law and Policy, 7*(1—2), 9—29.

Langholz, J.A., & Lassoie, J.P. (2001). Perils and promise of privately owned protected areas. *BioScience, 51*(12), 1079—1085.

Lincoln, Y.S., & Guba, E.G. (1985). *Naturalistic inquiry.* Newbury Park, CA: Sage.

Liu, J., Ouyang, Z., & Miao, H. (2010). Environmental attitudes of stakeholders and the perceptions regarding protected area-community conflicts: A case study in China. *Journal of Environmental Management, 91*, 2254—2262.

Marín, G., & Marín, V.M. (1991). *Research with Hispanic populations.* Newbury Park, CA: Sage.

Meza, L.E. (2009). Mapuche struggles for land and the role of private protected areas in Chile. *Journal of Latin American Geography, 8*(1), 149—163.

Moore, B. (1978). *Injustice: The social basis of obedience and revolt.* White Plains, NY: M.E. Sharpe.

Núñez, D., Nahuelhual, L., & Oyarzún, C. 2006. Forests and water: The value of native temperate forests in supplying water for human consumption. *Ecological Economics, 58*(3), 606—616.

Petty, N.J., Thomson, O.P., & Stew, G. (2012). Ready for a paradigm shift? Part 2: Introducing qualitative research methodologies and methods. *Manual Therapy, 17*(5), 378—384.

Ponce, I. 2007. *Informe preliminar levantamiento de información socio-cultural para el sitio prioritario Curiñanco, en las localidades de Pilolcura, Bonifacio, Curiñanco, Las Minas y Los Pellines.* [Preliminary socio-cultural information gathering report for the priority site Curiñanco, in the towns of Pilolcura, Bonifacio, Curiñanco, Las Minas y Los Pellines] (Unpublished paper). Valdivia, Chile.

Sarkki, S., Rantala, L., & Karjalainen, T.P. (2015). Local people and protected areas: Identifying problems, potential solutions and further research questions. *International Journal of Environment and Sustainable Development, 14*(3), 299—314.

Scott, J.C. (1985). *Weapons of the weak. Everyday forms of peasant resistance.* New Haven, CT: Yale University Press.

Sepúlveda, C., & Villarroel, P. (2012). Swans, conflicts, and resonance: Local movements and the reform of Chilean environmental institutions. *Latin American Perspectives, 39*(4), 181—200.

Serenari, C., Peterson, M.N., Leung, Y.-F., Stowhas, P., Wallace, T., & Sills, E.O. (2015). Private development-based forest conservation in Patagonia: Comparing mental models and revealing cultural truths. *Ecology & Society,* DOI:org/10.5751/ES-07696-200304

Silva, J.A., & Motzer, N. (2015). Hybrid uptakes of neoliberal conservation in Namibian tourism-based development. *Development and Change, 46*(1), 48—71.

Sims-Castley, R., Kerley, G.I., Geach, B., & Langholz, J. (2005). Socio-economic significance of ecotourism-based private game reserves in South Africa's Eastern Cape Province. *Parks, 15*(2), 6—18.

Singh, T., Slotkin, M.H., & Vamosi, A.R. (2007). Attitude towards ecotourism and environmental advocacy: Profiling the dimensions of sustainability. *Journal of Vacation Marketing, 13*(2), 119—134.

Stolton, S., Redford, K., Dudley, N., Adams, W., Corcuera, E., & Mitchell, B. (2014). *The futures of privately protected areas.* Gland: IUCN.

Taylor, J.E. (2010). *Pilgrims of the vertical: Yosemite rock climbers and nature at risk.* Cambridge, MA: Harvard University Press.

The Nature Conservancy. (2015). *Avoiding planned deforestation and degradation in the Valdivian Coastal Reserve project, Region XIV, Chile*. Valdivia: Author.

Thompson, C. (1999). Qualitative research into nurse decision making: Factors for consideration in theoretical sampling. *Qualitative Health Research, 9*(6), 815–828.

Vidal, F., Smith-Fleuck, J.A.M., Flueck, W.T., & Arias, E. (2011). Patagonian huemul deer (*Hippocamelus bisulcus*) under captive conditions: An historical overview. *Animal Production Science, 54*(4), 340–350.

Wakild, E. (2009). Purchasing Patagonia: The contradictions of conservation in free market Chile. In W.L. Alexander (Ed.), *Lost in the long transition: Struggles for social justice in neoliberal Chile* (pp. 112–125). Lanham, MD: Lexington Books.

Wilson, K., Newton, A., Echeverría, C., Weston, C., & Burgman, M. (2005). A vulnerability analysis of the temperate forests of south central Chile. *Biological Conservation, 122*(1), 9–21.

World Bank. (2001). *World development report 2000/2001: Attacking poverty*. New York, NY: Oxford University Press.

Tourism concessions in National Parks: neo-liberal governance experiments for a Conservation Economy in New Zealand

Valentina Dinica ⓘD

ABSTRACT
This paper analyses the use of concessions for tourism business in protected areas, including National Parks (NPs), in New Zealand, with specific reference to their financial and in-kind contributions to conservation work. A holistic approach is taken, examining the design and implementation of policies affecting concessions, and their interactions with wider governance arrangements for NPs. The case study considers a range of legal and policy questions, informed by 42 non-structured interviews with concessionaires, park staff and other stakeholders. It explores neo-liberal government policies implemented since 2009, as part of a Conservation Economy vision, aiming to stimulate economic growth by opening-up business access to natural heritage and resources in NPs. More and longer contracts have been granted against the promise of conservation, environmental and infrastructural gains from concessionaires. The research found that, so far, there is no evidence of improvements in the latter aspects, and expectations of gains from donations and voluntary business action have remained only symbolic. Powers to impose such contractual responsibilities are unused. Other significant, and related, neo-liberal governance changes have been implemented or under consideration, raising concerns that a tipping point in NPs' governance towards unsustainable tourism development may be on the horizon.

Introduction: governance and unsustainable tourism in the context of National Parks

National Parks (NPs) constitute major attractions for tourism worldwide. They can be accessed independently or through authorized tourism-related companies. Authorizations include permits (typically for low impact activities), licenses or leases (when infrastructural works are included). They are often referred to in the academic literature and by practitioners as concessions. Examples of activities under concessions are hiking/walking/cycling businesses, and a range of boating, kayaking, climbing, skiing, caving, fishing, hunting, hospitality and accommodation enterprises. Some activities are not allowed in some NPs, such as jet-skiing in New Zealand's park waters inhabited by protected marine wildlife. However, motorized sightseeing or thrill-seeking activities may be allowed (sometimes subject to conditions and zoning arrangements), such as helicopter tours, dune buggy rides and vehicle use on beaches. Examples of facilities/infrastructures are roads, tracks, bridges, huts, camping and picnic sites, signage, toilets, shelters.

Businesses contribute to the increased accessibility of natural areas and help diversify the types of nature which can be enjoyed. They are enabled to operate in NPs because governmental organizations do not have sufficient expertise, or human and financial resources, to deliver such serv-ices themselves. In some countries, ideological/political beliefs are such that legal frameworks do not allow state agencies to manage commercial operations in NPs. Nevertheless, the increasing supply of activities and facilities often raises concerns regarding NPs' commercialization beyond ecologically and socially acceptable levels. An increasing body of literature documents the negative impacts of tourism in NPs, and the widely variable financial contributions from tourism to the budgets of park agencies. For example, writing on the relationship between parks and tourism, Buckley observes that "Some park agencies earn up to 80% of total revenue by charging individual visitors directly. Partner-ships with tourism developers, however, have incurred higher costs, brought few visitors and minimal revenue (<6%), earned no net revenue for conservation, and reduced benefits for private recrea-tional visitors" (2009, p. 1). The distribution of tourism impacts, costs and benefits for NPs – environ-mental (including biodiversity conservation), economic, social and cultural – is important, as tourism is a major commercial user of protected areas. If governed properly, tourism has potentially lower environmental impacts than, for example, mining and hydropower production. And other writers have long noted that in a world of declining state funding for protected areas, NPs need to tap into tourism's cash flow to support their conservation work (Eagles, 2002, 2014).

In this context, a useful approach to the study of the sustainability of tourism development in NPs is to focus on the design and implementation of tourism concessions: what do they say about con-cession fees; environmental obligations; social and cultural heritage responsibilities; criteria for con-tract renewal; monitoring and enforcement? What are the opportunities for public engagement in concession decisions, as a safeguard that public interests are properly considered? These are impor-tant questions, as tourism concessions are exceptionally powerful instruments in the toolbox that governments have available to find a balance between the often conflicting environmental, eco-nomic and social objectives legally set for NPs' governance.

However, studying tourism concessions in isolation is unlikely to be sufficient, as they are part of wider governance arrangements affecting the achievement of the hierarchy of objectives for NPs. From the standpoint of legally available instruments, concessions represent the last layer of tools for the implementation of the governance arrangements pertaining to NPs. "Governance" will be used in this paper to refer to all legal instruments, policies, strategies, administrative arrangements, public–private collaborations and societal processes (such as public participation in policy processes) rele-vant for the study area/unit (Dinica, 2013, p. 664–665), in this case the NP of interest. Governance arrangements for NPs may be visualized as a pyramid, as suggested in Figure 1, at the top of which

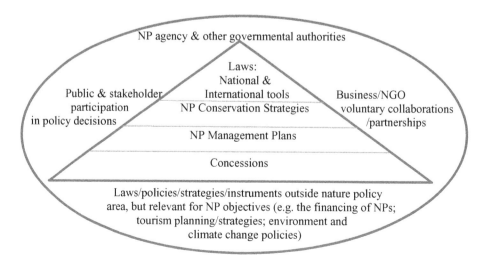

Figure 1. A holistic approach to concessions' study, as part of NP governance arrangements.

we find the guiding national laws and ratified international conventions, implemented by national and regional policy instruments and processes (including business partnerships, and volunteering programmes); the bottom layer is represented by concessions, which operationalize the governmental approach to managing the respective NPs under discussion.

Across all layers there are public and stakeholder engagement processes, ensuring that citizens and stakeholders have input in the design of legal and policy frameworks, and collaborations with businesses/NGOs/communities for the implementation of objectives and instruments. This structure is embedded in the wider institutional framework, where the NP agency works with other governmental departments and regional/local bodies to implement policies that require coordination with the legal-policy frameworks applicable for NPs. Therefore, a holistic governance approach is needed when studying the role of concessions in the sustainability of tourism development in NPs, as their design is influenced by the instruments and processes surrounding them.

In the New Zealand context studied in this paper, the relevant legal framework is substantial, but the two key legal tools are the 1980 National Parks Act, and 1987 Conservation Act. The latter appointed the Department of Conservation (DOC) as the governmental authority charged with managing NPs (and the public Conservation Estate, covering 33% of the country; Treasury, 2014a, p. 58). Key regional policy instruments are the NP Conservation Strategies that may span several NPs, and are implemented through NP Management Plans. Since 1993, the country uses one of the oldest visitor planning frameworks, the Recreation Opportunity Spectrum (ROS, see below), which is operationalized in each NP Plan, in terms of zones with different volumes and types of tourism activities and facilities/infrastructures (Taylor, 1993). Concessions can only be given based on the ROS zoning reflected in individual Park Plan provisions. The Conservation Minister or delegates may exceptionally approve concessions for activities/facilities/timings not included in the Plan. At the same time, the 1987 Conservation Act regulates many aspects of concessions' approval process, content, monitoring and enforcement, and the discretion available to regional officials.

Clearly, the sustainable development of tourism is also influenced by the volume and behaviour of independent visitors to NPs. This paper focuses, however, on how tourism that takes place through tourism concessionaires is governed. It examines how governance arrangements pertaining to concessions in NPs are changing in New Zealand, following a neo-liberal turn towards a "Conservation Economy", aiming to increase the commercial utilization of natural resources and heritage, and the possible consequences in terms of the sustainability of tourism in NPs. These analyses are placed on the background of existing deficiencies in the wider governance framework within which DOC operates and the concessions' design, focusing on environmental and biodiversity responsibilities.

Unsustainable developments can typically be traced back to governance weaknesses or gaps. In this paper, unsustainable tourism development in NPs will be conceptualized as the patterns of development that are likely to undermine the existing hierarchy of objectives for NPs' protection in (1) the national legislation (see section below for New Zealand) and (2) a NP's definition and management objectives, as defined by the International Union for the Conservation of Nature and Natural Resources (IUCN), and represented in Figure 2.

Whenever key governance elements are changed, new opportunities and threats emerge; existing unsustainable patterns of development may be reversed or accelerated, and new ones may emerge. For example, the hierarchy of legally set objectives for NPs may change or be undermined by new legal provisions, or new ways of implementing legal-policy frameworks, with changes in the discretion available to officials. Regulatory inconsistencies and gaps may emerge, which may be exploited by colluding commercialization-minded actors. Other changes may have an open radical nature, such as enabling long-term concessions for environmentally damaging mass tourism in NPs.

Some use the term tipping points (Vormedal, 2012, p. 258), to refer to cumulative governance changes that cannot be regarded anymore as suitable to ensure the initially envisaged objectives. Schelling defines a tipping point as "a point of discontinuity on a trajectory, a disruption that marks the beginning of a cumulative process of change deflecting from the previous path" (paraphrased in Vormedal, 2012, p. 258). In the NP context, a tipping point would represent a set of governance

Primary objective
To protect natural biodiversity along with its underlying ecological structure and supporting environmental processes, and to promote education and recreation.
Other objectives
- To manage the area in order to perpetuate, in as natural a state as possible, representative examples of physiographic regions, biotic communities, genetic resources and unimpaired natural processes;
- To maintain viable and ecologically functional populations and assemblages of native species at densities sufficient to conserve ecosystem integrity and resilience in the long term;
- To contribute in particular to conservation of wide-ranging species, regional ecological processes and migration routes;
- To manage visitor use for inspirational, educational, cultural and recreational purposes at a level which will not cause significant biological or ecological degradation to the natural resources;
- To take into account the needs of indigenous people and local communities, including subsistence resource use, in so far as these will not adversely affect the primary management objective;
- To contribute to local economies through tourism.

Figure 2. The hierarchy of objectives for National Parks, based on the IUCN classification of protected nature areas. (Source: http://www.iucn.org/about/work/programmes/gpap_home/gpap_quality/gpap_pacategories/gpap_pacategory2).

changes resulting in the de facto and/or de jure reversal of the hierarchy of objectives for NPs envisaged in national and international IUCN-backed biodiversity agreements.

This paper presents some of the results of a long-term research project in New Zealand investigating whether recent governance changes surrounding concessions, aiming to implement a vaguely defined political idea of a "Conservation Economy" (see section below), can be assessed as a potential tipping point for NP tourism sustainability. The paper starts with a literature review reflecting on what we know so far about concessions in NPs. Later sections present the research methods, and the empirical analyses based on the New Zealand case-study.

Literature review

In a recent special issue of the *Journal of Sustainable Tourism* dedicated to protected areas and sustainable tourism planning, Eagles (2014, p. 538–539) argues that tourism concessions should be seen as one of the top 10 research priority areas on sustainable tourism in nature areas. In an earlier publication, Eagles, Baycetich, Chen, Dong, Halpenny, Kwan, Lenuzzi, Wang, Xiao, and Zhang list 17 research questions they view as "the most important gaps of knowledge in concession management in park agencies" (Eagles et al, 2009, p. 93–94). Among them are questions including:

1. What are the major problems identified by park agencies in their;
    (a) Selection of concessions, and
    (b) Management of concessions?(...)
5. What are the contract lengths?
6. What criteria are used by park agencies for concessionaire choice?(...)
15. Are independent compliance auditors used?(...)
17. How often do concession contract conflicts go to the courts for resolution?

Despite the importance of this theme, literature on NP concessions is remarkably scarce. Several guidelines are available, with recommendations and illustrations on how to design tourism concessions, to serve sustainability objectives (Eagles et al, 2009; Eagles, Mc Cool, & Haynes, 2002; IUCN, 2000; UNEP-WTO, 2005). However, only a few focus comprehensively on tourism concessions in protected areas. Eagles et al. (2009) offer detailed recommendations for NP managers on aspects to be included in concession contracts, including pricing policy; capital/infrastructure investments; environmental guiding principles; visitor education; monitoring procedures and costs, contract suspension and termination; benefits for local communities, etc. The authors also encourage NP managers to insert best practice conditions like biodiversity conservation measures; renewable energy/fuels use; application of the most environmentally friendly methods for wastes' management, wastewater treatment and transportation (p. 48–60).

Some guidelines offer examples of concession design. UNEP-WTO guidelines consider that governments should use the opportunity to influence the behaviour of tourism concessionaires by attaching contractual conditions that "require compliance with the sustainability agenda" (2005, p. 94). Some Australian states have linked environmental certification programmes to the concession process/criteria. Tourism Queensland has been consolidating this link by means of small subsidies to businesses for steps such as in-house training or auditing (UNEP-WTO, 2005 , p. 134). In South Africa, concession conditions often require tourism businesses to implement particular measures for community benefit. The South African concession regime has also been successful in securing funds for conservation, beyond the regular concession fees (UNEP-WTO, 2005, p. 134).

A more elaborated treatment of concessions can be found in several academic studies focusing exclusively on the design and/or performance of the NP concessions' regime from the standpoint of sustainability. This literature stream is important but seriously under developed; it is here that this paper aims to contribute. Only two studies were identified as relevant. Wyman, Barborak, Inamdar, and Stein (2011) offer a short "birds-eye view evaluation" of tourism concession agreements in 22 countries, drawing exclusively on document analysis. This has clear limitations, given that monitoring and enforcement could be a problem, as discussed below. Their study includes New Zealand and concludes that its concession regime is best designed on several concessionaire qualification aspects: financial capacity, tourism experience and educational level. However, New Zealand concessions are found to be lacking prescriptions on social responsibilities, and satisfy only two environmental responsibility criteria (out of six): monitoring plan and risk analysis.

To understand the governance arrangements for NP concessions properly, one needs to start with a level of in-depth analysis such as that provided by Fearnhead (2007) on commercial tourism development in South African National Parks (SANParks). Featherhead describes the key elements included in concession contracts, such as the required concessionaire qualifications, financial and legal arrangements, environmental responsibilities and social empowerment measures. Each concessionaire is subject to Environmental Guidelines, which are to be observed in addition to the applicable environmental legislation; they include site-specific Environmental Impact Assessments, Environmental Management Plans with "site carrying capacity" studies and "the appointment of an environmental control officer" (p. 307). Nevertheless, a larger picture of the concessions' governance arrangements in SANParks and their sustainability performance cannot be elicited from Featherhead's publication. Featherhead does not reflect on the wider context in which tourism concessions are embedded: that of visitor planning frameworks, NP Management Plans and higher order policies.

Setting the stage: the legal framework and financial challenges for DOC's operations

The main legal framework for NP management in New Zealand consists of the 1980 National Parks Act, the 1987 Conservation Act and all legal revisions of these Acts. In addition, the 1953 Wildlife Act is relevant for biodiversity conservation, and some provisions of the 1991 Resource Management Act are relevant for several environmental sustainability aspects of human impacts on NPs. Based on the 1980 National Parks Act, DOC is required to preserve NPs

in perpetuity..., for their intrinsic worth and for the benefit, use, and enjoyment of the public, areas of New Zealand that contain scenery of such distinctive quality, ecological systems, or natural features so beautiful, unique, or scientifically important that their preservation is in the national interest.

The hierarchy of objectives for NP management by DOC, emerging from the legal framework, is clear and, so far, has remained unaltered since the 1980s: (1) conserving nature; (2) educating the public; (3) fostering recreation; (4) allowing for tourism whenever compatible with nature conservation.

Based on the 1987 Conservation Act, Section 6, DOC's first responsibility is

(a) to manage for conservation purposes, all land, and all other natural and historic resources ...and... to preserve so far as is practicable all indigenous freshwater fisheries, and protect recreational freshwater fisheries and freshwater fish habitats.

While paragraph (a) refers to all natural resources, emphasis is given to land and ecological functions; water resources require protection especially from the standpoint of ensuring healthy indigenous fisheries and recreational fishing. It is important to note that no reference is made to air quality or climate stability, which can affect the quality and productivity of soils, the quality and availability of water resources and the health of many types of terrestrial and aquatic fauna. From this standpoint, New Zealand's legal framework on nature protection is dated; it does not incorporate global policy developments on climate change and air pollution mitigation, and lacks a holistic approach to ecosystem health. Second, DOC has education and advocacy responsibilities towards the public and tourists (Sections 6.b and 6.d). Further, Section 6.e of the 1987 Conservation Act envisages that

to the extent that the use of any natural or historic resource for recreation or tourism is not inconsistent with its conservation, to foster the use of natural and historic resources for recreation, and to allow their use for tourism (emphasis by author).

Consequently, in the legal hierarchy, recreation is clearly a third legal responsibility for DOC, while the support for tourism activities and infrastructures only comes fourth, provided that the highest ranked objective is not compromised – nature conservation.

DOC's ability to manage tourism in NPs sustainably has been, however, historically affected by two governance constraints: its inability to influence tourism planning; and the persistent lack of political willingness to introduce NP entry fees or other financial instruments, for conservation funding and visitor management. Since the countrywide deregulation of most economic sectors in the 1980s and early 1990s, tourism planning has been slowly abandoned at all levels (Shone and Memon, 2008). Visitor numbers have increased steadily over the past decades across NPs, especially from international tourism (DOC, 1996; DOC, 2011a).

The country's tourism marketing agency, Tourism New Zealand, and the Ministry for Business, Innovation and Employment (MBIE, hosting the government's Tourism Policy Group) have ambitious targets to increase the share of international visitors using NPs, from 46% in 2007/2008 (1.1 million people) to 50% by 2016/2017 (DOC, 2011a, p. 30). This has led DOC into the difficult position of having to meet increasing demands for tourism facilities/infrastructures in the context of decreasing budgets. Table 1 shows that recently, the share of DOC's budgets dedicated to tourism-recreational objectives (third and fourth objective – second row from the top) is approaching that dedicated to its primary objective of natural heritage conservation (top row).

Some may argue that the hierarchy of legal objectives need not be mirrored in the budgetary spending, but this argument is weak in the context of disappointing conservation outcomes. By 2004, only 38% of New Zealand's 268 ecological districts were surveyed for its Protected Natural Area Programme (Wildlife Consultants, 2004), while "less than 25% of conservation land receives interventions on key threats, with around 8% receiving possum, rat and stoat control" (DOC, 2014a, p. 2). New Zealand's native biodiversity is largely unique and vulnerable to introduced fauna, having evolved in the absence of predators, after its islands split from the ancient continent of Gondwana. All native species of reptiles, frogs and bats can only be found there, while the same can be said of 90% of the insects and birds, and 80% of the plants forming New Zealand's ecosystems (DOC, 2014a, p. 2). However, as DOC often argues "New Zealand has one of the highest proportions of threatened species and one of the highest extinction rates in the world" (DOC, 2014a, p. 2).

In 2013, the concession fees collected by DOC from commercial users represented 3.15% of DOC's annual budget (based on Treasury 2014a, p. 6 and 2014b, p. 69). It is unclear how much tourism contributed to this, but the overall picture is clear: New Zealander taxpayers foot most of the bill for DOC's expenses. Table 2 shows the revenue generated by all business sources, between 2010 and 2013. Treasury did not offer numbers on revenue from partnerships/sponsorships/donations in the revenue graphs, but for 2013, this appears around 10 million NZ$, i.e. around 2.8% of DOC estimated budget. Considering

Table 1. Budgetary allocations for DOC. (Sources: Treasury, 2014a, p. 2 and 3).

Allocations per (some) activity/Budgeted for, in Million NZ$	2013/2014	2014/2015
management of natural heritage including the maintenance/restoration/protection of ecosystems, habitats and species	164.936 (37.7%)	160.303, (37%)
recreational (including tourism-related) facilities and services, and the management of business concessions	148.564 (34%)	144.993 (34%)
Payment of Rates to local councils on Properties for Concessionaires (limited to conservation areas that are used for private or commercial purposes; and for services provided by a local authority)	839	839
the protection and conservation management of historic heritage	5.565	5.996
working with communities to protect natural and historic resources	25.500	24.346
Crown Contribution to Regional Pest Management	3.614	3.292
Policy Advice, Statutory Planning, and Services to Ministers and Statutory Bodies	6.698	6.021
Impairment of Public Conservation Land (this appropriation is limited to the impairment in value of Public Conservation land to be transferred to iwi as part of Treaty of Waitangi Settlements)	3.059	50
Non-Departmental Output Expenses	15.236	6.979
Various other smaller expenses	60.302	77.993
Total	437.313	430.812

the three revenue sources in Table 2, business (including tourism-recreation) revenue seems to have represented 6.5% of DOCs allocated budget for 2013 (based on Treasury 2014a, p. 6 and 2014b, p. 69), while DOC spent a third of its budget in tourism-recreation facilities, in the context of chronic underfunding of activities under the primary NP objective.

Two questions emerge: what are the environmental responsibilities of concessionaires in New Zealand, and do concession contracts require direct/in-kind contributions to biodiversity conservation? The above context sketches a situation of financial unsustainability for DOC's operations, considering New Zealand has only 4.5 million people, while the taxpayer base is even smaller. The financial structure and tourism policy associated with the governance arrangements put in place in the 1980s fails to enable DOC to deliver adequately on the hierarchy of legal objectives allocated by law-makers. This governance background is important when assessing the impact of changes emerging from the Conservation Economy vision.

Political drivers for governance changes: the conservation economy idea

The post-2009 governments of neoliberal orientation (the National Party and its coalition partners) have been implementing a business growth agenda, which has a programme building on natural resources (New Zealand Government, 2013). The Resources Program is led by the ministers of seven Ministries/Departments that are either responsible for the management of natural resources (marine and terrestrial) or are managing economic sectors that depend on such resources. This group includes the Conservation Minister, the Environment Minister and the MBIE Minister. The programme aims to increase the contribution of all economic sectors relying on natural resources to 40% of

Table 2. DOC's revenue from businesses, 2010–2013 (Based on: Treasury 2014b, p. 60).

Revenue NZ$ million/year/type of tourism-related revenue	2010	2011	2012	2013
Concessions	13.91	14.37	14.31	13.79
Recreational charges	10.25	11.29	11.23	12.65
Retail sales	2.89	2.71	2.31	2.20

national GDP by 2025; this is referred to as "greening growth" or "sustained growth from natural resources" (New Zealand Government, 2012, p. 5 and7). For the tourism sector, the government aims to "Grow the number of new business opportunities on public conservation land in order to deliver increased economic prosperity and conservation gain" (New Zealand Government, 2012, p. 23).

Soon after the introduction of this programme, in 2009, the term 'Conservation Economy' began to be used by political decision-makers and DOC officials, in relation to this economic objective. Interestingly, the term has been predominantly used in the spoken media in New Zealand, in political discussions broadcasted through radio or television. Very limited explanations are offered in governmental documents. So far, it has only been spelled out by a Conservation Minister in DOC's Statement of Intent for 2009–2012, when he wrote:

> In its totality, conservation plays a critical role in validating the "clean pure" brand that is the market advantage on which our producers rely. It is increasingly clear that sound management of our natural areas produces the life-sustaining ecosystem services on which our lifestyle and prosperity depend. These are services such as freshwater yield and storage, soil fertility and stability, and carbon storage. Tourism is New Zealand's largest single foreign exchange earner, and the destinations for both domestic and international visitors are primarily around public conservation lands and waters. The businesses that support and complement tourism are major contributors to our regional economies and local communities. Once we recognise these interdependencies, we can start to capitalise on them to achieve social, economic and conservation gains. This gives meaning to the term, "the conservation economy." (DOC, 2009, p. 5)

This text is rather unclear as to the governance arrangements that should underpin the Conservation Economy idea, and the full range of expected outcomes. Conservation Economy seems to be the "short name" for the part of the Program of Building Natural Resources that refers to the Conservation Estate, or a way of more efficiently communicating such changes to the New Zealand public. It has become a convenient and eye-catching term. However, a closer inspection of governmental and DOC documents indicated that there are no new environmental, climate or biodiversity policies, or governmental funding schemes for at least some project-level initiatives, to support the "greening growth" ambitions, or the 40% objective for natural resources' contribution to national GDP by 2025. The underlying philosophy seem to be that, in exchange for access to conservation land and natural resources, businesses will increase their contributions towards environmental and biodiversity objectives. Governmental documents also fail to clarify which natural resources envisaged by authorities to "sustain growth" are renewable, at what rates of exploitation, and how do they plan to sustain growth based on non-renewable resources. They also fail to specify whether business contributions will be requested de-jure, or voluntarily, or both. In the same 2009 DOC Statement, the Conservation Minister wrote that the document "sets out how the Department of Conservation will contribute to the wellbeing and prosperity of New Zealanders over the medium term". (DOC, 2009, p. 5).

This is a surprising statement, as it has nothing in common with the hierarchy of legal objectives for DOC, summarized above. Furthermore, he wrote:

> I have directed my Department to investigate ways in which it can evolve its approach to tourism. This includes working to streamline the statutory processes in the granting of concessions. It is also about planning and developing its recreation infrastructure in ways and in places that are most likely to stimulate and support tourism, including by shifting the focus to more heavily populated areas. (2009, p. 5–6)

Therefore, the idea seems to be that DOC should move from "allowing tourism" to "enabling" it (DOC, 2010), while commercial concession fees should increase their contribution to DOC's budgets (Treasury, 2012). Governmental strategies and policies suggest, therefore, a reshuffling of the de facto hierarchy of objectives for DOC, whereby tourism is lifted to the second rank, at least. This is inconsistent with the 1987 Conservation Act, 1980 National Parks Act and also inconsistent with the IUCN definition and hierarchy of management objectives for NPs.

This vision requires more and longer concessions in NPs, as drawing cards for tourism. It is unclear whether higher concession fees are being considered by DOC, as such information is kept confidential. If concession fees increase significantly and indiscriminately, this may push out of the market, small locally owned and operated businesses. As New Zealand's legislation permits overseas companies to apply for NP concessions, this may undermine the last IUCN objective for NPs: "To contribute to local economies through tourism". It seems that the government prefers to use the option of increasing tourism volumes (DOC, 2011a, p. 30), rather than introducing NP entry fees for (at least) international visitors or other economic instruments that could enable DOC to raise more revenue from the same tourism volumes.

The valid question emerges whether mass tourism in NPs (through radical increases in the number of concessionaires or the number of tourists they are allowed to involve) and significant infrastructural developments (with significant landscape impacts) are considered politically desirable types of tourism development. Key governance changes have already been implemented that could possibly make this a reality, such as the 2010 and 2013 changes in the Conservation Act provisions regarding concessions' design and decision-making processes. There are concerns that – if the guidance role of NP Management Plans for concession decisions is eroded – and if public input in decision-making continues to be undermined, through further legal revisions – the ongoing governance changes may converge to generate a tipping point, meaning that insufficient safeguards will be left in place to ensure that the increased commercialization of NPs remains ecologically and socially sensitive.

It is difficult to see how the political vision to increase DOC's tourism revenues from concessions, and the new demand oriented approach to tourism, will not affect the legally imposed guiding role of NP Management Plans for concession allocations. NP Plans are underpinned by the ROS visitor planning framework, which uses a supply approach to NP access. Designed in the 1970s in the United States, ROS

> was offered as a means of identifying and determining the diversity of recreation opportunities for a nature area. ROS assumes that by providing diversity the adverse effects of increasing levels of use both on natural environment and visitors' experience would be mitigated. These effects would be reduced in large part by allocating high impact activities to more resilient sites and low impact activities to less resilient locations. (Newsome, Moore & Dowling, 2002, p. 157)

The Conservation Economy vision is demand-driven, with respect to access and infrastructural development, and can be expected to generate significant tensions during its implementation in a governance structure that at this stage has remained supply-oriented, through its regional planning tools. What will be the outcome of such tensions?

Furthermore, the Conservation Chapter of the 2013 Treasury Budget Report states that, in exchange for the new business opportunities, those obtaining DOC concessions are *expected* to start contributing to DOC's work on biodiversity conservation (such as pest trapping, native flora and fauna breeding/repopulation) and the maintenance of facilities and infrastructures used by tourism across the country, which are deteriorating fast and are insufficiently extensive to meet the current demand from international tourists (Treasury, 2013, p. 3;12–13). Additionally, it is also *hoped* that concessionaires, communities, recreational user groups and individuals will increase their contributions through donations and volunteer work to implement DOC's legal objectives (Treasury, 2012, 2013; New Zealand Government, 2012).

However, given the gradual weakening of the legal-policy frameworks regarding concessions, it is important to obtain more insight into the legal and policy status of such hopes and expectations. Three questions emerge as important if we are to understand whether NP's governance in New Zealand is heading towards a tipping point:

- What does the new concession regime look like, since 2009, and what are the implications in terms of public participation in decision-making?
- How do the current NP Management Plans guide the approval and management of tourism concessions? Do we see changes to their guidance role, following the shift to a Conservation Economy?
- Do the existing and new concession contracts deliver on the expectation that tourism businesses will generate biodiversity and environmental gains in NPs?

Research methods

Data collection was carried out between November 2013 and February 2015, as part of a large project on Governance Challenges and Opportunities for Sustainable Tourism Development in New Zealand. Four research methods underpin the empirical findings reported here. First, a review was performed of the New Zealand literature relevant for concessions, including academic publications and reports for/by DOC. Second, an in-depth legal and policy analysis was carried out of the past and revised legislative frameworks regarding NP management, and DOC's policies on concessions. Third, the Management Plans and Strategies for a selection of three NPs – Westland Tai Poutini, Mount Cook Aoraki and Mount Aspiring were examined, together with a selection of concession contracts in those parks.

The three NPs are located in the South Island (for a map of all NP and information on each park, please see: http://www.doc.govt.nz/parks-and-recreation/national-parks). They were selected because, together with Fiordland, they are part of the UNESCO South-West Wilderness World Heritage Site and enjoy some of the highest visitation numbers of all NPs (while they have been less researched than the Fiordland Park). In addition, they are prominent in the 'operation areas' for which many concessions are issued (the list of concessionaires and their operation areas/products can be viewed at http://www.doc.govt.nz/parks-and-recreation/activity-finder/activity-finder/guides-and-commercial-tourism-providers).

Twelve concession contracts for tourism in these NPs were reviewed, to appraise the incorporation of biodiversity and environmental objectives/measures. Concession contracts are publicly available only upon request and the selection was made by DOC. The selection ensured good representation of older and newer contracts, and a diversity of activities and facilities. Furthermore, 12 concession applications were considered that were publicly notified at the website of the DOC in the period July 2013–March 2015, and regarded tourism operations throughout the whole Conservation Estate.

Finally, the qualitative method of semi-structured interviews was used. Interviews were carried out with research participants from DOC regional centres, tourism concessionaires, Conservation Boards (representing communities, advisory for DOC), and NGOs. Forty-two interviews were carried out but evidence from only 11 interviews is invoked in this paper, due to size limitations. The response rate for interview invitations was 38%. Given that, for some interviews, confidentiality was requested, all interviews were treated as such. The interview schedule was structured around the themes included in the last three research sub-questions mentioned above.

Changes in the legal framework for concessions and implications for public participation

The 1987 Conservation Act dedicates Part 3 to concessions, distinguishing among (Art. 2.1):

- permits, granting the "right to undertake an activity that does not require an interest in land";
- licenses, offering a "nonexclusive interest in land or a grant that makes provision for any activity on the land that the licensee is permitted to carry out";
- leases, "granting an interest in land that (A) gives exclusive possession of the land and (B) makes provision for any activity on the land that the lessee is permitted to carry out"; easements are also envisaged but less used for tourism.

DOC operationalized these forms by considering also the expected nature impacts and the length of the requested concessions. These aspects influenced whether concession applications were to be publicly notified. If concessions activities/facilities were assessed as *high impact* and/or were requested for *a longer term*, the public had to be notified. In 2009, the Conservation Minister triggered a revision of the concession approval process, invoking the need to offer businesses more certainty regarding their investments in the Conservation Estate and "the need to improve the timeliness and efficiency of decision-making" (DOC, 2010, p. 7). The main changes to the concession regime regard the terms highlighted in italics above.

In the pre-2009/2010 regime, 'longer term' was operationalized in the Act as 5 years, while in the new one, this has become 10 years (Revised Sections 17T[4] and[5]. This term was recommended in a 2006 submission from the Tourism Industry Association New Zealand to DOC (TIANZ, 2006, p. 20-21). The revised Act requires the Minister to publicly notify, before granting, any "lease or a license with a term (including all renewals) exceeding 10 years", according to the revised Section 17T(4). However,

> Before granting a license with a term (including all renewals) not exceeding 10 years, or a permit … the Minister may give public notice of the intention to do so if, having regard to the effects of the license permit or easement, he or she considers it appropriate to give the notice

according to the new Section 17T(5). While permits may not be longer than 10 years and are not renewable (Section 17Z[2]), leases and licenses "may be granted for a term (which shall include all renewals of the lease or license) not exceeding 30 years or, where the Minister is satisfied that there are exceptional circumstances, for a term not exceeding 60 years" (Section 17Z[1]). These quotes show that the Minister has been given discretion on the notification of the decision to grant a permit or license for less than 10 years, for which the likely impacts are assessed as high. The new system was implemented in July 2010.

Another change regards the time available to the public to make submissions, when they are entitled to be consulted. "Sufficient time" for public submission was considered in the past to be 40 working days. In the new system, DOC internal procedures only allows for 20 working days (DOC, 2010). It is widely accepted that DOC took a long time to process concession applications, largely due to internal operational processes. In 2010, DOC wrote that

> throughout the organisation there are approximately 100 concession applications being processed that have been in the system for over 2 years. Many of these are for low impact activities. (DOC, 2010, p. 22)

This begs the question: was the shortening of the time allowed for public reaction by 20 days an absolutely necessary legal measure, to address the problem of processing delays by DOC?

The reduction in the opportunities for public engagement in concession decisions raises questions regarding the rate and number of approved concessions. Table 3 presents DOC's statistics on the annually active tourism concessions. Since 2009/2010, the trend towards fewer accommodation concessions has been reversed, with an almost 17% increase by 2013/2014. There are also significant increases in the numbers of guiding, vehicle and aircraft concessions; only the number of boating concessions is in decline. However, these statistics do not provide insight into the number of tourists allowed under each concession (or actually using) these tourism products.

NP Management Plans and the planning framework: emerging cracks in the system

De jure, New Zealand follows a supply-oriented system whereby concession contracts are used as tools for implementing the ROS zoning/planning framework set in NP Management Plans. Based on

Table 3. The number of annually active tourism concessions, 2006–2014.

Service	2006/2007	2009/2010	2013/2014
Accommodation	436	419	504
Aircraft	150	197	240
Boating	88	116	89
Guiding	408	468	604
Skifields	25	25	27
Vehicle	46	72	119

Source: http://www.doc.govt.nz/about-doc/concessions-and-permits/concessions/concession-statistics/activity-statistics.

Table 4. Operationalizations of the ROS framework (Sources: Canterbury Conservancy, 2004, p. 215; West Coast Conservancy, 2008).

Mount Cook Aoraki NP	Westland Tai Poutini NP
1. Backcountry remote;	1. Remote Experience
2. Backcountry walk in;	2. Backcountry with Facilities
3. Backcountry accessible – motorized;	3. Front-country with Facilities
4. Front-country – short-stop;	4. Highways, Roadside Opportunities & Visitor Service
5. Highways, roadside opportunities and visitor service sites	5. Intense Interest Sites

Section 17W of the 1987 Conservation Act, concessions can only be issued within the development limits, and under the terms and consultation processes, specified in NP Management Plans and higher order legal/policy tools. In their turn, NP Plans cannot derogate from regional Conservation Management Plans and Strategies that typically covers several NPs. At the national level, the latter are guided by the 2005 General Policy for National Parks, which in turn must be consistent with the legal framework (Controller and Auditor General, 2012). According to DOC (2003, p. 38), "the term 'recreation opportunity' is used to describe the mix of settings at the places where people visit and the recreation activity they undertake there".

In each NP, various zones are distinguished that require different levels of nature protection. The less protection is needed, the higher the number of facilities and activities (in terms of both number and diversity) that may be approved (for details see Taylor, 1993). Table 4 shows how the framework was implemented with respect to two NPs (Mount Aspiring Park is larger and uses a more complex system). Each NP Management Plan needs to specify what kinds of facilities and activities are allowed, to what extent, in each ROS zone. Management Plans also specify which activities and facilities may be commercially offered.

There are signs that the Conservation Economy idea generates changes beyond the level of concession (see Figure 1) through: (1) the potential watering down of the guiding role of current NP Plans, which can be changed to accommodate Minister approved concessions; (2) regulatory change proposals on the boundaries of Conservation Management Strategies (CMS), and DOC restructuring; (3) the introduction of a different management framework for visitors.

To illustrate the first point, in 2011–2012, two major infrastructural tourism projects were proposed to the Conservation Minister: a monorail-based project trying to link the popular resort of Queenstown to Milford Sound (by boat, monorail and bus); and a road with tunnel through the Parks' backcountry mountains linking the same areas. Both proposals received positive evaluations based on internal DOC assessments, and were open to public submissions based on notification of the intention to grant the concession. While eventually both were declined (the former in an election year), they attracted major public interest, seen in hundreds of public submissions on the two projects (mostly in opposition[1]). If approved, any of these projects would have triggered a change in the Fiordland Park Management Plan, to accommodate the concessions (based on Sections 17W[3] and [4], 1987 Conservation Act). Similar attempts to change a NP Plan, to accommodate new/extended concessions for more helicopter hikes and a road project, for a faster access to a retreating glacier, have been noted in the Westland NP (see West Coast Conservancy, 2012).

Of relevance for the second point is that, following the 2013 radical restructuring of DOC, the number of concession allocation services has been reduced to four (see: http://www.doc.govt.nz/get-involved/apply-for-permits/contacts/). This has the effect of reducing the number of decision-making points and increasing the distance between monitoring rangers and decision-makers. In a 2013 media release, DOC, wrote regarding its restructuring:

> The proposed structure involves: the removal of DOC's existing 11 regional conservancy boundaries and replacing them with six new regions; (...) the creation of a Conservation Partnerships Group; the disestablishment of 118 regional management and administration roles; the disestablishment of 22 asset management, planning and inspection positions. (DOC, 2013a)

Following the restructuring, countrywide there only seem to be 31 full time positions involved in concessions work (Parliamentary Financial Select Committee, 2011, p. 9). As the number of concession applications is expected to increase, one may wonder about the consequences of time pressures on staff, and limited familiarity with the regions for which concessions are requested, for the proper evaluation of concession effects and mitigation/remedy plans.

In a 2013 Conservation Board Review led by the Associate Conservation Minister, it has been already proposed to lower the number of CMS to 3, from 11, invoking efficiency arguments in drafting and updating them (DOC, 2013b). Reducing the number of regional conservancies and CMS generates new concerns. To be relevant for a larger number of NPs, they will need to be formulated in much broader terms, weakening the planning guidance role they offer for the design of NP Plans. This would also increase the discretion available to officials to decide on the zoning for NPs and concession allocations across zones and NPs. This moves the politics of access to NPs to the local level, where there are de facto fewer/weaker public and non-commercial stakeholders, to balance out commercial interests, including those held by some DOC officials.

To make this point, it is helpful to invoke here a recent case, where DOC has shown its interest to "govern by exception", to advance commercial interests. In 2014, DOC approved a concession request to increase the number of overnight visitors allowed to a monopoly concessionaire for the popular Routeburn Track in the Mount Aspiring Park. Many submissions in opposition were received from the local public and Conservation Board (representing stakeholders and public interest), on grounds that the recently adopted NP Plan disallowed an increase in overnight visitors in that zone. The delegated DOC official decided that exceptional circumstances applied, and approved the application. A Conservation Board member complained to the Ombudsman, who concluded that DOC misused the 'exceptional circumstances' provisions and that the decision "makes a 'mockery' of the process of public consultation in the development of the Plan and undermines public participation" (Ombudsman, 2014).

The Department was asked to cancel the concession and to apologize to the Conservation Board member, and it did so (see DOC, 2015).

Finally, on the third point, a new demand-oriented framework was introduced for tourism-recreation: the 2012–2017 Destination Management Framework, replacing the older supply-oriented 1996 Visitor Strategy (DOC, 2011b). In this document, DOC writes that

> DOC is a significant provider of tourism and outdoor recreation opportunities in New Zealand. This framework aims to ensure that the delivery of these opportunities is focused, fit for purpose, demand-driven and affordable" and that "Destinations can be described in terms of required infrastructure and in terms of promotion. (DOC, 2011b, p. 5; author's emphasis)

This framework regards the whole Conservation Estate and in NPs is not restricted to zones like 4 and 5 in Table 4. A list of proposed icon sites and destinations was published, with some icons/destinations crossing many types of ROS zones (DOC, 2011b, p. 44–53). This politically required approach is at odds with the currently applicable hierarchy of objectives for NP (where DOC as asked to *allow* for tourism rather than *promote* it), and with the supply-oriented approach of the legal framework for visitor planning. The latest DOC Statement of Intent confirms progress along a demand-focused approach:

> DOC is changing how it provides access to the lands and waters in its care to ensure current and future generations continue to enjoy New Zealand's outdoors to the fullest. The resource commitments of maintaining a 'demand-driven' network of opportunities is being confirmed, and informed by growing knowledge about people's recreation preferences. (2014b, p. 18)

Notable here is the use of the term "recreation" to refer also to tourism, masking the prioritization of the infrastructure predominantly used by international visitors, at the expense of the backcountry facilities cherished by New Zealanders.

Unfulfilled hopes for environmental benefit from tourism concessions

Several observations emerge regarding the provisions on environmental and nature management included in tourism concession contracts and the concession applications received over the past years (see Research Methods). First, DOC's contractual requirements are typically formulated in terms of don'ts rather than do's: the concessionaire should not break applicable laws, not light fires, "not cut down or damage any vegetation; or damage any natural feature or historic resource on the Land" (DOC, no date, p. 12); not dispose of toilet wastes near water, etc. The requirements of do's are the normal ones, expected in any commercial contract: do deal with weeds, rodents and pest insects, "keep all structures, facilities and land alterations and their surroundings in a clean and tidy condition"; and "make adequate provision for suitable sanitary facilities for the Land if directed by the Grantor and for the disposal of all refuse material" (p. 12).

A suitable requirement in the standard concessions uploaded at DOC's website was found in the Guiding Permit only, asking businesses and their clients to adhere to the international "Leave No Trace" principles at all times (www.leavenotrace.org.nz). Requirements to provide environmental and cultural interpretation to clients are also included in Permits, Licenses and Leases. However, one may wonder how the neo-liberal government is going to achieve "greening growth" or "sustained growth from natural resources" (New Zealand Government, 2012, p. 5 and7) with such concession provisions? Section 17 ZG(2) of the 1987 Conservation Act gives the Conservation Minister good tools: the Minister may

> include in any concession provisions for the concessionaire to carry on activities relating to the management of any conservation area on behalf of the Minister or at any time enter into any agreement providing for the concessionaire to carry out such activities.

Second, there is little evidence of meaningful environmental requirements included in concessions, of the type recommended in international guidelines, such as concrete measures taken regularly to achieve biodiversity conservation outcomes; the use of renewable energy/fuels and or environmentally friendly methods for wastes' management, wastewater treatment and transportation (Eagles et al, 2009, p. 48–60). The current situation in New Zealand makes one wonder: if a government cannot afford to ask companies carrying out business in NPs – a country's most precious lands –to use the best available environmental practices, technologies and renewable resources, and to be proactive on biodiversity management, then who can that government ask to do so?

Interviews with concessionaires indicated they are unlikely to implement such measures unless required through concession agreements, because their priority is making a living, and the market in the NPs is already too competitive to afford voluntary measures (Interviewees A, B, 2013). The culture so far in New Zealand has been that – if concessionaires avoid, rectify and mitigate environmental effects, then all will be fine. This is clear when examining concession contracts and the reports accompanying public notifications on concession applications. However, as the Youth Parliament argued:

> Businesses could do more than just rectify damage caused by their own commercial activities. As well as protecting the conservation estate, they could enhance it (... .). In terms of behavioural change, it is more desirable to have businesses commit to carrying out conservation action themselves, rather than just giving funding to DOC to do it on their behalf, as this is likely to result in a more meaningful commitment to environmental values by the business, its staff, and its customers. (2013, p. 5)

A DOC participant admitted that:

> we accept we have not had the skills needed to negotiate and work with companies so that we can have a win-win on conservation; we are now working with the new commercial partnership unit, but it is still very early days in terms of the major concessionaires literally buying into and understanding that concept. (Interviewee C, 2013)

The appointed interviewee for Tourism Industry Association New Zealand was not aware, by October 2014, of any initiatives that would aim to achieve "conservation gains" from tourism concessionaires through concession contracts or formal public–private partnerships (Interviewee D, 2014).

Interviewees from the Department's Head Office could also not point towards any specific implementation plans, suggesting that it is too early for that (Interviewees E, F, 2013). Reference was made to a DOC webpage (http://www.doc.govt.nz/about-us/our-partners) listing some sponsoring/volunteering businesses, few of which are, however, tourism concessionaires. The main approach, detectable so far, is to rely on voluntary initiatives, which primarily concentrate on saving iconic species like kiwi and tuatara, as these have higher impacts on the marketing of sponsoring businesses.

The government seems to be expecting environmental initiatives and biodiversity gains to come in the form of donations and voluntary measures by concessionaires. However, these are very low (see Table 2), and government departments cannot plan work based on donation expectations; they may survive on them (or not) from one day to another. One Conservation Board member raised the issue of tourism concessionaires trying to negotiate lower concession fees, in exchange for some voluntary biodiversity measures, arguing that "in other countries this would be seen as corruption; in New Zealand this is seen as good business sense" (Interviewee G, 2013). The use of the term corruption is rather inadequate here, but it indicates the respondent's frustration with the situation. The main point raised is, however, a serious one: if the Department feels under pressure to negotiate lower concession fees (as others believe as well, e.g. Interviewees A, H, I, 2013), and the voluntary projects are not meaningful enough for the work that needs to be done, how can the current arrangements help address DOC financial sustainability problem for biodiversity conservation?

Concluding reflections

Neoliberal governments have presented the Conservation Economy vision as an advantageous exchange for tourism businesses in protected areas: relaxing the rules and approval processes around concessions, in exchange for contributions from concessionaires to conservation, environmental and infrastructural challenges facing DOC, given budgetary constraints. This is reiterated in the latest DOC Statement of Intent, for 2014–2018. While the Conservation Economy term has technically not reappeared in Statements after the 2009 one (see DOC, 2009, p. 5–6), the basic ideas have remained the same:

The Department is contributing to Government's Better Public Service results by:

- Working with businesses to achieve conservation gains in ways that deliver environmental, social and economic benefits to New Zealanders. (...)
- Putting more emphasis on partnerships, relationship building, sharing skills and knowledge, and involving others.
- Working with all Natural Resources Sector agencies to implement medium term priorities agreed by Government for the sector and described in the Building Growth from Natural Resources Progress Report. (DOC, 2014b, p. 5)

However, the vision for a Conservation Economy with environmental and biodiversity gains is being implemented at a disheartening slow pace, particularly by the tourism sector. The empirical analyses provided here suggest that, while we have seen indeed a softening of rules and approval processes around concessions, there is no evidence at this stage of the latter part of the equation – neither in terms of concession contract provisions, incorporating meaningful environmental/biodiversity responsibilities, nor in terms of significant voluntary contributions from concessionaires. Moreover, significant governance changes have been implemented or are under preparation that were not initially envisaged in the 2010 "Building Natural Resources" Program and other early statements.

The holistic conceptualization of NPs governance, summarized in Figure 1, suggests that interventions at the concessions level may trigger/require changes for other governance arrangements, to restore the internal coherence, assuming the preservation of the incumbent hierarchy of objectives for NPs. In the New Zealand context, the question is whether such changes can be seen as merely restoring coherence. The analyses presented here suggest that this is not the case and that the observable changes could potentially cumulate into a tipping point, which may shift tourism development in NPs towards unsustainability. The following developments are of concern, on the background of incumbent governance weaknesses:

- DOC's inability to use effective financial instruments to raise revenue from tourism, limiting its ability to deliver on its highest order objective: conservation of natural heritage in NPs (incumbent weakness)
- DOC's inability to plan/influence tourism volumes, other than through concessions (incumbent weakness)
- an outdated legal framework that fails to stimulate a holistic environmental approach to NP management, including air and climate protection (incumbent weakness)
- the absence of meaningful environmental and biodiversity conservation responsibilities in concessionaires' contracts, despite the legal provisions in the Conservation Act enabling the Minister to impose such conditions (incumbent weakness)
- a de facto, politically prescribed, change in the hierarchy of objectives to be pursued by DOC, lifting tourism development from fourth to second rank; tourism-related expenses are on the increase while tourism revenues have not increased much since 2010/2011 (primarily a "post-2009 issue");
- changes to the concession approval process and rules, implying: (1) fewer opportunities and less time for public input, with concerns that commercialization attempts will be exposed to less public interest scrutiny; (2) the exacerbation of existing problems around the synchronization and content harmonization between NP Plans, CMS and concessions (Interviewees J, K, 2013); formerly, all these tools were envisaged for (maximum) 10 years; synchronization is needed to ensure consistency, and preserve the guiding role of CMP and NP Plans in concession allocation; (3) less time available to DOC staff to assess applications (some: post-2009 issue; others: incumbent);
- the radical restructuring of DOC with many redundancies, resulting in (1) fewer staff available to deal with higher concession-related workloads; (2) an increased distance between application and processing regions, due to the significant centralization of concession services (post-2009 issue);
- the tourism industry attempts at watering down the guidance role of NP Plans, seen in concession applications for major infrastructural projects that would necessitate the amendment of NP Plans designed following years of public consultations (mainly post-2009 issue);
- the proposal to reduce the number of regional CMS, guiding NP Plans and concessions' allocation; CMSs will then be formulated in broader terms and increase the discretion of regional officials (post-2009 issue);
- the demand-oriented approach to visitor planning advanced by the Conservation Economy idea, creates tensions with the still de jure supply-oriented approach embedded in the ROS planning framework; the government would like DOC to invest in infrastructure where (especially international tourist) demand is high/increasing, but the ROS framework requires the prioritization of natural and cultural heritage objectives, which is consistent with the hierarchy of NP objectives envisaged by IUCN and in national legislation (a post-2009 issue);
- the over-reliance on donations and volunteering by concessionaires regarding environmental, biodiversity and infrastructural gains, which so far has not materialized in terms of meaningful commitments and outcomes (more emphasis since post-2009).

These incumbent and post-2009 issues can be seen as potentially contributing to a tipping point towards unsustainability for tourism development in New Zealand's NPs. The de jure and de facto hierarchy of objectives pursued by DOC for NPs still has natural and cultural heritage and landscape quality in the lead. But tourism development priorities are catching up fast, claiming increasing financial and human resources from DOC, while conservation outcomes remain disappointing.

To avoid the tipping point, the government needs to enable DOC to use more effective financing instruments than small concession fees and donations; to use a planning approach that can ensure the prioritization of natural heritage objectives, in line with national and IUCN objectives for NPs; to refresh the legal framework for a holistic environmental management of the Conservation Estate; to foster public engagement in all types of decisions for NP governance; and to use the legally available mechanisms to insert environmental and biodiversity responsibilities in concessionaires' contracts.

Note

1. See http://www.doc.govt.nz/get-involved/have-your-say/all-consultations/2012/dart-passage-tunnel-milford-dart-ltd/ and http://www.doc.govt.nz/get-involved/have-your-say/all-consultations/2012/fiordland-link-experience-monorail-riverstone-holdings-ltd/#report

Acknowledgments

The author would like to thank the two reviewers involved, for the inspiring feedback, very constructive comments, and clear suggestions for improvement.

Disclosure statement

No potential conflict of interest was reported by the authors.

List of interviews in the period 15 November 2013–30 November 2014

Interviewee A, concessionaire operating on a Westland glacier
Interviewee B, concessionaire in Mount Aspiring NP
Interviewee C, DOC staff appointed for interview on Mount Aspiring NP
Interviewee D, employee for the Tourism Industry Association New Zealand
Interviewee E, DOC staff appointed for interview, Wellington Head Office
Interviewee F, DOC staff appointed for interview, Wellington Head Office
Interviewee G, member of Westland Conservation Board representing public interests
Interviewee H, concessionaire Westland NP
Interviewee I, member of Mount Aspiring Conservation Board representing public interests
Interviewee J, DOC staff appointed for interview on Westland NP
Interviewee K, DOC staff appointed for interview on Mount Cook NP

ORCID

Valentina Dinica (iD) http://orcid.org/0000-0003-0302-7893

References

Buckley, R. (2009). Parks and tourism. *PLOS Biology, 7*(6), e1000143.
Canterbury Conservancy. (2004). *Aoraki/Mount Cook national park management plan.* Christchurch: Canterbury Conservancy.

Controller and Auditor General. (2012). *Department of conservation: Prioritising and partnering to manage biodiversity.* Wellington: Controller and Auditor General.

Dinica, V. (2013). International sustainability agreements: Are they politically influential for tourism governance innovations? *Tourism Analysis: An Interdisciplinary Journal, 18*(6), 663–676.

DOC. (1996). *Visitor strategy.* Wellington: Author.

DOC. (2003). *Towards a better network of visitor facilities.* Wellington: Author

DOC. (2009). *Statement of intent for 2009-2012.* Wellington: Author.

DOC. (2010). Concessions Reviewing Report. Retrieved on 30 March 2015 at from http://www.doc.govt.nz/Documents/about-doc/concessions-and-permits/concessions/concessions-processing-review-report.pdf

DOC. (2011a). *Statement of intent for 2012-2017.* Wellington: Author.

DOC. (2011b). *Destination management framework – a new approach to managing destinations.* Wellington: Author.

DOC. (2013a). The department of conservation's strategic direction. Retrieved 31 March 2015 from http://www.doc.govt.nz/about-us/our-policies-and-plans/strategic-direction.

DOC. (2013b). *Conservation boards review.* Wellington: Author.

DOC. (2014a). *Briefing to the incoming minister of conservation.* Wellington: Author.

DOC. (2014b). *Statement of intent for 2014-2018.* Wellington: Author.

DOC. (2015). DOC to develop new guidelines. Wellington. Retrieved 28 January 2015 from http://www.doc.govt.nz/news/media-releases/2015/doc-to-develop-new-guidelines/

DOC. (no date). Concessions contract - lease and licence agreement. Retrieved 5 June 2014 from http://www.doc.govt.nz/about-doc/concessions-and-permits/concessions/applying-for-a-concession/pre-application-requirements/#contracts

Eagles, P.F.J. (2002). Trends in park tourism: Economics, finance and management. *Journal of Sustainable Tourism, 10*(2), 132–153.

Eagles, P.F.J. (2014). Research priorities in park tourism. *Journal of Sustainable Tourism, 22*(4), 528–549.

Eagles, P.F.J., Baycetich, C.M., Chen, X., Dong, L., Halpenny, E., Kwan, P.B., ... Zhang, Y. (2009). *Guidelines for planning and management of concessions, licenses and permits for tourism in protected areas.* Canada: University of Waterloo. Retrieved 2 June 2014 from http://www.areasprotegidas.net/sites/default/files/documentos/Park%20Tourism%20Concession%20Guidelines.pdf

Eagles, P.F.J., Mc Cool, S.F., & Haynes, C.D. (2002). Sustainable tourism in protected areas: Guidelines for planning and management. Guidelines for IUCN. Retrieved on 1 June 2014 from http://cmsdata.iucn.org/downloads/pag_008.pdf

Fearnhead, P. (2007). Concessions and commercial development: Experience in South African national parks. In R. Bushell, & P. Eagles. (Eds.), *Tourism and protected area: Benefits beyond boundaries* (pp. 301–314). Wallingford: CABI International.

IUCN. (2000). *Financing protected areas: Guidelines for protected area managers.* Gland: Author.

Newsome, D., Moore, S.A., & Dowling, R.K. (2002). Visitor planning. In *Natural area tourism: Ecology, Impact and Management.* Clevedon: Channel View Publications. Retrieved 17 November 2015 from: http://www.treasury.govt.nz/downloads/pdfs/b13-info/b13-2528847.pdf

New Zealand Government. (2012). *The business growth agenda progress report 2012: Building natural resources.* Wellington: New Zealand Government.

New Zealand Government. (2013). *The business growth agenda progress report 2013.* Wellington: New Zealand Government.

Ombudsman. (2014). Investigation of DOC renewal of Routeburn Track concession (Reference 361523). Wellington. Retrieved 24 September 2015 from http://www.ombudsman.parliament.nz/system/paperclip/document_files/document_files/925/original/361523_-_grant_of_routeburn_track_concession.pdf?1420418674

Parliamentary Financial Select Committee. (2011). Department of Conservation responses to questions relating to the briefing to the incoming minister of conservation in 2011 (Document No. 139-225120305). Retrieved 6 June 2014 from http://www.parliament.nz/resource/0000180894.

Shone, M.C., & Memon, P.A. (2008). Tourism, public policy and regional development: A turn from neoliberalism to the new regionalism. *Local Economy, 23,* 290–304.

Taylor, P.C. (1993). *The New Zealand recreation opportunity spectrum: Guidelines for users.* Wellington: DOC.

TIANZ. (2006). *Concession allocation by the Department of Conservation where supply is limited: Submission.* Retrieved 1 June 2007 from http://www.tianz.org.nz

Treasury. (2012). *The Treasury Budget 2012 information release document.* Wellington: New Zealand Treasury.

Treasury. (2013). The 2013 four year plan framework and vote conservation (forwarded by the Minister of Conservation). In *The Treasury Budget 2012 information release document.* Wellington: New Zealand Treasury. Retrieved 17 November 2015 from http://www.treasury.govt.nz/downloads/pdfs/b13-info/b13-2528847.pdf

Treasury. (2014a). The estimates of appropriations 2014/15: Environment Sector B.5 Vol.3. Vote Conservation. Retrieved 24 August 2014 from http://www.treasury.govt.nz/budget/2014/estimates/v3/est14-v3-conser.pdf

Treasury. (2014b). Budget 2014 information release document – Department of conservation. July 2014. Wellington. Retrieved 23 August 2014 from http://www.treasury.govt.nz/downloads/pdfs/b14-info/b14-2810032.pdf

UNEP-WTO. (2005). *Making tourism more sustainable: Guidelines for policy-makers*. Madrid: UNEP-WTO.

Vormedal, I. (2012). States and markets in global environmental governance: The role of tipping points in international regime formation. *European Journal of International Relations, 18*(2), 251–275

West Coast Conservancy. (2008). Westland *Tai Poutini* National Park Management Plan 2001-2011. Amended June 2008.

West Coast Conservancy. (2012). Notification on partial review of Westland *Tai Poutini* national park management plan. Retrieved 20 August 2014 from http://www.doc.govt.nz/Documents/getting-involved/consultations/current-consulta tions/west-coast/partial-review-westland-tai-poutini-factsheet.pdf

Wildlife Consultants. (2004). *A review of the use and management of Protected Natural Area Programme (PNAP) survey reports*. Wellington: Wildlife Consultants.

Wyman, M., Barborak, J.R., Inamdar, N., & Stein, T. (2011). Best practices for tourism concessions in protected areas: A review of the field. *Forests, 2*(2), 913–928.

Youth Parliament. (2013). *Inquiry to parliament committee on local government and environment into whether government should restrict or permit private business profiting from conservation activities*. Wellington. Retrieved 2 June 2014 from http://www.myd.govt.nz/documents/youth-parliament-/local-government-and-environment-committee-back ground-paper.pdf

The health and well-being impacts of protected areas in Finland

Riikka Puhakka, Kati Pitkänen and Pirkko Siikamäki

ABSTRACT

Following the growth of nature-based tourism, national parks and other protected areas have become important tourist attractions and tools for regional development. Meanwhile, research on the impact of nature on human health and well-being is increasing and taken into account in park management. This study examines health and well-being benefits perceived by visitors to Finland's protected areas. It is based on survey data from five national parks and one strict nature reserve in 2013–2015: an on-site visitor survey ($N = 3152$) and an Internet-based health and well-being survey ($N = 1054$). The study indicates that visitors' perceived benefits to their well-being were highly positive. Visits to protected areas promoted psychological, physical, and social benefits. In particular, park visits were found to provide strong and multi-faceted, long-lasting, embodied and sensory well-being experiences as well as escape from everyday life and work. Overnight visitors reported more well-being benefits than day visitors, and different types of park had different well-being benefits. The study suggests that the potential benefits of protected areas for public health are significant, emphasizing the need to integrate health and well-being arguments into the neoliberalist politics assessing the economic benefits of protected areas and their role in regional development.

Introduction

Following the growth of nature-based tourism, national parks and other protected areas have become important tourist attractions and tools for regional development. According to recent modeling, protected areas are globally visited c. 8 billion times annually: 80% of those visits are in Europe and North America (Balmford et al., 2015). Visitor numbers in protected areas have also increased in northern Europe, including Finland (Puhakka & Saarinen, 2013). In particular, the national park label has been shown to increase the attractiveness of protected areas (Wall Reinius & Fredman, 2007). Therefore, the global tourism industry has become a significant user, stakeholder, and element of change in protected areas. Coordinating conservation and the utilization of nature is often considered advantageous for both conservation and regional development goals. The touristic attractiveness of natural areas is seen as offering potential income for local peripheral communities struggling with economic restructuring (Hammer, Mose, Siegrist, & Weixlbaumer, 2007; cf. Byström & Müller, 2014; Mayer, 2014). Thus, protected areas increasingly justify their existence by local and regional economic gain and by satisfying visitor need. This discursive policy shift reflects the rise of neoliberalist politics in which nature conservation has become more instrumental and market oriented than

---

ⓑ Supplemental data for this article can be accessed ⓩ here.

before (Puhakka & Saarinen, 2013). These developments have important and far-reaching management implications.

Although recreation has been an integral part of national parks since the beginning, interest in the potential role of parks in human health and well-being is relatively new. Meanwhile, there are growing efforts to develop nature-based well-being tourism (e.g. Hjalager et al., 2011). Empirical evidence is mounting that contact with nature promotes mental and physical health. Direct physical and emotional health benefits arise from the opportunity to observe nature and be in a natural environment (Maller et al., 2008). Natural settings activate people to move, which produces indirect health benefits (Björk et al., 2008). Nature also promotes people's mutual interaction and their sense of community (Health Council of the Netherlands and Dutch Advisory Council for Research on Spatial Planning, Nature and the Environment, 2004). It has been argued that current health care practices alone cannot deal with the growing stress and other problems connected with urban living and contemporary work practices (Karjalainen, Sarjala, & Raitio, 2010). The economic implications of the benefits of natural environments to health and well-being have been considered substantial. Utilizing green spaces effectively in health promotion could reduce public health care budgets and create new sources of income (see Nilsson, Baines, & Konijnendik, 2007).

National parks and other nature reserves with recreational value can thus be seen as a fundamental health resource, particularly in terms of disease and illness prevention (Maller et al., 2008; Stolton & Dudley, 2010). While research on the health and well-being benefits of nature has a long history, previous studies have focused on the effects on physical health, psychological well-being, and cognitive ability (Keniger, Gaston, Irvine, & Fuller, 2013). And, research has largely focused on urban and suburban parks while the perceived health benefits of visitors to national parks and nature reserves have not been widely studied (see Bowler, Buyung-Ali, Knight, & Pullin, 2010). Also, most of the research has been carried out in North America (e.g. Lemieux et al., 2012, 2015), and comparisons between different kinds of protected areas have been rare (see Weber & Anderson, 2010). Accordingly, more research on the perceived health and well-being outcomes associated with visiting different types of protected areas is needed.

This study examines the health and well-being benefits perceived by visitors to national parks (IUCN Category II) and strict nature reserves (IUCN Category IA) in Finland. First, the article analyzes the type and strength of health and well-being benefits. Second, the article explores if different demographic characteristics or characteristics of a park visit lead to different perceived health and well-being outcomes. The study is based on two types of survey data collected in five national parks and one strict nature reserve: an on-site visitor survey ($N = 3152$) and an Internet-based health and well-being survey ($N = 1054$). Data were collected during years 2013–2015. The study aims to discuss the new and additional mandate of protected areas as "the fountains of health and well-being" in urbanized societies.

The impact of nature on human health and well-being

Psychological and cognitive benefits

Previous studies on the psychological benefits of interacting with nature can broadly be divided into psychological well-being benefits and those associated with cognitive performance, i.e. positive effects on mental processes and cognitive ability or function (Keniger et al., 2013; see Bowler et al., 2010).

Interaction with nature has been shown to increase self-esteem and mood (Kuo & Sullivan, 2001), reduce anger (Moore, Townsend, & Oldroyd, 2006), and improve general psychological well-being with positive effects on emotions and behavior (Kaplan, 2001). The positive impacts of nature on children's self-esteem and mental well-being has also been discovered (Maller, 2009).

Several studies have focused on the psychological well-being effects of exercising in a natural environment (Keniger et al., 2013). The multi-study analysis by Barton and Pretty (2010) showed that acute short-term exposures to facilitated "green exercise" improved both self-esteem and mood irrespective of duration, intensity, location, gender, age, and health status (see also Pretty et al., 2007). Considerable improvements in well-being, mood, relaxation, joy, and other health and well-being indicators were also perceived by hikers, runners and walkers in national parks (Wolf & Wohlfart, 2014). Research suggests that exercise is more beneficial, leading to the relief of anxiety and depression, when it occurs in natural rather than urban settings (Hartig, Mang, & Evans, 1991). A national study in Finland indicated that repeated exercise in nature was, in particular, connected to better emotional well-being (Pasanen, Tyrväinen, & Korpela, 2014). A green environment is considered to encourage people to exercise more often and for longer periods than a non-natural environment which has positive benefits for both mental health and physical fitness (Health Council of the Netherlands and Dutch Advisory Council for Research on Spatial Planning, Nature and the Environment, 2004; Kaczynski & Henderson, 2007). Besides exercise, intentional interactions with nature, such as watching wildlife, have been shown to increase psychological well-being (Curtin, 2009).

In terms of cognitive benefits, it has been hypothesized that green spaces are restorative, contributing to attentional recovery and reducing mental fatigue (Björk et al., 2008; Hartig, Evans, Jamner, Davis, & Gärling, 2003; Kaplan, 2001; Tyrväinen et al., 2014). In a Finnish study, Korpela, Borodulin, Neuvonen, Paronen, and Tyrväinen (2014) discovered that the longer the time in nature-based recreation associated with restorative experiences, the better emotional well-being is perceived. Many of the health and well-being benefits from outdoor activities can be ascribed to the restorative capacity of natural environments (Wolf, Stricker, & Hagenloh, 2015). Research findings suggest that exposure to nature in urban and wilderness settings has positive effects on academic performance and the ability to perform mentally challenging tasks (Berman, Jonides, & Kaplan, 2008). For instance, Hartig et al. (1991) concluded that a prolonged wilderness experience had restorative effects. Van den Berg, Koole, and van der Wulp (2003) showed that, compared with urban environments, natural environments have positive impacts on the ability to concentrate. Performing activities in green areas has been found to reduce the symptoms of attention-deficit/hyperactivity disorder (AD/HD) in children (Kuo & Taylor, 2004). In addition, positive, restorative experiences in natural environments may promote greater ecological behavior (Hartig, Kaiser, & Strumse, 2007).

According to Han (2010), natural environments differ in their restorative potential. Many studies have explored, for instance, the positive contribution of the forest environment on psychological health and well-being (see Karjalainen et al., 2010; Shin, Yeoun, Yoo, & Shin, 2010). Barton and Pretty (2010) concluded that spending time near waterside (e.g. beach or river) or participating in water-based activities may give a greater benefit, although all green environments improved their participants' self-esteem and mood. Fuller, Irvine, Devine-Wright, Warren, and Gaston (2007), in turn, found that the restorative benefits to urban park users increased with plant species richness in urban green spaces.

Physiological benefits

Research has identified a broad range of physiological benefits from interacting with nature, i.e. positive effects on physical function and/or physical health. Research findings suggest that contacts with green space alleviate the negative physiological effects of various stressors in urban environments (Lee et al., 2012; Tsunetsugu et al., 2007). Compared with urban environments, natural environments produce positive changes in human physiology after stressful or attention-demanding situations (Hansmann, Hug, & Seeland, 2007; Hartig et al., 2003; van den Berg et al., 2003). For instance, blood pressure, heart rate, skin conductivity, and muscle tension are at lower levels in natural environments than in urban settings. Forest visits also reduce salivary cortisol levels (stress hormone), suppress sympathetic nervous activity, and enhance para-sympathetic nervous activity (Hartig et al., 2003; Lee et al., 2012; Tyrväinen et al., 2014).

Natural settings activate people to move and, thus, increase energy expenditure and produce indirect physical health benefits. For instance, Björk et al. (2008) found that recreational values for nearby natural environments were positively associated with physical activity and also with a normal or low body mass index (BMI) for tenants. Wolf and Wohlfart (2014) observed that although hiking in national parks was performed with the primary motivation to experience nature and not to exercise, hikers burned more energy than runners and walkers as they preferred more difficult tracks with greater slopes.

Visits to green areas may also strengthen the human immune system by increasing natural killer (NK) cell activity. NK cells can kill tumor cells by releasing anticancer proteins and, thus, nature visits may have a preventive effect on cancer generation and development (see Lee et al., 2012). Furthermore, interactions with the natural environment influence the composition of the human commensal microbiota and its immunomodulatory capacity. The "biodiversity hypothesis" proposes that reduced contact with natural environment and biodiversity, including environmental microbiota, leads to poor human microbiota, immune dysfunction and finally to chronic inflammatory diseases (Hanski et al., 2012).

Social and spiritual benefits

Nature can also have a beneficial effect on health by promoting social contact (Health Council of the Netherlands and Dutch Advisory Council for Research on Spatial Planning, Nature and the Environment 2004). Natural environments and shared nature experiences provide opportunity for social interaction and strengthen bonds within families and communities (Wolf et al., 2015). Research suggests that provision and access to natural environments may ameliorate or even reverse some of the social challenges in urban areas (Keniger et al., 2013). Natural environments foster social empowerment, enhance interracial interaction, and promote social cohesion and support (e.g. Kuo & Sullivan, 2001; Maller, 2009). Nature can help in personal and community identity formation, social activity, and social participation (Irvine & Warber, 2002). In comparison with urban areas with limited greenery, significantly lower levels in crime rates and violent behavior have been observed in urban areas with surrounding green space or vegetation (Kuo & Sullivan, 2001; Moore et al., 2006).

A small number of studies have focused on the spiritual benefits of interacting with nature. Spiritual benefits identified include the increased inspiration and feelings of connectedness to a broader reality (e.g. Curtin, 2009; Fredrickson & Anderson, 1999; Humberstone, 2011). Participation in rural tourism has also been found to elicit a deeper, emotional or spiritual experience (Jepson & Sharpley, 2015).

Study areas

Parks & Wildlife Finland (formerly Natural Heritage Services) manages state-owned protected areas in Finland, including 39 national parks and 19 strict nature reserves. Parks & Wildlife Finland is a unit of the state-owned enterprise *Metsähallitus* which runs business activities on state-owned land and water areas while also fulfilling public administration duties, such as nature conservation, facilities and services for outdoor recreation, and protected area management planning. While national parks are important destinations for recreation, strict nature reserves are primarily reserved for the purposes of nature conservation and research (see Metsähallitus, 2016a).

In 2010, Parks & Wildlife Finland launched the *Healthy Parks, Healthy People Finland* program that aims to improve public health by activating people to get out into natural settings, enjoy positive and genuine experiences, and improve their physical health through outdoor activities. The key objective is to effectively monitor and measure the health benefits of protected areas so that the findings can be used to enhance services.

This study is based on survey data collected by Parks & Wildlife Finland in six protected areas: Kurjenrahka, Patvinsuo, Repovesi, Pyhä-Luosto, and Syöte National Parks as well as Kevo Strict Nature Reserve. These areas were selected to represent parks located in different parts of Finland and with different geographical and visitor characteristics (see Figure S1, in Supplemental Data in the online

Table 1. Characteristics of study areas.

Name of the protected area	Classification of nature[a]	Recreational facilities[a]	Tourism services outside the park[a]	Population density in the surrounding area[b]	Description of the surrounding area[b]	Number of visits (year)
Kevo Strict Nature Reserve				0.3	Countryside	5000 (2013)
Kurjenrahka National Park	Mire	Low	High	94.9	Population centre	32,100 (2013)
Patvinsuo National Park	Mire	Moderate	High	3.1	Countryside	12,900 (2013)
Pyhä-Luosto National Park	Fell	High	Moderate	1	Tourist centre	115,100 (2015)
Repovesi National Park	Water and scenery	Moderate	Moderate	26.8	Population centre	93,200 (2013)
Syöte National Park	Forest	High	Moderate	2	Tourist centre	40,300 (2015)

[a]Puustinen et al. (2009).
[b]Inhabitants per km^2; Statistics Finland and Parks & Wildlife Finland (2011).

version of this paper). The results of Puustinen, Pouta, Neuvonen, and Sievänen (2009) indicated that the natural characteristics of national parks, recreation services and the tourism services in the surrounding municipalities were associated with the number of park visits in Finland (see Table 1). The number of visits is highest in northern parks characterized by fells (i.e. relatively mountainous formations) and abundant recreation and tourism services (e.g. Pyhä-Luosto), while in parks dominated by bog land or mires (e.g. Patvinsuo), the recreation service level has not affected the number of visits. Biodiversity has also been shown to be linked with the perceived attractiveness of Finnish national parks – parks with high biodiversity values are more attractive for visitors than parks with lower biodiversity values (Siikamäki, Kangas, Paasivaara, & Schoderus, 2015).

Kevo Strict Nature Reserve in the northernmost Finland was established in 1956 and extended in 1982. Kevo is the largest strict nature reserve in Finland with an area of over 712 km^2. As Kevo is located in a sparsely populated area and far away from the major tourist centers of northern Finland, the number of yearly visits is only about 5000 (Table 1). Kevo is popular especially for long-distance wilderness hikes; the 40-km-long canyon-like valley of the River Kevojoki forms the core of the area. Inside the strict nature reserve, visitors are allowed to walk only on marked trails.

Pyhä-Luosto and Syöte National Parks are located near middle-sized skiing resorts in sparsely populated northern Finland. Pyhätunturi National Park was established in 1938, but the entire Pyhä-Luosto area, with an area of 142 km^2, was designated as a national park in 2005. Pyhä-Luosto National Park combining two popular fell areas attracts both day trippers and long-distance hikers; the number of yearly visits is over 115,000 (Table 1). Syöte National Park was established in 2000 and it covers an area of 299 km^2. In Syöte, landscape changes from broad aapa mires[1] to wilderness-like hills growing spruce forests. There are marked trails for day trips as well as for longer hikes. According to visitor studies, in Pyhä-Luosto, Syöte, and Kevo approximately 90% of the visitors stay overnight in the surrounding area or in the park and the share of local visitors is low.

Patvinsuo National Park, established in 1982, is located in sparsely populated eastern Finland, and it covers an area of 105 km^2. There are both raised bogs and open aapa mires in the park. The park is suitable for one or two-day hikes and observing the natural environment of the wilderness. Patvinsuo is the least visited national park in this study, and over half of its visitors come from the surrounding area or nearby cities (Table 1).

Kurjenrahka and Repovesi National Parks are located in more densely populated southern Finland, and they are much smaller than the other study areas. Kurjenrahka National Park, 29 km^2, was designated as a national park in 1998. Kurjenrahka includes the largest raised bogs of southwestern Finland, which are in their natural state. Within an easy access from the city of Turku with 186,000 inhabitants, Kurjenrahka is the only close-to-home-recreation area in this study. Kurjenrahka is a destination for day trips and also for longer excursions. Repovesi National Park was established in 2003, and it covers an area of 15 km^2. Rugged forests in Repovesi are dotted with lakes and ponds, and it is located within a two hour drive from the capital city Helsinki. Repovesi is one of the most popular hiking areas in southern Finland and suitable for both day trippers and overnight hikers. The number of yearly visits has grown fast to over 140,000 (Metsähallitus, 2016b; see Table 1).

Materials and methods

This study was inspired by the study of Lemieux et al. (2012) exploring the perceived health and well-being benefits of two protected areas in Canada. In their study, a questionnaire was developed to reflect a comprehensive suite of health and well-being indicators (or attributes): physical, psychological/emotional, social, intellectual, spiritual, ecological, environmental, cultural, occupational, and economic well-being. The perceived benefits received from protected area experiences were substantial; the greatest well-being benefits were psychological/emotional, social, cultural, and environmental.

In this study, two types of survey data were used (Table 2). In the first phase, data were collected through an on-site visitor survey ($N = 3152$). Visitor surveys are conducted by Parks & Wildlife Finland on state-owned conservation and recreational areas every five years. The survey questionnaire includes questions on the length of a visit, traveling to the destination, motives, activities, service demand and rating as well as spending of money. To find out about health and well-being benefits, the visitors were asked to what extent they thought that the visit increased their social, psychological and physical well-being respectively (1 = strongly disagree to 5 = strongly agree) and to estimate the monetary value of perceived well-being benefits. Social well-being was specified as "e.g. improved working capacity, strengthened social relations, enjoyed doing things alone or together", psychological well-being as "e.g. satisfaction with life, improved the mood, recovery from mental stress, learned something new", and physical well-being as "e.g. enjoyed sensing the nature, maintained the fitness, learned new skills, perceived physical well-being". In addition, visitors were prompted to leave their email address to participate in the more detailed second survey.

In the second phase, those respondents who had left their contact details were emailed a web questionnaire approximately one week after their visit ($N = 1054$). The web survey included 36 statements of different well-being effects assessed with a five-point Likert scale (1 = strongly disagree to 5 = strongly agree). The statements covered different aspects of physical well-being benefits (activities, sensations), psychological well-being (restoration, relaxation, being creative, intellectual stimulation), and social well-being (interaction, togetherness, bonding, occupational well-being). The statements were designed based on the Canadian study (Lemieux et al., 2012) but adapted to the Finnish context. Questions related to economic, cultural, and ecological well-being were left out of the survey. In addition, the questionnaire included a structured question on the visitors' estimation on the duration of the positive impacts ("for a long time", "for some time", "during the visit", "no positive impacts"), the monetary valuation of the impacts (open-ended), respondents' relationship with the place of a visit (protected area), their physical exercise habits and physical characteristics, their relationship with nature, and if they traveled with children, the impacts of the visit to the children. The visitor survey was available in several languages, while the web survey was available only in

Table 2. Data used in the study.

Name of the protected area	On-site visitor survey N (year)	Web survey N (year)
Kevo Strict Nature Reserve	524 (2013)	290 (2013)
Kurjenrahka National Park	413 (2013)	132 (2013)
Patvinsuo National Park	213 (2013)	50 (2013)
Pyhä-Luosto National Park	760 (2015–2016)	109 (2015–2016)
Repovesi National Park	902 (2013–2014)	399 (2013–2014)
Syöte National Park	375 (2015)	74 (2015)
Total	3152	1054

Table 3. Demographic characteristics of survey respondents.

		On-site visitor survey		Web survey	
		N	%	N	%
Gender	Male	1409	44.7	453	43.3
	Female	1743	55.3	590	56.7
Age	Under 30 years	650	20.8	214	20.7
	30–44 years	945	30.2	337	32.6
	45–64 years	1225	39.2	411	39.7
	65 years or older	305	9.8	73	7.1
Education	Less than bachelors	1537	49.4	465	45.0
	Bachelors or higher	1572	50.6	569	55.0
BMI	Normal or underweight			594	57.3
	Overweight			443	42.7

Finnish (Kaikkonen et al., 2014). On average, the web survey was responded three weeks after the on-site survey.

Key demographic characteristics of both studies are presented in Table 3. The share of respondents with higher education is larger in the web survey, which indicates that especially the educated respondents of the visitor survey left their contact details and answered the follow-up survey.

Data gathered through the questionnaires were combined, stored and analyzed using SPSS software. The results are reported in the following sections via the descriptive statistics of different variables. Differences between respondent groups are compared using the nonparametric Mann–Whitney U and Kruskal–Wallis tests since the data were not normally distributed.

Factor analysis was conducted to detect the different dimensions of well-being studied with the extensive web survey and to see how they compared with those of the Canadian study (Lemieux et al., 2012). Exploratory factor analysis via the principal component method and using Varimax-rotation (orthogonal) with Kaiser normalization resulted in eight factors with eigenvalues greater than 1. Four variables that did not have a strong loading (below 0.5) at any of the factors were removed from the analysis after repeating the procedure twice. Two variables ("Weather conditions felt unpleasant", "I found insects (mosquitos, elk flies, wasps, mites etc.) disturbing") were removed from the analysis as weather and insect conditions are highly changeable and dependent on the time of the visit. The remaining 30 variables had six factors with eigenvalues above 1 and explain together 59.5% of the variance. The factors were turned into composite variables by calculating an average of the variables assigned to each of the factor. These six variables were used as dependent variables in the analysis.

Results of the on-site visitor survey

Type and strength of well-being benefits

Visitors' perceived benefits to their health and well-being were highly positive. All three dimensions of well-being (social, physical, psychological) had means above 4 on the five-point Likert scale (Table 4). Physical and psychological well-being scored mean values above 4.4 and social well-being 4.2.

As another indicator of the strength of the perceived well-being effects, the monetary value of the well-being benefits of the visit was surveyed in an open-ended question. As a reference, the respondents were given examples of prices of commercial wellness services ranging from a gym visit (€5) to a trip abroad (€3000). Responses ranged from €0 to €100,000, the median being €150. In the on-site survey, 31.9% of respondents gave values under or equal to €50 and 16.1% between €51 and €100. Forty-two percent of respondents estimated the value to be in the range of €101–€500, 1.6% between €501 and €999, and 8.4% above or equal to €1000.

Impact of background and visitor variables

The impact of background variables on the perception of well-being benefits was studied comparing respondent groups of different gender, age, and education. The impact of the characteristics of the visit, in turn, was studied comparing visitors to different protected areas, day and overnight visitors, those who traveled alone vs. in a group, and first-time vs. repeat visitors (Table 4).

In general, women rated all well-being benefits higher than men. The oldest (65 and older) and the youngest (under 30) age group rated the three types of well-being benefits the lowest, while those between 30 and 64 years of age gave higher ratings. The youngest and the oldest differed significantly from others in terms of social and psychological well-being, while in terms of physical well-being, only the difference between the youngest and middle age group (45–64 years old) was significant.

Differences between the visitor ratings of different areas were significant, especially in relation to social and physical well-being. The visitors to Kevo, Syöte, Pyhä-Luosto, and Patvinsuo in northern and eastern Finland rated higher the different forms of well-being benefits they had gained during the visit. In turn, Kurjenrahka and Repovesi visitors in more densely populated southern Finland gave the most moderate ratings.

In terms of length of stay, overnight visitors perceived the impacts more positively than those who stayed a shorter time. Especially, psychological well-being was rated higher by overnight respondents. Travel companion affected people's perceptions so that those who traveled alone rated physical well-being higher than others, whereas those who had company rated social well-being higher than others. There were no significant differences between first-time and repeat visitors.

Results of the web survey

Type and strength of well-being benefits

The six latent factors identified by factor analysis from the extensive web survey gave a more detailed understanding about the different dimensions of health and well-being benefits than the three dimensions used in the on-site survey (Table 5). The first factor includes variables related to restoration and relaxation ("My vitality and energy increased", "I calmed down", "My concentration improved") as well as self-esteem ("My self-confidence increased", "I got better hope for tomorrow", "My life was put into perspective"). These have often been listed as the key psychological well-being benefits of nature (Keniger et al., 2013) as well as protected areas (Lemieux et al., 2012). In this study, psychological well-being was associated with improved work motivation and motivation for everyday life, that is, occupational well-being as defined by Lemieux et al. (2012). Therefore, the first factor was named as *Psychological and occupational well-being*.

Variables relating to cognitive skills and opportunities to engage in creative and stimulating activities (see Lemieux et al., 2012) were loaded into the second factor ("I learned new skills", "I learned more about nature", "My interest towards nature increased"), which was named as *Intellectual well-being*. Also, the variable "I enjoyed meeting new people during the visit" had the highest loading to the second factor.

Related to psychological well-being, a separate third factor was formed by variables "I forgot everyday worries", "I had a chance to get away from work" and "I had a chance to get away from everyday life". These are all variables related to finding a counterbalance to stressful everyday life and work: this factor was named as *Escape*.

The fourth factor was called *Social well-being*. It consists of variables promoting togetherness, social contact, participation, and bonding that have been found to be important social well-being benefits from nature ("I enjoyed spending time with people I cherish", "I enjoyed shared activities with people I cherish", "I found it easier to talk about personal matters in nature", "Being in nature fostered my relationship with people I cherish") (see Lemieux et al., 2012). In addition, the variable "Having company increased my feeling of security" loaded to the factor, which suggests that common

Table 4. Descriptive statistics and tests of significance for the ratings of health and well-being benefits between different groups of the on-site visitor survey.

	Social well-being	Mental well-being	Physical well-being
N	3116	3115	3118
Mean	4.23	4.41	4.43
Std. deviation	0.80	0.70	0.69
Gender (p-value)	<0.01**	<0.01**	<0.01**
Male (N)	1384	1382	1384
Mean	4.10	4.32	4.32
Std. deviation	0.82	0.72	0.72
Female (N)	1721	1722	1723
Mean	4.32	4.47	4.51
Std. deviation	0.77	0.67	0.65
Age (p-value)	<0.01**	<0.01**	0.024*
Under 30 years (N)	647	646	647
Mean	4.13	4.34	4.37
Std. deviation	0.85	0.77	0.69
30–44 years (N)	938	940	937
Mean	4.28	4.48	4.44
Std. deviation	0.77	0.64	0.69
45–64 years (N)	1205	1203	1208
Mean	4.26	4.43	4.46
Std. deviation	0.78	0.66	0.67
65 years or older (N)	292	292	292
Mean	4.10	4.22	4.37
Std. deviation	0.81	0.74	0.73
Education (p-value)	0.063	0.008*	0.887
Less than bachelors (N)	1519	1515	1518
Mean	4.20	4.37	4.42
Std. deviation	0.80	0.71	0.71
Bachelors or higher (N)	1553	1556	1556
Mean	4.25	4.44	4.43
Std. deviation	0.81	0.68	0.68
BMI (p-value)	0.120	0.600	0.746
Normal or underweight (N)	587	587	588
Mean	4.33	4.47	4.47
Std. deviation	0.76	0.65	0.67
Overweight (N)	438	439	439
Mean	4.27	4.46	4.46
Std. deviation	0.74	0.61	0.64
Park	0.009*	0.081	0.001**
Kevo Strict Nature Reserve (N)	516	514	514
Mean	4.22	4.47	4.47
Std. deviation	0.77	0.64	0.64
Kurjenrahka National Park (N)	404	404	405
Mean	4.13	4.38	4.33
Std. Deviation	0.88	0.71	0.72
Kurjenrahka National Park (N)	404	404	405
Mean	4.13	4.38	4.33
Std. deviation	0.88	0.71	0.72
Patvinsuo National Park (N)	210	210	210
Mean	4.37	4.49	4.47
Std. deviation	0.72	0.65	0.68
Patvinsuo National Park (N)	210	210	210
Mean	4.37	4.49	4.47
Std. deviation	0.72	0.65	0.68
Pyhä-Luosto National Park (N)	743	746	746
Mean	4.17	4.38	4.43
Std. deviation	0.81	0.72	0.68
Repovesi National Park (N)	882	880	882
Mean	4.27	4.37	4.38
Std. deviation	0.78	0.72	0.72

(continued)

Table 4. (*Continued*)

	Social well-being	Mental well-being	Physical well-being
Syöte National Park (*N*)	361	361	361
Mean	4.26	4.43	4.52
Std. deviation	0.79	0.66	0.65
Overnight (*p*-value)	0.342	0.004*	0.370
Overnight visitors (*N*)	1161	1160	1161
Mean	4.25	4.45	4.44
Std. deviation	0.78	0.68	0.68
Day visitors (*N*)	1777	1777	1779
Mean	4.21	4.38	4.41
Std. deviation	0.81	0.71	0.70
Alone or group (*p*-value)	0.004*	0.084	0.042*
Alone (*N*)	214	216	216
Mean	4.05	4.50	4.53
Std. deviation	0.92	0.63	0.61
Group (*N*)	2902	2899	2902
Mean	4.24	4.40	4.42
Std. deviation	0.79	0.70	0.70
First-time visitor (*p*-value)	0.178	0.879	0.760
First-time visitor (*N*)	1212	1211	1212
Mean	4.20	4.40	4.42
Std. deviation	0.82	0.73	0.69
Repeat visitor (*N*)	1904	1904	1906
Mean	4.24	4.41	4.43
Std. deviation	0.79	0.68	0.69

*Result significant at 0.05 level.
**Result significant at 0.01 level.

experiences in somewhat challenging natural environments may strengthen bonds with family members and friends.

Variables related to the perceived benefit to *Physical well-being* were loaded to the fifth factor. These included variables related to exercise ("During my visit to the area I exercised more than in everyday life", "I felt the nature exercise improved my physical condition"), being able to test one's physical strength ("I was able to test my physical strength"), and in general the feeling of physical well-being ("I felt my physical well-being improved") (see Lemieux et al., 2012).

Separate from physical fitness, strength, and exercise, the sixth factor was formed by variables emphasizing the different sensations provided by the visit. These were related to sound ("I enjoyed sounds of nature", "I enjoyed silence"), sight ("I enjoyed beautiful nature"), smell ("I enjoyed the fragrance of nature", "It felt good to breathe fresh air"), and feel ("The feel of nature was pleasant (wind on my face, soft moss, the shapes of different surfaces"). These variables are related to the often instantaneous and momentary feelings provided by nature, and the factor was named as *Sensory satisfaction*.

Out of these six factors, *Sensory satisfaction* (factor 6) and *Escape* (factor 3) scored the highest mean values, above 4.5, while *Intellectual well-being* (factor 2) as well as *Psychological* and *occupational well-being* (factor 1) received the lowest ratings with mean values below 4. The web survey thus deepens the result of the on-site visitor survey by suggesting that, in terms of physical well-being and psychological well-being, sensory satisfaction and escape from everyday life are especially important.

In line with the visitor survey, the web survey included a question on the monetary value of the perceived well-being benefits. The estimations of respondents on monetary value ranged from 0 euros to a billion with the median of €150. Thirty-one percent gave values under or equal to €50 and 15.6% between €51 and €100. In turn, 40.1% estimated the value of well-being impacts between €101 and €500, 2.2% between €501 and €999, and 10.9% gave values equal to or above €100.

From those respondents who replied to both surveys, 28.1% gave higher values in the on-site visitor survey. Similarly, 27.9% of those respondents gave higher values in the web survey. In both cases,

Table 5. Factor loadings.

	Factors					
Scale items	Psychological and occupational	Intellectual	Escape	Social	Physical	Sensory satisfaction
Total variance explained	59.5					
Cronbach's (alpha)	0.86	0.70	0.70	0.82	0.78	0.83
My vitality and energy increased	0.614					
I got better hope for tomorrow	0.766					
My concentration improved	0.764					
My self-confidence increased	0.679					
I calmed down	0.568					
My life was put into perspective	0.751					
My work motivation improved	0.523					
My motivation for everyday life improved	0.600					
I learned new skills		0.709				
I learned more about nature		0.731				
My interest towards nature increased		0.565				
I enjoyed meeting new people during the visit		0.601				
I forgot everyday worries			0.686			
I had a chance to get away from work			0.775			
I had a change to get away from everyday			0.700			
I enjoyed spending time with people I cherish				0.875		
I enjoyed shared activities with people I cherish				0.873		
I found it easier to talk about personal matters in nature				0.555		
Having company increased my feeling of security				0.659		
Being in nature fostered my relationship with people I cherish				0.746		
During my visit to the area I exercised more than in everyday life					0.823	
I was able to test my physical strength					0.730	
I felt that nature exercise improved my physical condition					0.800	
I felt my physical well-being improved					0.593	
I enjoyed silence						0.689
I enjoyed sounds of nature						0.830
I enjoyed the fragrance of nature						0.828
It felt good to breathe fresh air						0.759
I enjoyed beautiful nature						0.629
The feel of nature was pleasant (wind on my face, soft moss, shapes of different surfaces)						0.653

the median for the value change was €100. Forty-four percent of respondents did not change their mind about the monetary value of well-being effects. Therefore, for the largest share of the respondents, the benefits of the park visit did not decrease after they returned home.

Web survey respondents were also asked to estimate the duration of the physical, psychological, and social well-being impacts of their visit. In general, respondents agreed that the impacts were not restricted only to the visit, but lasted longer. Psychological impacts were estimated to last the longest: 50.8% of the respondents answered that the mental impacts would last "for a long time after the visit", 45.8% estimated they would last "for some time after the visit". In turn, 3.3% responded the visit had psychological impacts only "during the visit" and 0.2% answered the visit "did not have positive impacts". The same percentages for social impacts were the following: "for a long time" – 40.3%, "for some time" – 46.5%, "during the visit" – 10.9%, "no positive impacts" – 2.3%, and for physical impacts: "for a long time" – 13.5%, "for some time" – 67.6%, "during the visit" – 17.4%, "no positive impacts" – 1.6%.

Impact of background and visitor variables

Similar to the on-site visitor survey, the perceived well-being benefits of groups of respondents of different gender, age, and education were compared. In addition, respondents' BMIs were calculated based on their height and weight asked in the web survey (Table S1, in Supplemental Data in the online version of this paper). The following equation was used:

$$BMI = \frac{weight}{height^2}$$

For weight and height, kilograms and meters were used, respectively. Twenty-five was used as the threshold value, with those of BMI equal to 25 or lower were categorized as normal or underweight, and those above 25 as overweight.

Similarly to the visitor survey, women ranked all well-being effects higher except from *Intellectual well-being* (factor 2). Both genders rated *Sensory satisfaction* as the highest (factor 5). In contrast to the visitor survey, age groups differed from each other only in terms of *Psychological* and *Occupational well-being* (factor 1) as well as *Escape* (factor 3). The well-being benefits of these were rated lower by those above 65 years of age. This is explained by the importance of work-related variables in both factors and the fact that most of those in the oldest age group are likely to be retired.

In terms of education, those with higher education rated *Intellectual well-being* (factor 2) higher than others. Thus, while in the visitor survey, psychological well-being was rated higher by the more educated, this might have been due to the cognitive dimensions of psychological well-being, not necessarily restoration, emotions, or stress relief.

Respondents who were overweight gave higher ratings to *Physical well-being* (factor 7) benefits than the rest. Those within the recommended or underweight category, in turn, appreciated higher the benefits of *Escape* (factor 3), *Social well-being* (factor 4), and *Sensory satisfaction* (factor 6).

When visitors to different protected areas were compared, the web survey gave even clearer proof that different types of parks have potentially different well-being benefits, partly due to different visitor profiles and characteristics of their visits. Similar to the on-site survey, visitors to Kevo gave the highest ratings, and Syöte and Pyhä-Luosto visitors also gave high scores. Kurjenrahka and Repovesi visitors, in contrast, gave the lowest ratings. Overnight visitors reported higher well-being with respect to all six dimensions of well-being. Travel companion affected people's perceptions so that those who traveled alone rated *Psychological and occupational well-being* (factor 1) higher, whereas those who had company rated *Social well-being* (factor 3) higher. While no differences between first-time and repeat visitors were found in the visitor survey, the first-time visitors gave higher scores to *Intellectual well-being* (factor 2), *Escape* (factor 3), *Social well-being* (factor 4) as well as *Physical well-being* (factor 5) benefits in the web survey.

Discussion

Our results confirm the positive effects of protected areas for visitors' psychological, physical, and social health and well-being. The perceived benefits of Finnish national parks and strict nature reserves were strong and long-lasting. Visitors highly rated their benefits during the visit as well as after returning home. Valuations in monetary terms were similar during and after the visit. On average, the health benefits of visiting a protected area were estimated to be equal to many popular commercial wellness services, but the range of monetary values was wide.

Six factors of well-being identified in the study corresponded to previous literature, but the study also revealed new aspects and deepen the understanding of different dimensions of well-being protected areas can offer. In terms of psychological well-being, three factors with a slightly different emphasis were distinguished. First, restorative and occupational well-being benefits formed one factor suggesting that a visit to a protected area can increase a person's psychological capacities and positive mood (see Barton & Pretty, 2010; Wolf & Wohlfart, 2014). Second, protected areas stimulate a person's brain and skills providing intellectual well-being benefits (see Berman et al., 2008; Hartig et al., 1991). In our study, also meeting new people in the park was associated with intellectual rather than social well-being, suggesting that, at least in the Finnish context, cognitive skills can equally be about nature, activities, and social relations.

Third, a visit to a protected area has a potential of providing a needed counterbalance to, and escape from, everyday life and routine environments. Escape is an aspect that has not been previously discussed as a separate well-being benefit of protected areas, but it is a common tourism motive (e.g. Iso-Ahola, 1982). Dunn Ross and Iso-Ahola (1991) suggested that the psychological benefits of recreational travel emanate from the interplay of escaping and seeking personal and interpersonal opportunities. The identification of escape as a separate well-being dimension indicates that a visit to a protected area has also potential of providing many well-being benefits looked for from other types of tourist trips. For instance, Gilbert and Abdullah (2004) showed that taking a holiday changed the sense of well-being and enabled individuals to enhance their sense of happiness (see Dolnicar, Yanamandram, & Cliff, 2012). When operationalizing and empirically testing Iso-Ahola's (1982) motivation theory, Snepenger, King, Marshall, and Uysal (2006) observed that motivational items for personal and interpersonal escape as well as for personal and interpersonal seeking were higher for the tourism experience (e.g. in a national park) than for similar recreation experience (e.g. in a local park). Weber and Anderson (2010), in turn, showed that escapism, either from personal or physical pressures, was a very important motivation for visitors in Australian urban and regional parks (see Wolf et al., 2015).

Similar to psychological well-being, physical well-being also turned out to be more multifaceted than expected. On the one hand, a visit to a protected area has positive impacts through encouraging physical activities and exercise as well as testing one's physical limits and strength (see Kaczynski & Henderson, 2007). On the other hand, and separately from physical activity, being out in nature increases perceived well-being through pleasant and satisfactory sensations. These sensations were stronger than other types of physical well-being, which indicates that embodied experiences of nature are memorable (see Humberstone, 2011). When studying park users' own reasons for, and benefits gained from, green space usage with open-ended interview questions, Irvine, Warber, Devine-Wright, and Gaston (2013) identified an important breadth to the "experience of nature"; fresh air, getting outside and sunshine emerged strongly suggesting that these intangibles were highly valued.

In general, the different health and well-being benefits were rated highly by most respondents and differences between the different groups of respondents were small. Women rated the well-being benefits higher than men, those aged between 30 and 65 higher than younger or older respondents, and differences were also found between normal and overweight respondents. Gender impact has also been noticed in previous studies (e.g. Lemieux et al., 2012, 2015). It may be a reflection of women showing a greater need for, or susceptibility to, experiences that foster well-being (Wolf & Wohlfart, 2014) or of women's better health literacy (Niemelä, Ek, Eriksson-Backa, & Huotari, 2012). This means that women are more motivated to obtain health information and are more able to reflect their own well-being.

Although differences between younger and older people have been observed in other studies (e.g. Barton & Pretty, 2010), the small difference between age groups in our study seemed to be related to different life phase and time-use patterns, rather than age or generational differences. The oldest and for the most part retired respondents gave lower ratings to the work-related benefits than others. However, this does not mean that these benefits were unimportant for the oldest age group; for instance, they rated escape with a mean value above 4. Although retired people seek less of a counterbalance to work, they may seek personal and interpersonal opportunities and escape from everyday life. For instance, Hunter-Jones and Blackburn (2007) observed that taking a holiday offered significant benefits to senior tourists in terms of personal health and social effectiveness (e.g. interaction with others, feelings of inclusion). Wolf et al. (2015) highlighted the role of guided activities in national parks in better integrating senior citizens into the community and providing opportunities to increase their well-being. Therefore, it might be more beneficial to compare groups of visitors with different motivations rather than gender, age, or education (see Konu & Kajala, 2012).

In general, our study suggests that protected areas have great potential for providing positive health and well-being experiences to a range of different groups despite their background, age, gender, or physical condition. In our study, the respondents categorized as overweight rated the physical well-being benefits higher than those who were normal or underweight. This result may signal that a visit to a protected area can function as a motivator to encourage exercise among groups that otherwise exercise too little (see Björk et al., 2008). Physical activity may be incidental to other activities such as sightseeing, socializing, and experiencing nature, as noticed in Wolf's and Wohlfart's (2014) study.

One of the most interesting findings of our study is the difference between the perceived well-being benefits of different types of protected areas. In general, visitors to the northern parks Kevo, Syöte, and Pyhä-Luosto gave higher well-being ratings. The differences are partly explained by the characteristics of the parks, visits, and visitor profiles. These areas are located relatively far away from the major population centers, and most visitors are domestic tourists. In Kevo, over 90% of the visitors are hikers who stay in the park overnight (in a tent, lean-to shelter or open hut), while in Syöte and Pyhä-Luosto, most visitors spend the night in the surrounding area. Since these visitors of the northern parks spend several days in the area, they are probably very motivated to travel to the area and prepared for the trip. Similarly, Kaikkonen and Rautiainen (2014) found that the well-being benefits perceived by fishermen and hunters were higher in the vast expanses of hunting areas in northern Finland. Lower rated Kurjenrahka and Repovesi parks, in turn, are within a day or weekend trip zone for city dwellers in southern Finland. Length of stay, and especially spending the night in the park, significantly increased the perceived well-being benefits in our study. Interestingly, first-time visitors gave higher scores to several dimensions of well-being than repeat visitors in the web survey. The differences are partly explained by the high share (71%) of first-time visitors in Kevo Strict Nature Reserve which received the highest ratings. As Kevo is a remote destination for long-distance hikes, there are fewer repeat visitors than in other study areas. Also, motivation-based visitor segments have been shown to be different in parks (Konu & Kajala, 2012).

The results emphasize the importance of both close-to-home recreation and nature-based tourism destinations for human health and well-being. Protected areas located further away from people's living environments encourage people to spend longer times in nature and have potential to provide stronger benefits, including escape from everyday life, than other types of green spaces. While political attention (e.g. in EU) is increasingly directed towards urban parks and nature-based solutions in producing healthy and sustainable living environments, the contribution of rural and more peripheral natural areas to population health should not be forgotten. Visitor rates in these areas and the popularity of nature-based tourism constantly increase (see Jepson & Sharpley, 2015; Lane & Kastenholz, 2015). This conclusion is echoed by Weber and Anderson's (2010) result of a greater attainment for several benefits by visitors in regional rather than urban parks, emphasizing the importance of maintaining or expanding existing low development conservation zones. Meanwhile, it is important to manage problems (e.g. crowding, littering or erosion) caused by increasing visitor numbers as they may also influence perceived well-being benefits.

While our study indicates the potential dimensions of well-being in protected areas and differences between the respondent groups, our study does not decipher what exactly are the intrinsic qualities of nature itself that originate these perceptions or which qualities make benefits stronger and last longer. The importance of considering biological diversity and complexity as opposed to loosely defined "nature" when investigating the benefits of interacting with nature has increasingly been emphasized as a solution to this problem (Fuller et al., 2007; Keniger et al., 2013). Although the highest rated northern parks are located in landscapes dominated by fells or hills, any further conclusions about the influence of topography, land cover, or biodiversity on the perceived benefits would require a more detailed study. Furthermore, while the importance of a range of different sensations and sensory satisfaction factors was highlighted in our study, they were also noted in a recent study of rural Portugal (Carneiro, Lima, & Lavrador Silva, 2015). Further research on the reasons for this and its management and marketing implications is needed. In northern countries such as Finland, these sensations may be very different in summer than in winter, which emphasizes the importance of studying perceived well-being benefits in different seasons. Also, Irvine et al. (2013) found in their qualitative study that when asked directly, users of urban green space gave more motivations and benefits than suggested by any existing theories or identified through the use of a closed-ended checklist drawn from previous research.

In Finland and other western societies, national parks and other protected areas have been increasingly assigned the new mandate as "the fountains of health and well-being" for the urbanized population. For instance, the global movement *Healthy Parks Healthy People* (see: http://www. hphpcentral.com/) harnesses the power of parks and public lands in contributing to a healthy civil society. Accordingly, health and well-being benefits are increasingly used to justify financial and political support for parks and committing to the preservation of biological diversity and ecosystem services (see Stolton & Dudley, 2010). Our findings suggest that this new mandate and objectives are well justified. At least in the Finnish context, protected areas have the potential of significantly contributing to public health. In 2013, the total value of the perceived health and well-being benefits of Finnish national parks was estimated €226 million (Vähäsarja, 2014) while the total maintenance cost of parks was over €6 million, according to Parks & Wildlife Finland. The value of the health and well-being benefits of all state-owned natural areas was estimated €1.1 billion while the total expenditure on Finnish health care was €17.1 billion in 2011 (Vähäsarja, 2014). However, the health and well-being impacts of protected areas should not be taken for granted, and the new mandate should not be assigned to parks without scientific evidence and appropriate indicators to monitor the benefits.

Our results indicate that different well-being impacts of protected areas are not yet sufficiently identified. Our respondents rated highest the benefits related to escape and sensory satisfaction. These types of well-being impacts have not been thoroughly identified or discussed separately in previous studies (e.g. Lemieux et al., 2012). However, it should be acknowledged that exploratory factor analysis used in the study is driven by the items included. Additional dimensions of well-being may exist that were not adequately covered by the item set used in the study. Similarly, our study revealed differences in the perceived well-being between respondents of different demographic groups that may partially be explained by different capacities to interpret and reflect the survey questions. Therefore, more research, including qualitative studies, is needed to deepen the understanding of the different dimensions of health and well-being provided by protected areas and for developing appropriate and demographically sensitive/equal indicators to monitor the health and well-being benefits. And, finally, the implications of this paper and of future studies need to be considered by Park and edge of Park planners in terms of infrastructure, and especially accommodation development. Much could be learned from the comparison between low-intensity and high-intensity park management discussed in Getzner, Lange Vik, Brendehaug, and Lane (2014).

Note

1. A broad wetland with an open area in its center (see http://eunis.eea.europa.eu/habitats/10154).

Acknowledgments

We gratefully acknowledge the staff of Parks & Wildlife Finland, especially Joel Erkkonen, Liisa Kajala, Veikko Virkkunen, and Matti Tapaninen.

Disclosure statement

No potential conflict of interest was reported by the authors.

Funding

This work was partly supported by TEKES funded project *ADELE – Autoimmune defense and living environment* [grant number 40333/14] and the Ecosystem services network funded by the Finnish Rural Policy Committee (Ministry of Agriculture and Forestry) [grant number 58/03.01.03/2016].

References

Balmford, A., Green, J.M.H., Anderson, M., Beresford, J., Huang, J., Naidoo, R., … Manica, A. (2015). Walk on the wild side: Estimating the global magnitude of visits to protected areas. *PLoS Biology, 13*, e1002074.

Barton, J., & Pretty, J. (2010). What is the best dose of nature and green exercise for improving mental health? A multi-study analysis. *Environmental Science & Technology, 44*, 3947–3955.

Berman, M.G., Jonides, J., & Kaplan, S. (2008). The cognitive benefits of interacting with nature. *Psychological Science, 19*, 1207–1212.

Björk, J., Albin, M., Grahn, P., Jacobsson, H., Ardö, J., Wadbro, J., … Skärbäck, E. (2008). Recreational values of the natural environment in relation to neighbourhood satisfaction, physical activity, obesity and wellbeing. *Journal of Epidemiology & Community Health, 62*, e2.

Bowler, D.E., Buyung-Ali, L.M., Knight, T.M., & Pullin, A.S. (2010). A systemic review of evidence for the added benefits to health of exposure to natural environments. *BMC Public Health, 10*, 456.

Byström, J., & Müller, D.K. (2014). Tourism labor market impacts of national parks. The case of Swedish Lapland. *Zeitschrift für Wirtschaftsgeographie, 58*, 115–126.

Carneiro, M.J., Lima, J., & Lavrador Silva, A. (2015). Landscape and the rural tourism experience: Identifying key elements, addressing potential, and implications for the future. *Journal of Sustainable Tourism, 23*, 1217–1235.

Curtin, S. (2009). Wildlife tourism: The intangible, psychological benefits of human–wildlife encounters. *Current Issues Tourism, 12*, 451–474.

Dolnicar, S., Yanamandram, V., & Cliff, K. (2012). The contribution of vacations to quality of life. *Annals of Tourism Research, 39*, 59–83.

Dunn Ross, E.L., & Iso-Ahola, S.E. (1991). Sightseeing tourists' motivation and satisfaction. *Annals of Tourism Research, 18*, 226–237.

Fredrickson, L.M., & Anderson, D.H. (1999). A qualitative exploration of the wilderness experience as a source of spiritual inspiration. *Journal of Environmental Psychology, 19*, 21–39.

Fuller, R.A., Irvine, K.N., Devine-Wright, P., Warren, P.H., & Gaston, K.J. (2007). Psychological benefits of greenspace increase with biodiversity. *Biology Letters, 3*, 390–394.

Getzner, M., Lange Vik, M., Brendehaug, E., & Lane, B. (2014). Governance and management strategies in national parks: Implications for sustainable regional development. *International Journal of Sustainable Society, 6*, 82–101.

Gilbert, D., & Abdullah, J. (2004). Holidaytaking and the sense of well-being. *Annals of Tourism Research, 31*, 103–121.

Hammer, T., Mose, I., Siegrist, D., & Weixlbaumer, N. (2007). Protected areas and regional development in Europe: Towards a new model for the 21st century. In I. Mose (Ed.), *Protected areas and regional development in Europe. Towards a new model for the 21*st *century* (pp. 233–246). Aldershot: Ashgate.

Han, K. (2010). An exploration or relationships among the responses to natural scenes: Scenic beauty, preference and restoration. *Environment and Behavior, 42*, 243–270.

Hanski, I., von Hertzen, L., Fyhrquist, N., Koskinen, K., Torppa, K., Laatikainen, T., … Haahtela, T. (2012). Environmental biodiversity, human microbiota, and allergy are interrelated. *Proceedings of the National Academy of Sciences of the United States of America, 109*, 8334–8339.

Hansmann, R., Hug, S., & Seeland, K. (2007). Restoration and stress relief through physical activities in forests and parks. *Urban Forestry Urban Greening, 6*, 213–225.

Hartig, T., Evans, G.W., Jamner, L.D., Davis, D.S., & Gärling, T. (2003). Tracking restoration in natural and urban field settings. *Journal of Environmental Psychology, 23*, 109–123.

Hartig, T., Kaiser, F.G., & Strumse, E. (2007). Psychological restoration in nature as a source of motivation for ecological behaviour. *Environmental Conservation, 34*, 291–299.

Hartig, T., Mang, M., & Evans, G.W. (1991). Restorative effects of natural environment experiences. *Environment & Behavior, 23*, 3–26.

Health Council of the Netherlands and Dutch Advisory Council for Research on Spatial Planning, Nature and the Environment (2004). *Nature and health. The influence of nature on social, psychological and physical well-being.* The Hague: Author. Retrieved September 20, 2016, from https://www.gezondheidsraad.nl/sites/default/files/Nature_and_health.pdf

Hjalager, A-M., Konu, H., Huijbens, E.H., Björk, P., Flagestad, A., Nordin, S., & Tuohino, A. (2011). *Innovation and re-branding Nordic wellbeing tourism.* Nordic Innovation Centre. Retrieved September 20, 2016, from http://www.nordicinnova tion.org/Global/_Publications/Reports/2011/2011_NordicWellbeingTourism_report.pdf

Humberstone, B. (2011). Embodiment and social and environmental action in nature-based sport: Spiritual spaces. *Leisure Studies, 30*, 495–512.

Hunter-Jones, P., & Blackburn, A. (2007). Understanding the relationship between holiday-taking and self-assessed health: An exploratory study of senior tourism. *International Journal of Consumer Studies, 31*, 509–516.

Irvine, K.N., & Warber, S.L. (2002). Greening healthcare: Practicing as if the natural environment really mattered. *Alternative Therapies in Health and Medicine, 8*, 76–83.

Irvine, K.N., Warber, S.L., Devine-Wright, P., & Gaston, K.J. (2013). Understanding urban green space as a health resource: A qualitative comparison of visit motivation and derived effects among park users in Sheffield, UK. *International Journal of Environmental Research and Public Health, 10*, 417–442.

Iso-Ahola, S.E. (1982). Toward a social psychological theory of tourism motivation: A rejoinder. *Annals of Tourism Research, 9*, 256–262.

Jepson, D., & Sharpley, R. (2015). More than sense of place? Exploring the emotional dimension of rural tourism experiences. *Journal of Sustainable Tourism, 23*, 1157–1178.

Kaczynski, A.T., & Henderson, K.A. (2007). Environmental correlates of physical activity: A review of evidence about parks and recreation. *Leisure Sciences, 29*, 315–354.

Kaikkonen, H., & Rautiainen, M. (2014). *Terveyttä ja hyvinvointia valtion mailta – tarkastelussa metsästäjät ja kalastajat* [Health and well-being in State-owned land – study of hunters and fisherman]. Nature Protection Publications of Metsähallitus. Series A 209. Vantaa: Metsähallitus. Retrieved September 20, 2016, from https://julkaisut.metsa.fi/assets/pdf/lp/Asarja/a209.pdf

Kaikkonen, H., Virkkunen, V., Kajala, L., Erkkonen, J., Aarnio, M., & Korpelainen, R. (2014). *Terveyttä ja hyvinvointia kansallis-puistoista – tutkimus kävijöiden kokemista vaikutuksista* [Health and well-being from Finnish national parks – a study on benefits perceived by visitors]. Executive summary in English. Nature Protection Publications of Metsähallitus. Series A 208. Vantaa: Metsähallitus. Retrieved September 20, 2016, from https://julkaisut.metsa.fi/assets/pdf/lp/Asarja/a208.pdf

Kaplan, R. (2001). The nature of the view from home: Psychological benefits. *Environment and Behavior, 33*, 507–542.

Karjalainen, E., Sarjala, T., & Raitio, H. (2010). Promoting human health through forests: Overview and major challenges. *Environmental Health and Preventive Medicine, 15*, 1–8.

Keniger, L., Gaston, K., Irvine, K.N., & Fuller, R. (2013). What are the benefits of interacting with nature? *International Journal of Environmental Research and Public Health, 10*, 913–935.

Konu, H., & Kajala, L. (2012). *Segmenting protected area visitors based on their motivations.* Nature Protection Publications by Metsähallitus. Series A 194. Vantaa: Metsähallitus. Retrieved September 20, 2016, from https://julkaisut.metsa.fi/assets/pdf/lp/Asarja/a194.pdf

Korpela, K., Borodulin, K., Neuvonen, M., Paronen, O., & Tyrväinen, L. (2014). Analyzing the mediators between nature-based outdoor recreation and emotional well-being. *Journal of Environmental Psychology, 37*, 1–7.

Kuo, F.E., & Sullivan, W.C. (2001). Aggression and violence in the inner city. Effects of environment via mental fatigue. *Environment and Behavior, 33*, 543–571.

Kuo F.E., & Taylor A.F. (2004). A potential natural treatment for attention-deficit/hyperactivity disorder: Evidence from a national study. *American Journal of Public Health, 94*, 1580–1586.

Lane, B., & Kastenholz, E. (2015). Rural tourism: The evolution of practice and research approaches – towards a new generation concept? *Journal of Sustainable Tourism, 23*, 1133–1156.

Lee, J., Li, Q., Tyrväinen, L., Tsunetsugu, Y., Park, B-J., Kagawa, T., & Miyazaki, Y. (2012). Nature therapy and preventive medicine. In J. Maddock (Ed.), *Public health – social and behavioral health* (pp. 325–350). InTech Open. Retrieved September 20, 2016, from http://www.intechopen.com/books/howtoreference/public-health-social-and-behavioral-health/nature-therapy-and-preventive-medicine

Lemieux, C., Eagles, P., Slocombe, D., Doherty, S., Elliott, S., & Mock S. (2012). Human health and well-being motivations and benefits associated with protected area experiences: An opportunity for transforming policy and management in Canada. *PARKS: The International Journal of Conservation and Protected Areas, 18*, 71–86.

Lemieux, C.J., Doherty, S.T., Eagles, P.F.J., Gould, J., Hvenegaard, G.T., Nisbet, E., & Groulx, M.W. (2015). *Healthy outside-healthy inside: The human health and well-being benefits of Alberta's protected areas - towards a benefits-based*

management agenda. Ottawa: CCEA Secretariat. Retrieved September 20, 2016, from http://scholars.wlu.ca/cgi/view
content.cgi?article=1024&context=geog_faculty

Maller, C.J. (2009). Promoting children's mental, emotional and social health through contact with nature: A model.
Health Education, 109, 522–543.

Maller, C., Townsend, M., Brown, P., St Leger, L., Henderson-Wilson, C., Pryor, A., … Moore, M. (2008). *Healthy parks healthy
people: The health benefits of contact with nature in a park context. A review of relevant literature* (2nd ed.). Melbourne:
Deakin University and Parks Victoria. Retrieved September 20, 2016, from http://www.deakin.edu.au/__data/assets/
pdf_file/0016/310750/HPHP-2nd-Edition.pdf

Mayer, M. (2014). Can nature-based tourism benefits compensate for the costs of national parks? A study of the Bavarian
Forest National Park, Germany. *Journal of Sustainable Tourism, 22,* 561–583.

Metsähallitus (2016a). *Principles of protected area management in Finland.* Nature Protection Publications of Metsähallitus.
Series B 217. Vantaa: Metsähallitus. Retrieved September 20, 2016, from https://julkaisut.metsa.fi/assets/pdf/lp/Bsarja/
b217.pdf

Metsähallitus (2016b). Kansallispuistojen, valtion retkeilyalueiden ja muiden virkistyskäytöllisesti merkittävimpien
Metsähallituksen hallinnoimien suojelualueiden ja retkeilykohteiden käyntimäärät vuonna 2015 [Number of visits in
national parks, national hiking areas and other conservation and recreation areas managed by Metsähallitus in
2015]. Retrieved September 20, 2016, from http://www.metsa.fi/documents/10739/3335805/kayntimaarat2015_fi.
pdf/0e3f3835-17e8-4a99-9b4c-51f1b1d1b9a5

Moore, M., Townsend, M., & Oldroyd, J. (2006). Linking human and ecosystem health: The benefits of community involvement in conservation groups. *EcoHealth, 3,* 255–261.

Niemelä, R., Ek, S., Eriksson-Backa, K., & Huotari, M-L. (2012). A screening tool for assessing everyday health information
literacy. *Libri, 62,* 125–134.

Nilsson, K., Baines, C., & Konijnendik, C.C. (2007). *Health and the natural outdoors* (COST Strategic Workshop, Larnaka.
Final Report). Retrieved September 20, 2016, from http://www.umb.no/statisk/greencare/general/strategic_work
shop_final_report.pdf

Pasanen, T., Tyrväinen, L., & Korpela, K.M. (2014). The relationship between perceived health and physical activity indoors,
outdoors in built environments, and outdoors in nature. *Applied Psychology: Health and Well-Being, 6,* 324–346.

Pretty, J., Peacock, J., Hine, R., Sellens, M., South, N., & Griffin, M. (2007). Green exercise in the UK Countryside: Effects on
health and psychological well-being, and implications for policy and planning. *Journal of Environmental Planning and
Management, 50,* 211–231.

Puhakka, R., & Saarinen, J. (2013). New role of tourism in national park planning in Finland. *The Journal of Environment
and Development, 22,* 411–434.

Puustinen, J., Pouta, E., Neuvonen, M., & Sievänen, T. (2009). Visits to national parks and the provision of natural and man-
made recreation and tourism resources. *Journal of Ecotourism, 8,* 18–31.

Shin, W.S., Yeoun, P.S., Yoo, R.W., & Shin, C.S. (2010). Forest experience and psychological health benefits. *Environmental
Health and Preventive Medicine, 15,* 38–47.

Siikamäki, P., Kangas, K., Paasivaara, A., & Schoderus, S. (2015). Biodiversity attracts visitors to national parks. *Biodiversity
and Conservation, 24,* 2521–2534.

Snepenger, D., King, J., Marshall, E., & Uysal, M. (2006). Modeling Iso-Ahola's motivation theory in the tourism context.
Journal of Travel Research, 45, 140–149.

Stolton, S., & Dudley, N. (2010). Vital sites: The contribution of protected areas to human health. The Arguments for Protection Series. Gland: WWF. Retrieved September 20, 2016, from http://assets.panda.org/downloads/vital_sites.pdf

Tsunetsugu, Y., Park, B.J., Ishii, H., Hirano, H., Kagawa, T., & Miyazaki, Y. (2007). Physiological effects of Shinrin-yoku (taking
in the atmosphere of the forest) in an old-growth broadleaf forest in Yamagata prefecture, Japan. *Journal of Physiological Anthropology, 26,* 135–142.

Tyrväinen, L., Ojala, A., Korpela, K., Lanki, T., Tsunetsugu, Y., & Kagawa, T. (2014). The influence of urban green environments on stress relief measures: A field experiment. *Journal of Environmental Psychology, 38,* 1–9.

Vähäsarja, V. (2014). Luontoympäristön terveys-ja hyvinvointivaikutusten taloudellinen arvottaminen [Assessment of the
financial value of the health and well-being benefits of natural environments]. Nature Protection Publications of
Metsähallitus. Series A 210. Vantaa: Metsähallitus. Retrieved September 20, 2016, from https://julkaisut.metsa.fi/
assets/pdf/lp/Asarja/a210.pdf

van den Berg, A.E., Koole, S.L., & van der Wulp, N.Y. (2003). Environmental preference and restoration: (How) are they
related? *Journal of Environmental Psychology, 23,* 135–146.

Wall Reinius, S., & Fredman, P. (2007). Protected areas as attractions. *Annals of Tourism Research, 34,* 839–854.

Weber, D., & Anderson, D. (2010). Contact with nature: Recreation experience preferences in Australian parks. *Annals of
Leisure Research, 13,* 46–69.

Wolf, I.D., Stricker, H.K., & Hagenloh, G. (2015). Outcome-focused national park experience management: Transforming
participants, promoting social well-being, and fostering place attachment. *Journal of Sustainable Tourism, 23,* 358–381.

Wolf, I.D., & Wohlfart, T. (2014). Walking, hiking and running in parks: A multidisciplinary assessment of health and well-
being benefits. *Landscape and Urban Planning, 130,* 89–103.

Operationalising both sustainability and neo-liberalism in protected areas: implications from the USA's National Park Service's evolving experiences and challenges

Susan L. Slocum

ABSTRACT

Following its 2013 Green Park Plan, the US National Park Service (NPS) has sought to achieve more sustainable operations at park level. However, a history of top-down governance, research aversion, contentious community relationships, and a move towards neoliberal governance has created numerous challenges to operationalising the sustainability agenda. Using comparative case study methodology, this paper approaches sustainability through a five-pillar framework that incorporates the economy, environment, society, good governance and sound science. Data from three very different parks, and from the newly formed Division of Climate Change and Sustainability at Park Headquarters, is presented. Using the National Mall in Washington, DC, Shenandoah in Virginia, and Big Bend in Texas, the paper highlights many successful sustainability initiatives. A number of serious problems remain, however, including lack of funding, the slow working budget cycle, the need for inter-park knowledge sharing, the need for behavioural change at both national and local management levels, issues surrounding the "whiteness" of the visitor mix, and the nature of park visitation, where driving remains the primary transportation mode. Overall, neo-liberalism is placing increased emphasis on concessionaires to encourage sustainable operations, distancing NPS from its core visitor base and encouraging rising visitation levels with their associated impacts.

Introduction

The US National Park Service (NPS) has been called "America's best idea" and has long been a source of pride for most Americans. NPS embodies the central values of American society, such as democracy and freedom (Weber & Sultana, 2013). Yet, stagnant visitation levels (NPS, 2016a) suggest that the NPS may be isolated from the changing expectations of the American public, especially now that sustainability is increasingly demanded by nature-based tourists seeking national park experiences (Schwartz & Lin, 2006). Furthermore, Slocum and Curtis (2016) beg the question "How proactive should park managements be in contrast[ing] traditional low intensity, reactive park management policies…with the high intensity, proactive, pro-local sustainable business marketing policies" (p. 164). In order to answer that, a more detailed look at the current role of sustainability in NPS operations is warranted. This paper looks at how the NPS is attempting to create sustainable conservation measures in an increasingly neo-liberal world. It also extends the discussion on behavioural change

---

🔓 Supplemental data for this article can be accessed 🔗 here.

necessary to achieve sustainable tourism from the market to the supplier, exploring the issues for NPS as an institution (Bramwell & Lane, 2013).

NPS has been archetypal for protected area management since its inception in 1916, setting standards and principles used by many other protected areas worldwide. Today, there are 411 protected areas, covering 84 million acres, located in every US state, the District of Columbia, American Samoa, Guam, Puerto Rico, and the US Virgin Islands (NPS, 2016a). These protected areas include not only spaces of abundant wildlife and unique ecosystems, but cultural and historical monuments, battlefields, lakeshores and seashores, and urban memorials. In 1962, NPS hosted the first World Parks Congress, followed by the World Heritage Convention, providing training for international park professionals which have resulted in the proliferation of protected areas worldwide (Sellers, 1997). In particular, NPS "provide(s) leadership in strategic global conservation initiatives, including the creation of the World Heritage program and initiatives to professionalize and systematize parks management through the World Commission on Protected Areas" (Mitchell, 2011, p. 8). While the influence of the NPS globally has been greatly reduced over time, their conservation philosophy remains a guide for park policies and management styles internationally. Protected area management on an international scale has now largely been turned over to international development agencies and conservation NGOs (Kline & Slocum, 2015), leaving NPS to address more domestic issues related to conservation and tourism. Those issues include its responses to the demands for more sustainable operations, including more sustainable forms of tourism.

The original NPS model, which has been replicated in many other countries, consisted of the removal of local inhabitants (as humans were not considered part of the environment), the prohibition of extractive economic activities, such as hunting, mining and forestry, and the preservation of landscapes in their "native" condition (Weber & Sultana, 2013). These practices often placed park officials at odds with local communities, resulting in long-standing conflicts that still impede sustainable development in the USA and worldwide. The isolation of national parks, both as environmental enclaves and socially detached administrative units, has limited the effectiveness of national parks in the overall management of protected areas, as many park impacts come from outside the park boundaries (Gimmi et al., 2011). This has left national parks vulnerable to public criticism (Gimmi et al., 2011).

Beginning early in NPS history, these isolated islands of ecological wonder have experienced both internal and external threats, including poaching, invasive species, excessive development, watershed quality and tourism impacts (Shafer, 2012). While national parks in other parts of the world are recognising the importance of stakeholder involvement and community-based initiatives to help counter these threats, the NPS has been reluctant to devolve power and influence to their conservation partners (Mitchell, 2011). Current literature suggests that the NPS "does not undertake major threats abatement activities unless first pushed by NGOs or the Congress" (Shaffer, 2012, p. 1107) and that "national parks in the United States have a well- documented history of indifference, if not hostility, to the support of basic research" (Parsons, 2004, p. 5). While there are a number of academic articles that approach the conservation science tactics of the NPS, sustainable tourism research in relation to the triple bottom line is surprisingly rare. Information on how NPS is approaching these challenges remains under-researched (Shaffer, 2012).

There has been international public outcry in relation to the increasing neo-liberalised management style of park administration that promotes tourism as a revenue source, often prioritising revenue management over ecological conservation (Park, Ellis, & Prideaux, 2010). Nyahunzvi (2016) claims that neo-liberalism in conservation is a direct result of reduced government budgets and inadequate state funding and is prevalent under the commercialisation policy of outsourcing that is common in national parks. In the US, parks are allowed to keep 80% of all entrance fees collected, as well as a percentage of concession revenue (Schwartz & Lin, 2006). While Park et al. (2010) claim that the highest support for fees structures by visitors occurs when this revenue remains in the park, visitors expect that this money will be spent to "improve informational signs, to protect resources, to improve wildlife habitat, to improve trailheads, and to increase visitor safety" (p. 205). Yet, there is limited research that addresses the changing management priorities of NPS and the impact this fee structure has on sustainability operations (Slocum & Curtis, 2016). Additionally, research has failed to comment on how neo-liberalism is operationalised when government agencies still possess relatively "socialist" planning cultures.

This paper recognises that each park operates in a unique ecosystem, works with heterogeneous communities with different cultures and needs, faces a variety of economic challenges, and potentially views sustainability through different lenses. It uses three distinctly different parks to represent the numerous sustainability implementation challenges facing the NPS. Through a qualitative study of the National Mall in Washington, DC, Shenandoah National Park in Virginia, and Big Bend National Park in Texas, this paper recognises that the lack of a holistic sustainability approach has fragmented the resources available to park administration and challenged the implementation of sustainability. It also explores the governance problems faced by a large, centrally controlled institution like the NPS as it seeks to tackle the potential conflicts between sustainable development and a neo-liberal approach.

Literature review: an evolving complexity

This paper uses the five key principles of sustainability as presented by the United Kingdom Government (2005) and Everett and Slocum (2013). This model includes not only the triple bottom line of the economy, environment and society, but also addresses good governance and sound science as vital elements within the sustainability dialogue (Figure 1). As Eagles (2014) writes, "Governance is the means for achieving direction, control, and coordination (and) is the process whereby organizations make their decisions" (p. 542). The framework provides a deeper exploration of sustainability theory and "underpins the discussion on the relationships between the principles and the challenges" (Everett & Slocum, 2013, p. 795) that is inherent in the operationalisation of sustainable

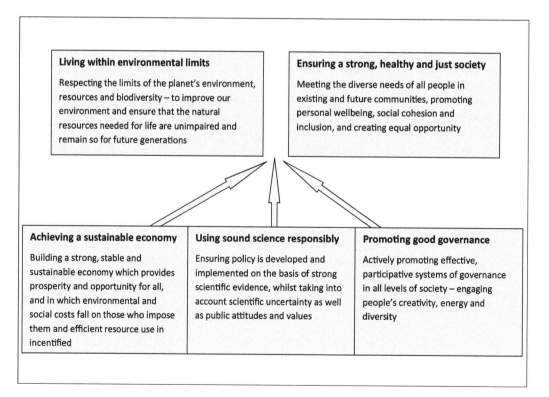

Figure 1. Five pillars of sustainability (Source: Everett & Slocum, 2013).

tourism. In particular, governance and science are relevant to an NPS study as neoliberal practices are both affected by and affect policy development related to sustainable tourism.

According to the National Park Service Organic Act (1916), the primary goal of the NPS is to "conserve the scenery and the natural and historic objects and the wild life therein and to provide for the enjoyment of the same in such manner and by such means as will leave them unimpaired for the enjoyment of future generations" (NPS, 1916). Nature-based tourism not only offers access and enjoyment, it provides economic opportunities for regional development as well as incentives and management resources for conservation (Eagles, 2002). While, traditionally, park entrance fees were kept low in order to ensure affordability for the American people (Wells, 1997), tourist spending inside the park was an uncaptured revenue source. Today, concessionaires provide a variety of services, including camping, showers and laundry, adventure activities, and basic supplies such as gasoline and groceries. These concessionaires, in turn, pay fees to the NPS to further protect the resources for future generations (Dinica, 2016).

National parks undoubtedly provide enormous economic opportunities for rural areas. The NPS reports the economic contribution of each park to their neighbouring communities, with, in 2015, over 307 million recreation visits resulting in $16.9 billion in visitor spending nationwide. The total economic contribution to the US (direct, indirect and induced) supported 295,000 jobs, creating $11.1 billion in labour income and $32 billion in economic output (Thomas & Koontz, 2016). Over 31% of visitor spending went to hotels, motels and bed-and-breakfasts, while over 20% went to restaurants and bars and almost 12% to gas and oil sales ((Thomas & Koontz, 2016). States where visitor spending was the highest included California ($1.8 billion), North Carolina ($1.2 billion) and Alaska ($1.2 billion). Parks with the highest visitor spending included Great Smokey Mountains ($873 million), Grand Canyon ($584 million) and Denali ($567 million) (NPS, 2016b).

The sheer volume of tourists and the economic impact they provide has been recognised as a viable funding source for a variety of government organisations. As federal agencies face pressure to work with reduced tax revenues, yet seek to enhance their service offer, neoliberal policies are supporting the delivery of public goods. Reduced government allocations for publically managed parks and the concept of "doing more with less" have changed the funding focus towards market structures that are related to tourism visitation (Eagles, 2014). Park budgets are moving to the provision of recreational activities that influence visitor satisfaction, in turn leading to increased usage of park resources, often beyond acceptable carrying capacities (Dinica, 2016). Jones (2012) believes that these emerging neoliberal policies encourage industrial development strategies that define "sustainability" and "conservation" (Jones, 2012).

As early as 1985, research was measuring "levels of acceptable change" within park boundaries by tourism visitation (Stankey, Cole, Lucas, Petersen, & Frissell, 1985). Wang and Miko (1997) reported that over 50% of park superintendents noted that visitation had adverse effects on water quality, 61% reported that visitation had a negative impact on air quality, and 84% reported adverse effects on vegetation within the parks. More recently, external research has addressed other environmental impacts associated with visitation, such as human–animal interactions (Taylor, Gunther, & Grandjean, 2014). Ament, Clevenger, Yu, and Hardy (2008) show that park officials consider traffic to be over capacity in more than half of the US national parks and that road-side mortality was impacting wildlife populations.

As government agencies rely more on market structures to meet operational objectives, their budgets become contingent not on public funding but on visitor spending, an important change. The simple economic impact numbers do little to explain how exogenous and endogenous factors affect visitor spending in these areas. Using a spatial lag model, Henrickson and Johnson (2013) claim that "increases in fuel costs, temperature increases of greater than 3°F, and restrictions on foreign inbound tourism all lower visitation to U.S. national parks, causing associated decreases in money spent by tourists, jobs created, and income generated" (p. 330). Poudyal, Paudel, and Tarrant (2013) measured changes in national park demand in the USA based on five recession variables, including unemployment rates, economic activity and output, business cycles, consumer confidence and consumer inflation expectations. All variables showed a significant effect on visitation, implying that national park tourism spending is adversely influenced by negative downturns in the economy. In 2006, Schwartz and Lin showed that increased park entrance fees and changes in revenue management by NPS resulted in a 22% decrease in visitation over a five-year period. Gabe (2016) looked at the effects of the October 2013 US government closure on Arcadia National Park in Maine. He found a "76% reduction in visitors to Acadia National Park in October 2013 and a 13% decrease in tourism-related spending" (p. 314). Additionally, the shutdown resulted in a 9% reduction in souvenir and gift shop sales and a 19% reduction in lodging revenue. Many gateway communities are increasingly becoming dependent on national park tourism and these economic returns.

The rise of new non-economic issues

The NPS began in conflict with local residents in areas that would become protected areas. The more remote parks, such as Yellowstone and Yosemite, were formed by forcibly removing native residents. In a later phase of park development, land was acquired through eminent domain (also known as compulsory purchase) and landholders were forcibly removed from their property if they refused to sell (Hamin, 2001). In many gateway communities, these clashes are still remembered and traditional industries, such as mining and forestry, are gone forever. Of course, these communities also reap the benefits of tourism development, and new businesses opportunities, in the form of gift shops and restaurants, have sustained these communities, even if visitation is often seasonal.

Tourism can also facilitate cultural change and leads to increased development in areas surrounding national parks. Visitors may grow attached to areas of outstanding beauty, value the recreational opportunities in an area and chose to move to gateway communities. This creates conflict between the long-time residents and the new immigrants. Hamin (2001) claims that the residents in Estes Park, Colorado (gateway to Rocky Mountain National Park) are composed of insiders, those who are long-time residents and dependent on tourism, and outsiders who are developers, telecommuters, second-home owners and retirees who are not dependent on tourism receipts. Frauman and Banks (2011) found that rising costs of living, a reduction in the number of farms, and the loss of open space outside the park boundaries were by-products of national park tourism development along the Blue Ridge Parkway in North Carolina.

Another issue that has plagued NPS in recent years is the lack of diversity in park visitation (Weber & Sultana, 2013). A number of studies show the recurring demographics of park visitors, who tend to be older, white, wealthy and educated (Benson, Watson, Taylor, Cook, & Hollenhors, 2013; Henrickson & Johnson, 2013; Slocum & Curtis, 2016). In a review of NPS visitor data since 2000, Weber and Sultana (2013) show that 93% of visitors were white, compared to 78% of the total national population, and that 13 parks reported no African American visitors at all. They claim this is because most parks are located in rural areas of the west where there are few African American residents. Additionally, there are few parks that celebrate African American, Asian American or Hispanic culture, which might appeal to a more diverse set of visitors. Up until the 1970s, "The official NPS view had previously been that any racial or ethnic history related to a site should be incidental to the site's importance… The fact that no African American site had previously been found to be nationally significant was presumably taken as simple evidence that African Americans had made no significant contributions" (Weber & Sultana, 2013, p. 450). Diversity issues are becoming a major threat to NPS as their staple visitation market is an aging population.

Yet, the social impact NPS has on the American people paints a very different picture. There is little doubt that America is proud of her national parks and the natural and cultural heritage that is protected through this agency (Weber & Sultana, 2013). Environmental education is a core mission of the NPS and is often credited with the growth in the environmentalist movement that has influenced development around the world (Mitchell, 2011). However, as external research has embraced the inclusion of local residents in the decision-making process, NPS has been slow to develop inclusive policies in gateway communities in the US, favouring visitors over residents (Hamin, 2001). And one by-product of neo-liberalism is a redefinition of stakeholders, towards stakeholders that are corporate customers and franchise holders seeking profitable market structures (Kline & Slocum, 2015). The role of community – and of conservationists – in park governance can potentially be weakened.

NPS was developed as a centralised management structure that operates within the Department of the Interior, which has privileged regulation over participation. As a federal agency, it must comply with a number of administrative authorities, including Congress, the Administrative Procedure Act, and the National Environmental Policy Act. Historically, all policies were top-down, relying on the expertise of agency staff, with stakeholder input limited to the public comment period as prescribed by law (Clarke & McCool, 1996). However, Cook (2014) shows how congressional gridlock has given way to "rule-making" as a form of policy and governance within some agencies, including NPS. Within the rule-making framework, the park superintendent is closely involved in the language of regulation and the process of community input. In a study of snowmobile regulation in Yellowstone National Park, Cook states "The fact that the NPS is not using a collaborative process may explain why the agency is having such a difficult time producing rule(s)…Moreover, it may also explain why the agency has begun to produce fewer successful rules than it has historically" (p. 1267).

Governance within NPS has historically centred on resource protection. But the ecological threats were often coming from outside the park boundaries, which forced park officials to look at regional land and other management partnerships and brought the NPS in direct contact with neighbouring communities (Hamin, 2001). Currently, there are four types of collaborative partnerships within national parks: contracting with concessionaires or local governments for tourism services; having a friends group run a unit's bookstore or provide other educational material; the NPS providing technical assistance to state or local parks; and sharing responsibility for resource ownership or management (Zube, 1992). It now appears that NPS is finding that the involvement of the general public is necessary in order to plan and accomplish their conservation and recreational goals (Baron, Theobald, & Fagre, 2000).

Along with a history of centralised governance, NPS has been accused of a long history of research avoidance (Sellers, 1997). Parsons (2004) explains, "The agency charged with protection of many of the nation's most valuable natural resources, evolved a culture that neither understood nor valued the importance of science to its management decisions" (p. 5). It was not until 1970, when the first university partnership was established, between NPS and the University of Washington, which brought applied research to the agency and in-park research centres, staffed by NPS employees, became more common. In 1993, the National Biological Survey (NBS) was developed where all federal conservation research was consolidated. Today, both the NBS and NPS work together to develop research-informed policy; however, the relationship between these agencies is still evolving (Parsons, 2004).

Shaffer (2012) shows that research-informed policy development is still a slow and cumbersome process within NPS. In his study of the development of buffer zones to protect parks from external threats, he acknowledges that while the first research study was published internally in the NPS in 1990, a clear policy agenda was not formalised until 2006. He blames a culture of stonewalling within the NPS as a primary cause. Miller, Carter, Walsh, and Peake (2014) write, "We endorse approaches that fit NPS studies to the ideals of sustainable development and good governance. We see a continuing need for multidisciplinary and interdisciplinary research that, in complementary fashion, protects the environment while improving the quality of human life" (p. 264). Research appears to still be a major hurdle to sustainable development within NPS.

In 2010, NPS published its Climate Change Response Strategy (NPS, 2010) leading to the establishment in 2012 of the Division of Climate Change and Sustainability to oversee implementation in an attempt to govern the operationalisation of policies in these areas. In 2012, it released the Green Park Plan, designed to increase sustainable operations at the park level (see NPS, 2012, available on the web), through nine strategic goals shown in Table 1. The Green Park Plan "defines a collective vision and a long-term strategic plan for sustainable management of NPS operations" (NPS, 2012, p. 3). Focusing primarily on environmental impacts, the plan prioritises "green" strategies, such as water, waste and energy reductions that *minimise the impact of facility operations on the external*

Table 1. The Green Park plan goals.

Goal	Description
Continuously improve environmental performance	The NPS will meet and exceed the requirements of all applicable environmental laws
Be climate friendly and climate ready	The NPS will reduce GHG emissions and adapt facilities at risk from climate change
Be energy smart	The NPS will improve facility energy performance and increase reliance on renewable energy
Be water wise	The NPS will improve facility water use efficiency
Green our rides	The NPS will transform our fleet and adopt greener transportation methods
Buy green and reduce, reuse, and recycle	The NPS will purchase environmentally friendly products and increase waste diversion and recycling
Preserve outdoor values	The NPS will minimise the impact of facility operations on the external environment
Adopt best practices	The NPS will adopt sustainable best practices in all facility operations
Foster sustainability beyond our boundaries	The NPS will engage visitors about sustainability and invite their participation

Source: NPS (2012).

environment. It also utilises their interpretive programmes to *engage visitors* and *invite their participation* within the sustainability dialogue. However, the focus on internal operations (NPS, 2012, p. 4) is counter to sustainability principles where the economy, environment, and society are dynamic and interrelated elements that require research-informed governance. As Eagles (2014) writes "The numerous park sites must be politically and societally relevant if their conservation mandate is to be accomplished" (p. 544).

Methodology

This research uses an independent instrumental case studies approach to assess how NPS senior park staff are implementing sustainability within park operations. Case studies allow examinations of specific bounded units of analysis, focus on "what" is to be studied rather than "how" it is to be studied, and can capture variable conditions through the reliance on multiple sources of evidence (Stake, 2003). The goal is to allow for the autonomy of the specific parks to reflect their unique cultural, social, and business networks within the research methodology (Slocum, Backman, & Baldwin, 2012) and to capture the shared experience of the participants, the researcher, the environment, and the study context (Stake, 2003). Instrumental case studies are used to develop theory, where the particular study site provides insight into a phenomenon or situation. By investigating multiple case study sites, "cross-case comparisons shed light on reoccurring issues, while simultaneously highlighting case-specific concerns" (Slocum et al., 2012, p. 524).

Three independent case studies were derived from a series of semi-structured qualitative interviews with senior park staff and headquarters staff from the Division of Climate Change and Sustainability. Interview questions were derived from current and past NPS research literature as a means to ground this study within the current challenges facing NPS officials. Convenience sampling was used to select the participating parks. These interviews were held on-site within each park and each lasted up to 2 hours. Six interviews were conducted in total, and all respondents' names are kept confidential. All interviews were recorded and transcribed and thematic coding was used. Additional information was gathered through participant observation, as the research team stayed in each site as tourists and watched the park operations over a three-day period in each park. Notes were taking throughout the process. Once the three case studies were developed, cross comparisons showed similarities and differences. Overall, four themes were derived from the cases, including operational/visitor sustainability, environmental sustainability, social sustainability, and knowledge sharing.

Research sites

This study selected three distinctly different research sites in order to find commonalities and difference in the challenges and operations of iconic, but different, national park in the USA (Weaver,

1998). Each is affected by national natural resource protection policy, but operates in different environments and within different historical and cultural milieus.

The National Mall is located in the centre of Washington, DC and contains key icons including the Washington, Lincoln and Jefferson Memorials, as well as war memorials and the White House. It stretches from the United States Capitol to the Potomac River and provides over 1000 acres of greenspace in the city centre, as well as park facilities, picnic areas, and public sporting facilities. The area was set aside as a unit of the national park system in 1987. It receives approximately 24 million visitors each year and hosts special events including veteran gatherings and First Amendment demonstrations (free speech) (see Figure S1, available in the online version of this paper, under the Supplemental Data tab).

Shenandoah National Park was conceived in 1926, but not designated until 1935 when the Commonwealth of Virginia seized the land through eminent domain and gave it to the federal government to be established as a national park. The 199,173 acres are traversed by the Blue Ridge Parkway as well as the Appalachian Trail and receive approximately 1.2 million visitors each year. It receives the majority of its visitors in the autumn when the leaves change colour. It borders eight counties in Virginia and over 40% of the park is designated as wilderness (see Figure S2 available in the online version of this paper, under the Supplemental Data tab).

Big Bend National Park is one of the most remote parks in the NPS system and was founded in 1944. Located in western Texas and bordering the Rio Grande River (and the country of Mexico), it is the largest protected area of Chihuahuan Desert topography in the US (see www.desertusa.com/chihuahuan-desert.html). The park encompasses 801,163 acres and faces some of the harshest climates, with temperatures over 100 °F (38 °C) for much of the spring and summer season. It receives approximately 314,000 visitors each year. The International Dark-Sky Association has recognised the park as one of only ten places in the world certified for dark-sky stargazing (see Figure S3, available in the online version of this paper, under the Supplemental Data tab).

Results

Four major themes emerged from this research: operational/visitor sustainability, environmental sustainability, social sustainability, and knowledge sharing. To set the stage for the interviews, each park official was asked to define sustainability as it relates to their park. All respondents used the NPS tag line which includes the phrase "future generations". However, they all struggled to relate sustainability to their park operations. One respondent recounted:

> You know, we are probably talking about natural resource sustainability, experience sustainability, the visitor experience sustainability, and the sustainability of all of the values that the (park) represents to everyone around our nation so that they can be enjoyed by future generations.

It appears that even though the NPS was established using the underlying concepts of sustainability, this was at a time when the notion of sustainability had not been defined, and is still, and ironically, a recent idea at the park level. All of the respondents went first to environmental sustainability and the visitor experience, which has been the focus of their ideology for many years. Thinking along the lines of sustainable operations has been a challenge for senior staff that have been with NPS for more than 15 years. Changing the outlook and behaviour of a major institution like the NPS is a complex process.

Operational/visitor sustainability

All the parks are implementing sustainable technology as a means to reduce the environmental impact of the visitors and their operations. For example, water and electricity usage is a major concern for parks, as municipal sources are expensive. However, each park is approaching this challenge differently. At the National Mall:

> We were the number one user of potable water and electricity in the entire National Park system and were using up to 50 million gallons of water a year. We had a lot of leaks. Now we are using non-potable water that has

come from the Potomac River through the tidal basin. It is being treated by an ozone system, then released back into the river, and for the first time we have a recirculation system in the Lincoln Memorial reflecting pool. So we treat the water and recycle it.

Big Bend faces the toughest water issues of all as they are located in the desert that generally gets only about 12–16 inches of rain a year. On a three-level scale of drought, the park has been at stage 2 for the past 5 years. In 1960, the park's wells ran dry and the park now mines water from an underground aquifer resulting in the use of a non-renewable resource threatening the water table for the entire region (UC Davis, n.d.). Their primary response has been to limit the amount of water staff can use in their homes, but they have not addressed water usage by the visitors. In contrast, Shenandoah is installing low-flow toilets and showerheads in all park facilities even though they have sufficient rain fall and do not have water shortages.

Examples of energy reduction include the use of LED lights at The National Mall and the installation of two solar panels on public bathrooms at Shenandoah. Big Bend, which is best suited for solar energy, has submitted a solar energy proposal to headquarters that has not yet been funded. The budget cycle is about 3 years, so they are hopeful funding will come through next year for energy-saving technologies.

Trash removal is another concern as many communities do not have adequate landfill facilities to support the waste from tourism, and zero waste is a policy goal at all national parks (NPS, 2012). Big Bend runs a recycling programme, not only for the park but the neighbouring communities. Twice each year they take a truck load of recyclables out of the desert, but they are finding that an expensive endeavour that may not continue. The concessionaire at Shenandoah has installed a propane recycling machine that operates using the residual propane in discarded canisters. The National Mall encourages visitors to take their trash home, or uses city recycling facilities. However, changing the practices of the visitors has been more difficult. Big Bend notes:

> You see cars pulling up and instead of having one 5 gallon container of water, they have 2 cases of plastic bottles. This is our biggest issue with the visitors. We have signs and water spickets (taps) so people can refill their bottles, but they don't. Two cases of bottles, the same as 5 gallons, equates to 48 discarded plastic bottles that we have to haul out of the desert.

NPS has also aggressively pushed their climate change initiative. Since 90% of national park visitors travel by car, greenhouse gas emissions are a top sustainability priority. Shenandoah has purchased natural gas lawn mowers. As a long, linear park that specialises in views across the Appalachian Mountains, managing wild grasses that block views is an ongoing challenge. They have also installed two electric car chargers. The National Mall has worked with a number of urban partners to promote bike sharing and public transportation and have also installed electric car chargers. Big Bend is just starting to look at climate-friendly operations, but has not implemented anything yet. Their initial suggestions have addressed transportation from lodges to trailheads. Additionally, the parks are discussing climate change in their interpretive programmes, such as at Shenandoah:

> The best example I can think of right now is if you were to go to the Henry F. Byrd Visitor Centre, we have a pretty nice exhibit about climate change. It's a touch screen exhibit where people can figure out what their own carbon footprint is and identify some way that they can reduce it. I think it's called the Changing World, and that's a relatively new exhibit for us.

Overall, three challenges emerged as parks try to move to more sustainably focused operations. First, the slow budget cycle and ageing infrastructure that limits the funding available for green technology. Second, the nature of park visitation, where driving is the primary transportation mode, and, third, the use of concessionaires who often have the largest impact on visitor sustainability, over which the parks have limited influence. One respondent noted:

> Every park has their own concessions contract. I think we have gotten a lot smarter as an agency about how to write those contracts to make those concessionaires much more sustainable. I keep coming back to concessions because they are in a lot of our structures and have the closest contact with our visitors.

Environmental sustainability

The theme of environmental sustainability addresses the protection of the natural resources within park boundaries. There are many challenges, and finding solutions to one problem often affects the sustainability of other natural systems. For example, invasive species are a common threat to the parks. Shenandoah has over 350 and has traditionally used pesticides to counter them, in turn affecting water quality and aquatic species.

> Where we can, we are doing individual treatments on every individual ash tree in the park. It's an insecticide injection around the base of the tree. So we will treat several thousand ash trees in the park. It is labour intensive. We are also starting to experiment with bio controls, another predator beetle from Asia that feeds only on the adelgid that causes the hemlock trees to die. So we are trying to shift away from the use of insecticides and more to the use of a natural predator for that species.

Yet, the demise of the hemlock trees has increased river temperatures, affecting the native brook trout, which has justified the use of pesticides that may seep into the riparian system. The Chihuahuan Desert was traditionally over-grazed, which eroded their natural grasses and provided good growing conditions for invasive species. These invasive grasses have affected wildlife, as they entangle native lizards and make them easy targets for the roadrunner predators.[1] They also increase the threat of range fires in the area; they burn faster and hotter than native grasses. Currently, the park is working with neighbouring protected areas in Mexico that face the same threats, yet there does not appear to be an easy solution. Big Bend has established a number of experimental plots to try to reintroduce native grasses.

Soil is the foundation of many ecosystems and has provided conservation challenges within the parks. The National Mall is addressing flood control as compacted soil within their park was a primary cause of urban flooding. All the soil has been replaced and underground cisterns have been installed. Water that is collected within these cisterns is used for irrigation around the park. Big Bend is attempting to address soil conservation in the same plots where they are growing native grass species. They are also trying to restore the riparian areas of the Rio Grande River in an effort to restore balance between the water, plant and animal life, and the soils of the region.

> We can eradicate one problem in the park, but that just leads to another issue. We need to work with other agencies. Our environmental problems occur all over the state, or even in far lying areas. What keeps an ecosystem healthy? Its soils, native plants, maintaining certain population levels of the animal species. It's both inside and outside the park because our ecosystem doesn't just lie within our borders and species don't understand our borders, they travel everywhere.

There is also a conflict between the natural environment and man-made heritage resources within the parks. For example, Shenandoah has a series of stone walls along the Skyline Drive, covered with plant species that erode the stone structures. One respondent claimed that he gets more complaints from tourists about the overgrown walls as the visitors value these cultural resources. However, spraying herbicides at the levels needed to abate this growth is costly to the environment. One solution is to maintain these artefacts by hand, but shrinking budgets have reduced the number of staff on-site. Big Bend is attempting to conserve both historical ranching buildings and Native American artworks, but struggles to justify the necessary finance. As the respondent at the National Mall remarks:

> When we do planning at the National Mall Plan we use an environmental impact statement, and then individual projects might get an environmental assessment. That also includes our historic resources. We work closely with the Smithsonian Institute, but we are competing for funds with all the other national parks on each individual project. The National Mall was never designed for use, and yet it gets used in so many different ways. So we have to take a look at what makes something able to withstand use, not just the design but how do we manage it differently. So it's a combination of goals, things that we are looking at, to make sure we are doing something that will be sustainable, as well as fiscally responsible.

Social sustainability

NPS prides itself on being a resource for all Americans, which has historically placed them at odds with local communities. For example, the National Mall must pay particular attention to the preservation of First Amendment rights and freedom of speech: demonstrations are common on the Mall. It is a place of reflection for war veterans or families who have lost loved ones in conflicts. When asked about the involvement of the local community during the master plan revisions for the park, the respondent remarked:

> I would say probably 25% of our public comments come from DC residents. You know the National Mall has a constituency that is way beyond. So that is what spreads things way beyond Washington DC. The National Mall has the iconic symbols of our nation that we want to make very sure that these places are here for the enjoyment of the public.

Both Shenandoah and Big Bend appear to be more actively involved in the local community. Both parks have representatives on the local tourism boards, although when probed, these contacts are often designed to promote events and activities within the park rather than using the park to promote regional activities or events. At Big Bend, the park administration is actively working with communities and protected areas in neighbouring Mexico, specifically providing training in resource management and general business skills with Mexican ferry boat operators. Because Big Bend was formed with the cooperation of their communities, they have a long-standing relationship. For example, they donated old fire trucks to a community when new ones were received. They also do joint wildfire and emergency response training and they work together on historic preservation. Shenandoah has a much more contentious history. The respondent from Shenandoah explains:

> We had tourism professionals on the east side of the park and tourism professionals on the west side of the park that had never spoken to each other. The park was a barrier to collaboration to the tourism community. So, the 75th anniversary was a catalyst for these people to get to know each other. At the end of the anniversary, they didn't want to disband and so they have now formed a separate non-profit called Celebrate Shenandoah which we are an active part of. They have 3 active committees, one works on branding and promotion, the other works on community relations, and the other works on land use issues surrounding the park.

The parks are also attempting to source locally in their gift shops and food service establishments. The NPS has initiated the Healthy Food Programme which calls for all concessionaires to source "a majority" of their supplies locally. While this is a voluntary programme, the response has been strong. The research team noted that the menu at Big Bend's restaurant highlighted local ranchers where meat was sourced, that the gift shop carried locally made processed food (salsas, jams, hot sauces) and that many of the crafts were made locally. At Shenandoah, only Virginian wine is sold and more ingredients are being sourced locally by the concessionaire. It was also noted that many of the park-related souvenirs are still made in China and a park t-shirt purchased was made in Haiti. In Shenandoah:

> We are certainly a global economy now. But, for example, the gift shop that the concessionaire operates at Big Meadows is about 99% made in the USA, and again they also sell a lot of local crafts and then they also use a fair number of local people for a variety of things that visitors enjoy, including local music. They typically have local bluegrass music in the concession lodges in the evening and almost all of those are local artisans.

Knowledge sharing

With sustainability being a new approach to management within the NPS, operationalisation has been problematic at the park level. This is due in part to a limited amount of information available and the inability to spread best practice from park to park. While the Green Park Plan lays out over 30 sustainability objectives, currently knowledge is disseminated via email and monthly newsletters. All respondents claim that they rarely find time to read these documents. Reduced budgets have prevented within agency travel, so staff experience less opportunity to interact with their counterparts in other parks. Therefore, most parks are making up sustainability as they go along. For example, the National Mall is looking into green roofs, but is searching for information outside of the NPS. The respondent at park headquarters explains:

> We use the term engagement, whether at the staff level, the gateway level or the visitor level. We have different tools at each level, obviously some are more successful than others. At the staff level we have the Green Park Plan and we communicate our performance agency-wide through an annual performance brief which has park and regional information. So we are feeding back information to the park so they know if they are trending upward or downward in energy conservation, for example. However, 90% of the footprint is the visitor, so if I don't get the visitor's participation, I don't meet my goal.

Since much of the sustainability agenda is influenced by the concessionaires, there is still little available data on visitor consumption patterns within the park. For example, the National Mall sells over 1 million hotdogs each year. Where those hot dogs come from is beyond the knowledge capacity of NPS staff and is not yet required to be captured. Furthermore, some of these concessionaires are Fortune 500 companies,[2] such as Aramark (Yosemite National Park), which isolates small local businesses from accessing the distribution channels. However, it should be noted that Xanterra Parks & Resorts, also a Fortune 500 company that operates in Yellowstone, Zion, Bryce Canyon, and Grand Canyon National Parks, was selected by the Environmental Protection Agency as the "2006 Performance Track Corporate Leader". Therefore, there are resources available through these concession contracts, but not always at the individual park level.

Furthermore, respondents claim that NPS has not been very good at showcasing their own successes in sustainability; therefore, tourists and local communities may be unaware of new policies or practices within the parks. The current practice is social media, press releases, youtube videos, and notices within park maps and brochures at the park level. As one respondent notes:

> That's not working. I am a big social network guy, but we don't have the platforms. We just recently got access to youtube, but we don't have facebook pages for sustainability. I don't know what the right social network platform is, and our supporters are aging. We don't even know if they use these social network platforms.

Park officials also know there have been tremendous successes at a number of parks, but how that information applies generally to the park service is under-researched. Many respondents discussed stories they had heard, such as "Zion is doing amazing things with their community", or "Yosemite and Arcadia have really embraced the local food movement". Yet, how these parks achieved success is unknown by the study participants. They are inundated with data about consumption, but often do not know what that information is telling them. As one respondent explains:

> We gather a great deal of information about how much energy we are using, how much fuel oil we are using, how much gasoline we are using, exactly how much waste we are taking to the landfill every year, how much lead we are introducing into the environment at the park (gun) range. Having said that, I am not sure we have been as sophisticated as we should be about using that information for decision making. I know what we consume, but that doesn't tell me how to reduce it.

Finally, the NPS is concerned as up to 25% of its workforce is slated to retire over the next 5 years. There has been a concerted effort to retain much of the knowledge that is possessed by these long-standing employees. There are processes in place to interview and document the institutional knowledge as a means to inform younger personnel on the systems and processes that have worked or not worked throughout a park's history.

Discussion

Classical economics recognises that public goods, goods that provide overall value to society as a whole, suffer from a lack of investment as profits are hard to capture (Morrell, 2009). Political pressure to balance budgets, reduce taxes, and meet the needs of an increasingly diverse society has pushed public agencies towards a neo-liberalised approach to funding and service provision. This move towards a market system is evident in many public goods, such as higher education, prisons, and national defence. NPS is facing similar issues as it strives to offer more services, accommodate an increasing number of tourists, and mitigate threats against the health of their parks, while simultaneously facing budget shortfalls (Eagles, 2014). The Green Park Plan provides a solid foundation within the NPS in their attempt to move towards a more sustainable agenda. This research showed that there have been many successes both at the institutional level and at the park level. However, many challenges still remain. In particular, the changing governance structure and the move towards neoliberal policies, within which the Green Park Plan must be implemented, require additional research-informed changes in relation to the five pillars of sustainability.

Park officials recognise their economic contribution to neighbouring communities, and some parks, such as Big Bend and Shenandoah, are implementing strategies to ensure that these economic impacts stay closer to home. Local sourcing is one avenue evident in the Green Park Plan. While these policies may counter exogenous factors that affect visitation and visitor spending (Poudyal et al., 2013), it may also increase dependency by the local communities who adapt agricultural practices to support national park visitors (Gabe, 2016). Shenandoah has taken locally sourcing to a new level by including local artisans, such as craft makers and musicians, which also may increase dependency.

NPS has always had a strong focus on environmental sustainability. Current challenges include transportation, invasive species and visitor impacts. In particular, NPS has attempted to address visitor carbon emissions through interpretive programmes, using both interactive technology and ranger lectures (Buckley & Foushee, 2012). However, the nature of national park visitation is heavily dependent on personal vehicles. While a few parks, such as Zion, have banned personal vehicles in the park, the primary motivation has been traffic congestion rather than air pollution or behavioural change (Ament et al., 2008). The National Mall has already supported public transportation for visitors, but this is due in part to their urban location and municipal partnerships. Big Bend is planning to implement shuttle services to trailheads, yet Shenandoah feels that the long linear layout of the park is not conducive to shuttle services. Invasive species are causing havoc in both Shenandoah and Big Bend, but is not a concern at the National Mall. Corrective practices bring forth new environmental impacts, such as the extensive use of herbicides and pesticides. The introduction of predator species is a risky option, and one hopes that appropriate research will inform these decisions (Parsons, 2004).

There is promise with the Healthy Food initiative that supports local agriculture and potentially offers an incentive to preserve open space in gateway communities (Gimmi et al., 2011). While none of the respondents commented on development issues outside the park boundaries, researcher observation noted extensive tourism capital in the form of hotels and tourism attractions (Shafer, 2012). Most notable were the number of restaurants and museums surrounding the National Mall and shopping centres around Shenandoah. While the Healthy Food initiative primarily focuses on providing the visitor with nutritious, tasty food, there is an element that promotes sourcing locally as a means to reduce the carbon footprint of imported food (Everett & Slocum, 2013). It also has potential to reduce the amount of farm land lost to development and to support recreational opportunities outside the park boundaries (Frauman & Banks, 2011), which in turn could lead to new economic opportunities for gateway communities.

The NPS still prescribes to a top-down governance style; however, it does appear that more power has been allocated to on-site park administrators than the historic centralised headquarter model. The practice of rule-making is still apparent, especially with the implementation of the Green Park Plan (Cook, 2014). While the implementation of this plan is still optional at the park level (Baron et al., 2000), the respondents showed a clear support towards sustainability. The National Mall plan highlights how a park with the highest usage of water and electricity can move quickly in finding sustainability alternatives. Big Bend, being remote and having substantially fewer visitors, is struggling to implement these same measures. With responsibility for sustainability located at the park level, the resources needed to implement changes are harder to come by, especially when the operationalisation focus is on visitor experiences that may or may not be affected by sustainability measures

(Schwartz & Lin, 2006). It is too early to fully understand the impact the Green Park Plan will have on park operations, but it is the first step in the sustainability dialogue and a foundation to counter climate change and impacts on the natural resources within the parks.

Each park representative expressed frustration with the process of achieving sustainable operations. The biggest constraint was the budget process, where proposals are submitted and it takes three to five years for these projects to be funded (Nyahunzvi, 2016). Furthermore, congressional gridlock is still apparent in the funding of NPS (Cook, 2014). Weaver (2012) argues that institutions must foresee a competitive advantage or a cost savings in order to embrace sustainability. Cost savings appear to be a primary motivator at NPS as the Green Park Plan focuses on reducing expenditures on energy, water, and waste disposal. Retrofitting current infrastructure is a viable capital investments strategy to realise savings in the long-run.

Research is still centralised at NPS Headquarters or through the National Biological Survey. Dissemination of this knowledge has been problematic as traditional modes of distribution do not fit with the increasing time commitments of park operational staff. Limited contact between parks and reduced travel budgets has restricted the availability of best practice sharing and centralised research is not always adaptable to the specific environments of each park (Parsons, 2004). Social scientists were notable absent at the park level. By having trained social science researchers, park-specific policies in relation to visitors' experiences and community engagement could support the rule-making process and encourage better social sustainability and governance (Miller et al., 2014). At present, there is little data that addresses the impact sustainability operations has on the visitor experience, which is one of the primary focuses of NPS policy. All respondents value research, yet need additional resources in order to utilise research to inform their sustainability efforts. There is potential to work with both local university researchers and communities at park level. The Blue Ridge Parkway has managed to do that, but is a lone example (see McGehee et al., 2013). Universities may also have a role in disseminating research findings through traditional extension work, or via webinars.

There is some debate on the changing nature of governance and the responsiveness of public institution in negotiating sustainable tourism (Büscher, 2008). Weaver (2012) recognises that a systematic change in values, mind-sets, and social organisations is required. This research shows that the Green Park Plan has brought sustainability and climate change to the forefront of park-level operations and appears to be instilling an organisational culture that supports a more holistic view of the role of parks in educating visitors. Yet, at headquarters and within the federal budgeting process (including Congress), organisational change is less apparent. Bramwell and Lane (2013) write "moving to radically more sustainable tourism will be very difficult as it is likely to involve reversing well-established and interlocking systems and social practices, countering powerful vested interests and fundamentally resetting policy agendas" (p. 2). Another point is that many managers are retiring over the next 10 years. While the loss of institutional knowledge is a concern, it also provides an opportunity to change the culture within the NPS as younger personnel may bring a different perspective on sustainable operations. One hopes that the Green Park Plan does not become another form of rule-making, with responsibility but limited resources, located at the park level (Cook, 2014).

The process of neoliberalism has increased the reliance on, and importance of, concessionaires within NPS, and on NPS policies. Dinica (2016) asserts that concessionaires are powerful tools in the provision of tourism services "because governmental organizations do not have sufficient expertise, or human and financial resources, to deliver such services themselves" (p. 2). Concessionaires are something that the NPS is increasingly utilising and which is supplementing their shrinking budgets (Nyahunzvi, 2016). For example, in 2015, the NPS received $92 million in concession fees and, in 2016, $97 million (NPS, 2016a, 2016b). However, it can be argued that the use of concessionaires is further distancing park administration and conservation from the visitor experience. Respondents in this study acknowledge that visitor impacts are the primary constraint to more sustainable operations. Yet, none of the respondents knew the details related to sustainability policies on the visitor experience or their effect on visitor impacts (Park et al., 2010). Interpretive programmes are furthering the sustainability dialogue, but more research on the efficiency and effectiveness of these messages is needed (see Babakhani, Ritchie, & Dolnicar, in press). There may be an opportunity to justify an increased fee structure or to require more responsible behaviour if visitors recognise how these policies impact the overall park experience (Schwartz & Lin, 2006). Further research into the knowledge-sharing opportunities between concessionaires and park administrators is also needed.

Kline and Slocum (2015) write "Key players are becoming more aware of their responsibility to mitigate human interactions… rather than policing human behaviour at odds with conservation science" (p. 13). The Green Park Plan is a commendable first step in recognising and mitigating visitor impacts. However, NPS has had difficulty showcasing their sustainability successes and promoting a more sustainable agenda to the travelling public (Slocum & Curtis, 2016). As neoliberal policy becomes the norm rather than the exception, further research is needed to explore and explain how the growing dependence on concessionaires is aiding sustainable tourism development. Moreover, a change in the institutional culture and the distribution of resources is needed to assist park administration in achieving these well-intended goals.

Limitations and future needs

This paper has presented a comparative case study of three distinctly different parks using the five pillars of sustainability model. It shows a promising move towards sustainable operational policies that can potentially have a large positive impact on the environment and host communities. However, NPS still maintains a culture that encourages top-down policies using limited research-informed implementation strategies, which prevents a holistic view of sustainability. There remain internal and external tensions between sustainability and neo-liberal policies in National Parks. Governance needs behavioural change perhaps as much as visitors.

This paper has a number of limitations and the author encourages further work to help inform protected area managers in the operationalisation of sustainability. Governance, staff support and training, and funding practices all need carefully researched attention if sustainability and neo-liberal practices are to live well together. Future research must take into account that each NPS park is unique and each faces different as well as common challenges. A larger study to incorporate more parks would be a beneficial starting point. Also, the visitor's voice has so far been notably silent in this examination. Understanding how sustainability affects the visitor experience is vital to reducing tourist impacts and the footprint of each park. Finally, studies that bring the local community to the table would be beneficial. Sustainability requires all stakeholder involvement, which is lacking in this research sector to date.

Notes

1. The roadrunner, also known as a chaparral bird, is a fast-running ground cuckoo that can fly, but normally runs.
2. Fortune 500 companies are the 500 largest companies in the USA by revenue.

Disclosure statement

No potential conflict of interest was reported by the author.

References

Ament, R., Clevenger, A.P., Yu, O., & Hardy, A. (2008). An assessment of road impacts on wildlife populations in US National Parks. *Environmental Management, 42*(3), 480–496.

Babakhani, N., Ritchie, B., & Dolnicar, S. (in press). Improving carbon offsetting appeals in online airplane ticket purchasing: Testing new messages, and using new test methods. *Journal of Sustainable Tourism*. doi:10.1080/09669582.2016.1257013

Baron, J.S., Theobald, D.M., & Fagre, D.B. (2000). Management of land use conflicts in the United States Rocky Mountains. *Mountain Research and Development, 20*(1), 24–27.

Benson, C., Watson, P., Taylor, G., Cook P., & Hollenhors, S. (2013). Who visits a national park and what do they get out of it?: A joint visitor cluster analysis and travel cost model for Yellowstone National Park. *Environmental Management, 52*, 917–928.

Bramwell B., & Lane, B. (2013). Getting from here to there: Systems change, behavioural change and sustainable tourism. *Journal of Sustainable Tourism, 21*(1), 1–4.

Buckley, L.B., & Foushee, M.S. (2012). Footprints of climate change in US national park visitation. *International Journal of Biometeorology, 56*(6), 1173–1177.

Büscher, B. (2008). Conservation, neoliberalism and social science: A critical reflection on the SCB 2007 Annual Meeting, South Africa. *Conservation Biology, 22*(2), 229–231.

Clarke, J.N., & McCool, D. (1996). *Staking out the terrain power differentials among natural resource management agencies.* Albany: State University of New York Press.

Cook, J.J. (2014). Are we there yet? A roadmap to understanding National Park Service rulemaking. *Society & Natural Resources, 27*(12), 1257–1270.

Dinica, V. (2016). Tourism concessions in National Parks: Neo-liberal governance experiments for a conservation economy in New Zealand. *Journal of Sustainable Tourism,* 1–19. doi:10.1080/09669582.2015.1115512

Eagles, P.F.J. (2002). Trends in Park tourism: Economics, finance and management. *Journal of Sustainable Tourism, 10*(2), 132–153.

Eagles, P.F.J. (2014). Research priorities in park tourism. *Journal of Sustainable Tourism, 22*(4), 528–549.

Everett, S., & Slocum, S.L. (2013). Food and tourism: An effective partnership? A UK-based review. *Journal of Sustainable Tourism, 21*(6), 789–809.

Frauman, E., & Banks, S. (2011). Gateway community resident perceptions of tourism development: Incorporating Importance-Performance Analysis into a Limits of Acceptable Change framework. *Tourism Management, 32*, 128–140.

Gabe, T. (2016). Effects of the October 2013 U.S. Federal government shutdown on National Park gateway communities: The case of Acadia National Park and Bar Harbor, Maine. *Applied Economics Letters, 23*(5), 313–317.

Gimmi, U., Schmidt, S.L., Hawbaker, T.J., Alcántara, C., Gafvert, U., & Radeloff, V.C. (2011). Increasing development in the surroundings of U.S. National Park Service holdings jeopardizes park effectiveness. *Journal of Environmental Management, 92*, 229–239.

Hamin, E.M. (2001). The US National Park Service's partnership parks: Collaborative responses to middle landscapes. *Land Use Policy, 18*, 123–135.

Henrickson, K.E., & Johnson, E.H. (2013). The demand for spatially complementary national parks. *Land Economics, 89*(2), 330–345.

HM Government (2005). *Securing the future: Delivering UK sustainable development strategy* (HM Government Publication No. Cmd 6467). Norwich: Crown. Retrieved 5 June, 2016, from https://www.gov.uk/government/uploads/system/uploads/attachment_data/file/69412/pb10589-securing-the-future-050307.pdf

Jones, C. (2012). Ecophilanthropy, neoliberal conservation, and the transformation of Chilean Patagonia's Chacabuco valley. *Oceania, 82*(3), 250–263.

Kline, C., & Slocum, S.L. (2015). Neoliberalism in ecotourism? The new development paradigm of multinational projects in Africa. *Journal of Ecotourism, 14*(2–3), 99–112.

McGehee, N., Boley, B.B., Hallo, J., McGee, J., Norman, W., Oh, C.-O., & Goetcheus, C. (2013). Doing sustainability: An application of an inter-disciplinary and mixed-method approach to a regional sustainable tourism project. *Journal of Sustainable Tourism, 21*(3), 355–375.

Miller, M.L., Carter, R.W., Walsh, S.J., & Peake, S. (2014). A conceptual framework for studying global change, tourism, and the sustainability of iconic national parks. *The George Wright Forum, 31*(3), 256–269.

Mitchell, B. (2011). Projecting America's best ideals: International engagement and the National Park Service. *The George Wright Forum, 28*(1), 7–16.

Morrell, K. (2009). Governance and the public good. *Public Administration, 87*(3), 538–556.

National Park Service (1916). *The National Park Service Organic Act.* Washington, DC: US Government. Retrieved 5 June, 2016, from https://www.nps.gov/grba/learn/management/organic-act-of-1916.htm

National Park Service (2010). *Climate change response strategy.* Retrieved 1 November, 2016, from https://www.nature.nps.gov/climatechange/docs/NPS_CCRS.pdf

National Park Service (2012, April). Green parks plan: Advancing our mission through sustainable operations. Retrieved 5 June, 2016, from https://www.nps.gov/greenparksplan/downloads/NPS_2012_Green_Parks_Plan.pdf

National Park Service (2016a). *Annual recreation visitation report by years: 2005 to 2015.* Retrieved 30 May, 2016, from https://irma.nps.gov/Stats/SSRSReports/National%20Reports/Annual%20Recreation%20Visitation%20By%20Park%20%281979%20-%20Last%20Calendar%20Year%29

National Park Service (2016b). *Visitor spending effects.* Retrieved 30 May, 2016, from https://www.nps.gov/subjects/social science/vse.htm

Nyahunzvi, D.K. (2016). The changing nature of national parks under neoliberalism. In J. Mosedale (Ed.), *Neoliberalism and the political economy of tourism* (pp. 99–116). Oxon: Routledge.

Park, J., Ellis, G.D., Kim, S.S., & Prideaux, B. (2010). An investigation of perceptions of social equity and price acceptability judgments for campers in the U.S. national forest. *Tourism Management, 31,* 202–212.

Parsons, D. (2004). Supporting basic ecological research in U.S. national parks: Challenges and opportunities. *Ecological Applications, 14*(1), 5–13.

Poudyal, N.C., Paudel, B., & Tarrant M.A. (2013). A time series analysis of the impact of recession on national park visitation in the United States. *Tourism Management, 35,* 181–189.

Schwartz, Z., & Lin, L. (2006). The impact of fees on visitation of national parks. *Tourism Management, 27,* 1386–1396.

Sellers, R.W. (1997). *Preserving nature in the national parks: A history.* New Haven, CT: Yale University Press.

Shafer, C.L. (2012). Chronology of awareness about us national park external threats. *Environmental Management, 50,* 1098–1110.

Slocum, S.L., Backman, K., & Baldwin, E. (2012). Independent instrumental case studies: Allowing for the autonomy of cultural, social, and business networks in Tanzania. In K. Hyde, C. Ryan, and A. Woodside (Eds.), *Field guide for case study research in tourism* (pp. 521–541). Bingley: Emerald.

Slocum, S.L., & Curtis, K.R. (2016). Assessing sustainable food behaviours of national park visitors: Domestic/on vocation linkages, and their implications for park policies. *Journal of Sustainable Tourism, 24*(1), 153–167.

Stake, R. (2003). Case studies. In N. Denzin and Y. Lincoln (Eds.), *Strategies of qualitative inquiry* (pp. 134–164). Thousand Oaks, CA: Sage.

Stankey, G.H., Cole, D.N., Lucas, R.C., Petersen, M.E., & Frissell, S.S. (1985). *The limits of acceptable change (LAC) system for wilderness planning* (General technical report INT-176). Ogden, UT: USDA Forest Service, Intermountain Forest and Range Experiment Station.

Taylor, P.A., Gunther, K.A., & Grandjean, B.D. (2014). Viewing an iconic animal in an iconic National Park: Bears and people in yellowstone. *The George Wright Forum, 31*(3), 300–310.

Thomas, C.C., & Koontz, L. (2016). 2015 National Park visitor spending effects: Economic contributions to local communities, states, and the nation (Natural Resource report NPS/NRSS/EQD/NRR—2016/1200). Retrieved 1 November, 2016, from https://www.nps.gov/subjects/socialscience/vse.htm

University of California Davis (n.d.). Mining water. Retrieved 15 October, 2016, from http://mygeologypage.ucdavis.edu/cowen/~gel115/115CH18miningwater.html

Wang, C.Y., & Miko, P.S. (1997). Environmental impacts of tourism on US national parks. *Journal of Travel Research, 35*(4), 31–36.

Weaver, D. (1998). *Ecotourism in the less developed world.* New York, NY: CABI.

Weaver, D. (2012). Towards sustainable mass tourism: Paradigm shift or paradigm nudge? In T.V. Singh (Ed.), *Critical debates in tourism* (pp. 28–34). Bristol: Channel View.

Weber, J., & Sultana, S. (2013). The civil rights movement and the future of the national park system in a racially diverse America. *Tourism Geographies, 15*(3), 444–469.

Wells, M.P. (1997). *Economic perspectives on nature tourism, conservation and development* (Environment Department Paper No. 55, Pollution and Environmental Economics Division). Washington, DC: World Bank.

Zube, E.H. (1992). Partnerships: New approaches to an old idea. *Cultural Resources Management, 15*(8), 9–10.

Visitor spending effects: assessing and showcasing America's investment in national parks

Lynne Koontz, Catherine Cullinane Thomas, Pamela Ziesler, Jeffrey Olson and Bret Meldrum

ABSTRACT
This paper provides an overview of the evolution, future, and global applicability of the U.S. National Park Service's (NPS) visitor spending effects framework and discusses the methods used to effectively communicate the economic return on investment in America's national parks. The 417 parks represent many of America's most iconic destinations: in 2016, they received a record 331 million visits. Competing federal budgetary demands necessitate that, in addition to meeting their mission to preserve unimpaired natural and cultural resources for the enjoyment of the people, parks also assess and showcase their contributions to the economic vitality of their regions and the nation. Key approaches explained include the original Money Generation Model (MGM) from 1990, MGM2 used from 2001, and the visitor spending effects model which replaced MGM2 in 2012. Detailed discussion explains the NPS's visitor use statistics system, the formal program for collecting, compiling, and reporting visitor use data. The NPS is now establishing a formal socioeconomic monitoring (SEM) program to provide a standard visitor survey instrument and a long-term, systematic sampling design for in-park visitor surveys. The pilot SEM survey is discussed, along with the need for international standardization of research methods.

Introduction

While the USA was the first country in the world to declare a National Park, at Yellowstone in 1872, the U.S. National Park Service (NPS), which is entrusted with the care of the Parks, was not set up until 1916. That service, now the largest of its kind in the world and a US Federal Agency, manages most aspects of the 340,000 square kilometers system, with its 417 separate areas, the great majority of which are in public ownership (see https://www.nps.gov). Those lands set a new visitation record in 2016 with 331 million recreation visits,[1] a 7.7% increase (up by 23.7 million visits) compared to the previous record of 307.2 million recreation visits in 2015 (Ziesler, 2017).

On vacations or on day trips, NPS visitors spend a considerable amount of time and money in the gateway communities surrounding those 417 land units. Many communities and regions near the national parks are highly dependent on economic activity generated by NPS tourism. The economic significance of national parks was underscored with the closure of all NPS lands during the 16-day October 2013 Federal government shutdown. In response to the negative economic impacts that the park closures were having on many communities and local businesses, the Federal government granted agreements for six state governments to fully fund NPS personnel to re-open national parks in their states. Each dollar funded by the six states was estimated to have generated an additional $10 in visitor spending in gateway regions surrounding these parks during the remainder of the shutdown (Koontz & Meldrum, 2014).

The mission of the NPS is to preserve unimpaired the natural and cultural resources and values of the national park system for the enjoyment, education, and inspiration of this and future generations. The NPS has many indicators to monitor natural resource conservation and public satisfaction, and estimates of the effects of NPS tourism on local and national economies serve as a key indicator of one of the many ways that parks benefit communities and the American public. To document this important benefit, the NPS has been measuring and reporting on the economic contribution of visitor spending for more than 25 years. The NPS estimates visitor spending and economic contributions for most park units, but there is a great need for better data across park types and geographic regions to improve the accuracy of visitor spending estimates. To improve visitor data, the NPS is establishing feasible approaches for implementing a long-term, systematic socioeconomic monitoring (SEM) program to better understand park visitors. This paper provides an overview of the new NPS visitor spending effects (VSE) framework and communication strategy for conveying the importance of national parks for jobs and local and regional economic vitality, and for helping to obtain political support for capital spending on, and the administration of, America's investment in national parks. These issues are not peculiar to the situation of the USA – see, for example, Mayer and Job (2014) for a European perspective. Therefore, the applicability of NPS methods and strategies for developing global visitor use monitoring standards is also discussed.

Tourism, and tourist expenditure, is becoming an increasingly important issue in the increasingly neo-liberal world in which most parks now function. As Font and McCabe (2017) point out, conservation and sustainable tourism management require detailed and accurate knowledge of markets and the trends in those markets if both private and public sector entities are to be managed successfully. Slocum (2017) examines the general reactions of the NPS to the twin challenges of sustainable development and a neo-liberal environment now surrounding US parks. This paper explains in detail how the NPS acquires market knowledge, and how it uses that knowledge, to adjust its management activities. It also points out the problems and the limitations in getting that knowledge, and the issues involved in trying to create a unified approach across a complex range of park types, sizes, and locations. At times, NPS methodology is explained in this paper in great detail, necessary because of the complexities of evaluating the economic impacts of the many types of park visitors.

Measuring the economic contributions of National Park Service visitation

Lands managed by the NPS are part of a diverse system of natural, cultural, and historic landscapes, representing 28 different designations including national parks, monuments, battlefields, historic sites, lakeshores, seashores, recreation areas, and scenic rivers and trails, in urban, rural, and sometimes remote settings. Visitors arrive and travel through parks in a variety of ways including cars, buses, cruise ships, bicycles, canoes, and on foot. The uniqueness of each NPS site requires a method of counting visitors and assessing their spending patterns that is customized to the site yet provides consistent metrics across the park system.

The NPS first started measuring and reporting the economic contributions of visitor spending with the development of the Money Generation Model (MGM) in 1990 (National Park Service, 1995). The initial MGM model was used on a limited basis to help individual parks estimate the local economic effects of visitor spending. An updated version of the Money Generation Model (MGM2) was used to produce the first system-wide estimates in 2001 (Stynes & Sun, 2003). Starting in 2005, NPS system-wide estimates have been published on an annual basis. In 2012, the MGM2 model was replaced by the VSE model. The VSE model builds on the framework developed by Stynes, Propst, Chang, and Sun (2000) for the historic MGM2, while increasing the ability of the NPS to update and improve input data over time (Cullinane Thomas, Huber, & Koontz, 2014). The Stynes methods have also been replicated and built upon by other U.S. Federal agencies including the U.S. Forest Service (White, Goodding, & Stynes, 2013), the U.S. Army Corps of Engineers (Chang, Propst, Stynes, & Scott Jackson, 2003), and the U.S. Fish and Wildlife Service (Carver & Caudill, 2013), as well as other countries including Germany (Mayer, Müller, Woltering, Arnegger, & Job, 2010), Finland (Huhtala, Kajala, & Vatanen, 2010), and Brazil (Souza, 2016).

Visitor spending effects model

VSE are defined as the jobs and business activity that result from the direct and "ripple effects" of NPS visitors' spending money within gateway economies surrounding parks. The ripple, or secondary effects, of NPS visitor spending include the indirect effects of local businesses purchasing inputs from local suppliers, and the induced effects of local employees purchasing goods and services within the local economy. Economic input–output models capture these complex interactions between producers and consumers in an economy and describe the secondary effects of visitor spending through regional economic multipliers. The sum of direct and secondary effects gives the total economic effect of visitor spending in a local economy. Figure 1 provides an overview of the framework for estimating VSE and describes the primary inputs and products of the VSE model. The VSE model utilizes three key data inputs: (1) the number of visitors who visit each park; (2) profiles of visitor spending patterns in local gateway regions; and (3) regional economic multipliers that describe the ripple effects of visitor spending throughout the economy. These inputs are used to produce two products: (1) estimates of total visitor spending; and (2) the resulting economic contributions to local gateway regions, states, and the nation. This section describes the methods and data sources of the VSE model inputs and the calculations required for estimating the VSE products.

VSE input 1: visitor use data

NPS visitor use records date back more than a century to 1904, when the national park idea was still new. In that first year, six national parks reported a total of 140,954 visitors (Secretary of the Interior, 1904). Early data collection efforts were informal and only intended to meet local park needs. As the number of NPS sites and visitation increased, so did the demand for consistent and reliable

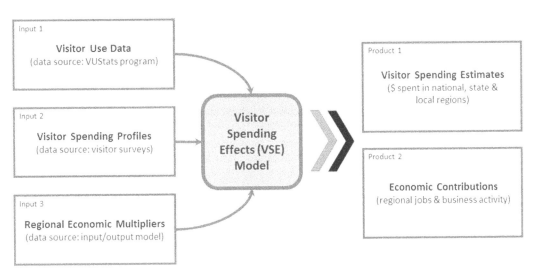

Figure 1. The Visitor Spending Effects (VSE) framework. The VSE model utilizes three key data inputs: (1) the number of visitors who visit each park; (2) profiles of visitor spending patterns in local gateway regions; and (3) regional economic multipliers that describe the ripple effects of visitor spending throughout the economy. These inputs are used to produce two products: (1) estimates of total visitor spending; and (2) the resulting economic contributions to local gateway regions, states, and the nation.

national-level visitor use data. In 2016, NPS visitation estimates grew to 378 parks reporting 330,971,689 recreation visits. Other parks are unable to report for a variety of reasons such as park administration by another agency or the park is too new to be open for regular, scheduled visitation. The responsibility for compiling and summarizing visitation records lies with the NPS Social Science Program. The Program relies on the conscientious efforts of NPS field staff to count, record, and report visitor use. Data and detailed, unit-specific instructions for the units that collect data are available on the Visitor Use Statistics (VUSTATS) website (https://irma.nps.gov/Stats).

NPS VUSTATS is the formal program for collecting, compiling, and reporting visitor use data. The objective of the VUSTATS program is to promote and design statistically valid, reliable, and consistent methods of collecting and reporting visitor use data for each park administered by the NPS. The VUSTATS program collects monthly recreation visit counts/visitor hours, non-recreation visit counts/visitor hours, and overnight stay counts for 378 of the 417 NPS units. VUSTATS staffs collaborate with individual parks to develop counting instructions that contain the procedures for measuring, compiling, and recording required visitor use data. Individual park counting procedures and calculations follow established guidelines and are continuously audited to ensure consistency across parks, to maintain the integrity of the data, and to ensure that the data and calculations are statistically sound.

While direct measurement of visitor use is possible at some parks, it is often cost prohibitive to count each individual entering a park each day. Some parks are able to use direct counts of visits at visitor centers, historic homes, or sites with ticket counts but most parks use estimates to form reasonable approximations of visitor use. Consistency among parks is maintained by implementing count systems that conform to a standard set of use definitions. The basic unit of measurement is a "visit": the entry of a person onto lands or waters administered by the NPS. The applicable rule is that one entrance per individual per day is countable (see Endnote 1).

The NPS utilizes a wide-variety of counting technologies in order to address the various ways visitors enter a park (cars, buses, cruise ships, bicycles, pedestrians, canoes, etc.). Most parks, particularly those with entrances on travel corridors (roads or trails), use automated counters, such as vehicular traffic counters, door counters, and trail counters. Vehicle counts are combined with person-per-vehicle multipliers to estimate the number of visitors. Person-per-vehicle multipliers are developed for each park using a study design in which observers record the number of vehicles and the number of passengers in each vehicle at prescribed times and days spread out over the year. Enough observations are collected to test for locational and seasonal differences in person-per-vehicle multipliers, which can vary by park (or location within a park), by time of year, and by type of vehicle (car, bus, plane, boat, canoe, etc.). Where necessary, mathematical relationships are used to estimate the number of visitors in a remote area based on the count of visitors in a more accessible area of the park. Duplicate reporting, also known as double-counting, is not permitted and the NPS works diligently to identify and correct situations that may lead to duplicate reporting. Situations that can lead to duplicate reporting include: commuter traffic going to and from work through the park on the same day; visitor traffic going to and from outside vendors during one day; visits to different areas of the same park on the same day that involve crossing non-park lands; and visitors staying outside the park and making morning and afternoon visits on the same day.

The NPS analyzes and publishes official visitation statistics on a monthly basis and produces annual summaries at the end of each calendar year. The various data and analyses capture changes in visitation over the decades, examine trends in park use, forecast future visitation, and provide inputs to other analyses outside and within the NPS. The results are easily accessible on the NPS VUSTATS website (irma.nps.gov/stats) and are used to support a variety of internal and external analysis efforts related to park visitation.

VSE input 2: visitor spending profiles

Visitor surveys are used to collect the essential visitor spending and trip characteristic data necessary for developing spending profiles to represent distinct visitor spending patterns for each park. Visitor spending profiles for the VSE model are derived from survey data collected through the NPS Visitor

Services Project (VSP). These surveys measure visitor characteristics and visitor evaluations of importance and quality for services and facilities.

VSP surveys follow established best practice guidelines for collecting visitor spending data, which include sampling travel parties, eliciting total party expenditures, and using mail-back surveys to gather data on each party's total spending within the local area for the full duration of their trip (Stynes & White, 2006). A party is defined as a group that is traveling together and sharing expenses (e.g. a family). The VSP surveys utilize a personally delivered self-administered mail-back technique in which sampled parties are given a stamped, addressed questionnaire to complete and mail back at the conclusion of their trip (Dillman, Dolsen, & Machlis, 1995; Rookey, Le, Littlejohn, & Dillman, 2012). Respondents are asked to report their party's expenditures during their time in the local area surrounding the park. Expenditure categories include spending on hotels, motels, and bed and breakfasts, camping fees, restaurants, groceries, fuel, local transportation, admission and guide fees, and souvenirs.

For development of visitor spending profiles, the best practice guidelines suggest estimating separate spending profiles for subgroups of visitors with distinct spending patterns (Stynes & White, 2006). White and Stynes (2008) found that segmenting by trip-type explains more of the variation in recreation visitor spending patterns than other strategies such as segmenting by recreation activity. Segmenting by trip-type is used in other countries too (Job, 2008; Mayer et al., 2010). The VSE model segments visitors by lodging-based trip-types including: local day trips, non-local day trips, overnight stays in lodging within the park, overnight stays in lodging outside of the park, overnight camping within the park, and overnight camping outside of the park. A seventh segment, "other", describes the spending patterns of visitors staying overnight in private homes, with friends or relatives, or in other unpaid lodging.

Visitor spending profiles are estimated for each visitor segment as spending per party per day for visitors on day trips, and spending per party per night for visitors on overnight trips. Total party days/ nights are defined as the sum of the number of days (for day trips) and the number of nights (for overnight trips) that parties spend visiting a park. To estimate total party days/nights, park visit data from the VUSTATS program are combined with trip characteristic information derived from the VSP surveys. The necessary trip characteristic data required to make these conversions are the average party size, entry rates (i.e. the average number of days parties enter the park over the course of a trip), and the length of stay (i.e. the average number of days that parties spend visiting the park) (see Cullinane Thomas & Koontz, 2017 for full details).

VSE input 3: regional economic multipliers

Economies are complex webs of interacting consumers and producers in which goods produced by one sector of an economy become inputs to another, and the goods produced by that sector can become inputs to yet other sectors. Thus, a change in the final demand for a good or service can generate a ripple effect throughout an economy as businesses purchase inputs from one another. Economic input–output models capture these complex interactions between producers and consumers in an economy and describe the secondary effects of visitor spending through regional economic multipliers.

The VSE model utilizes multipliers specifically developed for the geographic area of each park. Geographic information systems data were used to determine the local gateway region for each park unit by spatially identifying all counties partially or completely contained within a 60-mile radius around each park boundary. As an exception, the economic regions for parks in Alaska and Hawaii are defined as the State of Alaska and the State of Hawaii, respectively. Both of those states are distant from the remainder of the USA. Multipliers are derived from the IMPLAN software and data system (IMPLAN Group LLC, n.d.), a commercially available and widely used input–output modeling system available in the United States.[2] Four types of regional economic contributions are estimated:

- *Jobs* measure the annualized full and part-time jobs supported by NPS visitor spending;
- *Labor income* measures employee wages, salaries, and payroll benefits, as well as the incomes of sole proprietors that are supported by NPS visitor spending;

- *Value added* measures the contribution of NPS visitor spending to the gross domestic product of a regional economy. Value added is equal to the difference between the amount an industry sells a product for and the production cost of the product; and
- *Economic output* measures the total estimated value of the production of goods and services supported by NPS visitor spending.

The VSE analysis reports economic contributions at the park-level, state-level, NPS region-level, and national-level. Park-level contributions use county-level IMPLAN models comprised of all counties contained within the local gateway regions; state-level contributions use state-level IMPLAN models; regional-level contributions use regional IMPLAN models comprised of all states contained with the NPS region; and the national-level contributions use a national IMPLAN model. The size of the region included in an IMPLAN model influences the magnitude of the economic multiplier effects. As the area considered as the economic region expands, the amount of secondary spending that stays within that region increases, which results in larger economic multipliers. Thus, contributions at the national level are larger than those at the regional, state, and local levels.

VSE product 1: total visitor spending

VSE visitor spending estimation closely follows the methods developed by Stynes et al. (2000) for the MGM2 model. Table 1 details the data and the calculations required to estimate park-level visitor spending. Total park visit data provided by the VUSTATS program is converted to party days/nights using park-specific trip characteristic data from visitor surveys. Total park spending is estimated by multiplying the estimated party days/nights by the average spending profiles per party per day/night and then summing across visitor segments.

VSE product 2: economic contributions of visitor spending

Economic contributions describe the total economic activity associated with NPS visitor spending within a regional economy. The economic contributions of visitor spending to local economies are estimated by multiplying visitor spending estimates by regional economic multipliers. Direct spending for each of the defined spending categories (hotels, motels, and bed and breakfasts, camping

Table 1. Calculations required to estimate park-level visitor spending. Total park visit data provided by the VUSTATS program is converted to party days/nights using park-specific trip characteristic data from visitor surveys. Total park spending is estimated by multiplying the estimated party days/nights by the average spending profiles per party per day/night and then summing across visitor segments.

Total recreation visits are split into visits by visitor segment.	Let s represent visitor segments, and let p_s represent the share of park visits made by visitor segment s: Visit $s_s =$ Total Recreation Visits $\times p_s$, where $\sum_s p_s = 100\%$
For each visitor segment, visits are converted to visitor trips by dividing by average entry rates.	$\text{Visitor Trips}_s = \dfrac{\text{Visits}_s}{\text{Entry Rate}_s}$
Visitor trips are converted to party trips by dividing by average party sizes.	$\text{Party Trips}_s = \dfrac{\text{Visitor Trips}_s}{\text{Party Size}_s}$
Party trips are converted into party days and party nights based on the length of stay in the nearby area.	$\text{Party Days/Nights Nearby Area}_s = \text{Party Trips}_s \times \text{Days/Nights Nearby Area}_s$
Total park visitor spending is equal to the sum over all visitor segments of party days/nights multiplied by average spending per party per day/night.	$\text{Total Park Spending} = \sum_s \text{Party Days/Nights}_s \times \text{Avg. spending per party per day/night}_s$

fees, restaurants, groceries, fuel, local transportation, admission and guide fees, and souvenirs) are bridged to corresponding sectors of the economy, and IMPLAN multipliers are used to estimate the ripple effects of direct visitor expenditures. The VSE model inflates all dollar values to the analysis year, aggregates park-level visitor spending estimates to state and national totals, and multiplies visitor spending estimates by park-, state-, and national-level economic multipliers (Cullinane Thomas & Koontz, 2017).

Data improvements: socioeconomic monitoring program

Accurate estimation of VSE requires quality survey data that is representative of the variety of visitor uses and demographics from across the park system. Only 57 of the 417 NPS parks have conducted visitor surveys that include visitor spending questions. These parks were not randomly selected which has resulted in an over-representation of parks with high visitation, greater intensity of environmental compliance activities, and/or complex visitor services issues. Increased sampling rigor across park types and geographic regions could increase the accuracy of park-specific trip characteristic and spending data and thus improve the accuracy of future VSE analyses.

To address this issue, the NPS is in the process of establishing feasible approaches for implementing a formal SEM program that would provide a standard visitor survey instrument and a long-term, systematic sampling design for in-park visitor surveys. The NPS is currently piloting SEM visitor surveys in 16 parks. Results from the SEM pilot surveys will be finalized in 2018.

Recommended approaches for system-wide visitor monitoring will be communicated to the NPS leadership for further consideration after the completion of the pilot phase. This section describes anticipated data improvements that will be realized through the piloting and full implementation of the SEM program.

Improving park-level spending profiles

Currently, only 57 out of the 378 parks included in the VSE have conducted VSP surveys that included the spending questions required for developing spending profiles. The NPS has extrapolated data from these 57 surveyed parks to the remaining unsurveyed parks through the development of a set of generic profiles based on lodging availability and the amount of day use within parks. The generic profiles constructed from the available VSP data should be reasonably accurate for most park units; however, given the wide variety in the type of NPS parks across the nation, some parks are not well represented by the generic visitor spending and trip characteristic profiles developed from the limited VSP data. Full implementation of the SEM program would result in a greater number of parks having primary survey data updated regularly, and the SEM sampling design would ensure that the sampled parks were statistically representative of the system.

Updating and refining spending questions

In addition to improving sampling rigor, the new SEM survey instrument provides the opportunity to modify and add additional visitor spending-related questions to the original VSP survey questions. These new questions will help address several limitations with the currently available VSP data and will enable the exploration of improved visitor spending estimation methodologies. Specific updates and refinements include:

- Improved methods for estimating and attributing visitor spending based on primary trip purpose. The VSE model currently only counts expenditures for the number of days that visitors visit a park, but, due to data limitations, the model does not explicitly adjust the spending estimates for non-primary trip purposes. Many visitors come to local gateway regions primarily to visit NPS lands; however, some visitors are primarily in the area for business, visiting friends and relatives, or for some other reason, and their visit to an NPS unit is not the primary purpose for their trip. For these visitors, it may not be appropriate to attribute all of their trip expenditures to the NPS. Similarly, some visitors visit parks as part of a multi-destination trip or a longer vacation, and it can be difficult to determine how much of their total trip expenditures should be attributed to their trip to the national park. The pilot SEM survey includes a new set of questions about primary trip-purpose that will enable novel methods for attributing visitor spending for non-primary and multi-destination trips.

- Improved methods for estimating and attributing visitor spending for visitors who visit parks as part of tour packages or cruise trips. Many national parks, especially the large western-states parks and Alaska parks, are included as part of multi-destination tour packages or cruise trips, but the VSP surveys do not explicitly account for trip expenditures that are part of tour packages. The pilot SEM survey includes a new set of questions about expenditures on tour packages which will enable the inclusion of a new cruise or tour-package visitor segment for applicable parks.

- Refined expenditure categories that reflect recent trends in the tourism industry. Expenditure categories in the SEM survey remain consistent with the traditional spending categories used in the VSP surveys, but several expenditure categories have been divided in order to get better resolution on expenditures of interest. The original local transportation category is now divided into a rental car category and a taxis, shuttles, and public transportation category. Similarly, the original hotels, motels, and bed and breakfasts category is now divided into a hotels, motels, and resorts category and a specialty lodging category (e.g. B&Bs, hostels, cabins, vacation rentals). These new category divisions will enable improved bridging to IMPLAN economic sectors, and will provide data on shifts in transportation and lodging trends.

Communicating the economic contribution of national park visitation

Visitor use and economic contribution monitoring efforts must be visible and accessible to decision-makers and to the American public. The NPS has developed a highly effective strategy for communicating the economic importance of NPS tourism to the media and the US Congress. Every year, the new VSE estimates covering visitation from the previous calendar year are released in late April to coincide with the celebration of U.S. National Park week. The release events are typically initiated with a national media briefing by the US Secretary of the Interior and the Director of the NPS to discuss the findings and highlight the national importance of the report. The media briefing for the 2014 VSE results was elevated to the President of the United States in a speech highlighting the significance of climate change on national parks on 22 April 2015 (President Barack Obama, 2015). The national news release is shared with every Congressional office detailing the national and park level VSE estimates for parks within their home state or district.

Here is an excerpt from the April 2016 news release which highlighted the significance of America's return on investment on national parks:

> Spending by a record number of national park visitors in 2015 provided a $32 billion benefit to the nation's economy and supported 295,000 jobs, according to a report released today by National Park Service Director Jonathan B. Jarvis. 'The big picture of national parks and their importance to the economy is clear,' Jarvis said of the $16.9 billion visitors spent in communities within 60 miles of a national park. Each tax dollar invested in the National Park Service effectively returns $10 to the U.S. economy because of visitor spending that works through local, state and the U.S. economy.

> This is especially significant news to the gateway communities where national parks can be the community's primary economic engine,' Jarvis said. 'While we care for the parks and interpret the stories of these iconic natural, cultural and historic landscapes, our neighbors in nearby communities provide our visitors with important services like food and lodging and that means hundreds of thousands of local jobs. (National Park Service, 2016)

In the days following the national media release, many of the 378 individual parks included in the VSE report send out press releases on the park level VSE results for their local economy (see example https://www.nps.gov/grca/learn/news/econ-benefit-2016.htm). In the first month following the latest

VSE release in April 2017, there were more than 600 media stories about the economic benefits of NPS tourism in newspapers across the country.

To increase the usefulness of the VSE report series with the media, US Congress, and the public, the NPS collaborated with the U.S. Geological Survey to develop a web-based interactive data visualization tool. This interactive tool is available at http://go.nps.gov/vse. As shown in Figure 2, users can visually view year-by-year trend data and explore current year visitor spending, jobs, labor income, value added, and economic output effects by sector for national, state, and local economies.

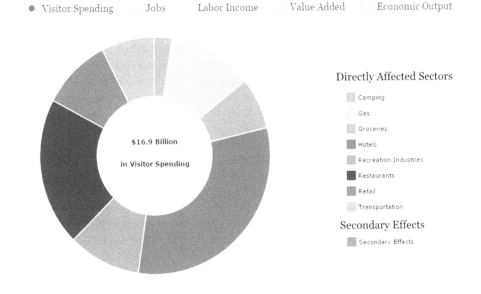

Figure 2. Example page from the visitor spending effects data visualization web-tool. This interactive tool is available at https://www.nps.gov/subjects/socialscience/vse.htm. Users can visually view year-by-year trend data and explore current year visitor spending, jobs, labor income, value added, and economic output effects by sector for national, state, and local economies.

Applicability of NPS strategies for global standards

Measuring visitor use, visitor spending, and the economic effects of visitor spending is not new, and there have been many studies of the economic impacts of protected-area tourism. Although these studies are common, their global application is minimal. There is generally a lack of international data on visitor use levels and economic impacts of protected areas, and there are no accepted international standard definitions or approaches for measuring visitation or visitor spending (Eagles, 2014; Eagles, McLean, & Stabler, 2000). Yet these data are increasingly important for promoting the conservation of natural and culturally significant sites and for justifying budgets for protected-area management. Eagles (2014) identifies visitor use monitoring and park tourism economic impact monitoring as urgent research gaps, and encourages international collaboration to build a set of methodological recommendations.

In recognition of the need for comprehensive and standardized data on visitor use and economic impacts, in September 2015, the German Federal Agency for Nature Conservation (BfN) convened an international meeting on the economic impacts of tourism in protected areas (Engels, Job, Scheder, & Woltering, 2015). The workshop concluded that there is a need for a global standard and suggested a multi-tiered approach that places monitoring and reporting of visitor use data as the minimum standard from which visitor spending and economic impact estimates could be developed. The experience and tools developed by the NPS have been replicated and adapted globally (see the case of Finland in Huhtala et al., 2010) and could be further leveraged as the international community works together to develop global standards and to promote an expanded collection of these important data.

Global visitor use monitoring standards need to be adaptable for a wide variety of protected areas both in terms of the type of protected area and in terms of the resources available for monitoring efforts. The great diversity of NPS sites provides a wide range of counting situations that could be used as examples for sites across the globe. The resources and counting technologies utilized by NPS sites range from basic to advanced, again providing a large array of example applications.

The NPS also has extensive experience conducting visitor surveys in a wide variety of park contexts. The VSP and SEM surveys satisfy the methodological, technical, and procedural requirements laid out in the BfN workshop, and provide examples that could be adopted and adapted by other countries. The SEM pilot surveys are being implemented in some of the most challenging parks in order to identify and address issues before the SEM program moves to an operational scale. Pilot parks include parks where multiple languages are spoken, parks with porous borders, parks in remote locations, geographically vast parks, and parks along cruise ship routes. The SEM pilot effort will yield a variety of visitor-related information through understanding some of the most challenging park contexts to conduct survey research.

As a general requirement, the BfN workshop stressed the need for clear communication of these data to media, policy-makers, park managers, and the public, and the need for communication to happen on local, regional, national, and international scales. The sophisticated communication program and the accessible data visualization tool developed by the NPS provide an example of communication success.

Conclusions

While it can be more challenging to estimate the economic activity generated by protected-area visitation as compared to extractive industries such as mining and forestry, incorporating these effects in policy and planning can help ensure wise trade-offs are made among sustainable development decisions. Over the past 25 years, the NPS has successfully developed methods and tools to estimate the VSE for a diverse range of parks and a successful communication strategy to increase the visibility of these estimates for decision-makers and the public. Efforts are currently underway to develop a formal SEM program to systematically survey park visitors across the National Park System. The procedures and experiences of the NPS are relevant to protected areas around the world and can be utilized for developing international visitor use and economic contribution monitoring protocols.

Notes

1. Recreation visitation data available from the NPS are measured in terms of visits, which are defined as the annual number of individuals who enter the NPS sites for recreation purposes. To count visits, the NPS uses the rule that one entrance per individual per day is countable. This measure of visitation is problematic for visitor spending estimation because a single visitor can count as multiple visits. For example, a family of four taking a week-long vacation to Yellowstone National Park and staying at a lodge outside of the park would be counted as 28 visits (four individuals who enter the park on seven different days). A different family of four, also taking a week-long vacation to Yellowstone National Park but lodging within the park, would be counted as four visits (four individuals who enter the park on a single day and then stay within the park for the remainder of their trip). These differences are a result of the realities of the limitations in the methods available to count park visits.

   FROM: 2016 National Park Visitor Spending Effects Economic Contributions to Local Communities, States, and the Nation Natural Resource Report NPS/NRSS/EQD/NRR—2017/1421

   Retrieved 31 July 2017 from: https://www.nps.gov/subjects/socialscience/vse.htm

2. Any use of trade, firm, or product names is for descriptive purposes only and does not imply endorsement by the U.S. Government.

Disclosure statement

The findings and conclusions in this article are those of the authors and do not necessarily reflect the views of the Department of the Interior or the United States Government.

References

Carver, Erin, & Caudill, James (2013). *Banking on nature – The economic benefits to local communities of national wildlife refuge visitation*. Washington, D.C.: U.S. Fish and Wildlife Service Division of Economics. Retrieved from https://www.fws.gov/uploadedFiles/Banking-on-Nature-Report.pdf

Chang, Wen-Huei, Propst, Dennis B., Stynes, Daniel J., & Jackson, R. Scott. (2003). Recreation visitor spending profiles and economic benefit to corps of engineers projects. (Report No. ERDC/EL TR-03-21). Washington, DC: U.S. Army Corps of Engineers.

Cullinane Thomas, C., Huber, C., & Koontz, L. (2014). *2012 National Park visitor spending effects – Economic contributions to local communities, states, and the nation* (Natural Resource Report No. NPS/NRSS/EQD/NRR – 2014/765). Fort Collins, CO: National Park Service. Retrieved from https://www.nature.nps.gov/socialscience/docs/NPSVSE2012_final_nrss.pdf

Cullinane Thomas, C., & Koontz, L. (2017). *2016 National Park visitor spending effects – Economic contributions to local communities, states, and the nation* (Natural Resource Report No. NPS/NRSS/EQD/NRR – 2017/1421). Fort Collins, CO: National Park Service. Retrieved from https://www.nps.gov/nature/customcf/NPS_Data_Visualization/docs/2016_VSE.pdf

Dillman, D. A., Dolsen, D. E., & Machlis, G. E. (1995). Increasing response to personally-delivered mail-back questionnaires. *Journal of Official Statistics-Stockholm, 11*, 129–129.

Eagles, P. F. J. (2014). Research priorities in park tourism. *Journal of Sustainable Tourism, 22*(4), 528–549.

Eagles, P., McLean, D., & Stabler, M. J. (2000). Estimating the tourism volume and value in protected areas in Canada and the USA. *George Wright Forum, 17*(3), 62–76.

Engels, B., Job, H., Scheder, N., & Woltering, M. (2015, September 25). *International workshop – Economic impacts of tourism in protected areas. Workshop proceedings*, Wadden Sea World Heritage Visitor Centre in Wilhelmshaven, Germany. Retrieved from https://www.bfn.de/fileadmin/BfN/sportundtourismus/Dokumente/Report_Workshop_Tourism_in_protected_Areas_bf.pdf

Font, X., & McCabe, S. (2017). Sustainability and marketing in tourism: Its contexts, paradoxes, approaches, challenges and potential. *Journal of Sustainable Tourism, 25*(7), 869–883.

Huhtala, M., Kajala, L., & Vatanen, E. (2010). Local economic impacts of national park visitors' spending in Finland: The development process of an estimation method (Working Papers of the Finnish Forest Research Institute). Vantaa: Metla. Retrieved from http://www.metla.fi/julkaisut/workingpapers/2010/mwp149.pdf

IMPLAN Group LLC. (n.d.). *IMPLAN System [data and software]*. Retrieved from www.IMPLAN.com

Job, H. (2008). Estimating the regional impacts of tourism to National Parks – Two case studies from Germany. *Gaia, 17*(S1), 134–142.

Koontz, L., & Meldrum, B. (2014). Effects of the october 2013 government shutdown on National Park Service visitor spending in gateway communities (Natural Resource Report No. NPS/EQD/NRSS/NRR – 2014/761). Fort Collins, CO: National Park Service. Retrieved from http://www.stgeorgeutah.com/wp-content/uploads/2014/03/DOI-Report-on-effects-of-government-shutdown-on-national-parks-and-gateway-communities.pdf

Mayer, M., & Job, H. (2014). The economics of protected areas – A European perspective. *Zeitschrift für Wirtschaftsgeographie, 58*(2–3), 73–97.

Mayer, M., Müller, M., Woltering, M., Arnegger J., & Job, H. (2010). The economic impact of tourism in six German National Parks. *Landscape and Urban Planning, 97*(2), 73–82.

National Park Service. (1995). *The money generation model*. Denver, CO: National Park Service.

National Park Service. (2016, April 21). National Park visitor spending contributes $32 billion to economy: Every public dollar invested in National Park Service returns $10. [press release], Washington, D.C. Retrieved from https://www.nps.gov/aboutus/news/release.htm?id=1821

President Barack Obama. (2015, April). *Remarks by the president on the impacts of climate change*. The Everglades, Florida. Retrieved from https://www.whitehouse.gov/the-press-office/2015/04/22/remarks-president-impacts-climate-change

Rookey, B. D., Le, L., Littlejohn, M., & Dillman, D. A. (2012). Understanding the resilience of mail-back survey methods: An analysis of 20 years of change in response rates to national park surveys. *Social Science Research, 41*(6), 1404–1414.

Secretary of the Interior. (1904). *Annual Reports of the Department of the Interior for the Fiscal Year Ended June 30, 1904*. Washington, DC: U.S. Department of Interior. 878 p.

Slocum, S. L. (2017). Operationalising both sustainability and neo-liberalism in protected areas: Implications from the USA's National Park Service's evolving experiences and challenges. *Journal of Sustainable Tourism*. doi: 10.1080/09669582.2016.1260574.

Souza, Thiago Do Val Simardi Beraldo. (2016). *Recreation classification, tourism demand and economic impact analyses of the federal protected areas of Brazil* (Unpublished doctoral dissertation). Graduate School of the University of Florida, Gainesville, FL.

Stynes, D. J., Propst, D. B., Chang, W., & Sun, Y. (2000). *Estimating national park visitor spending and economic impacts: The MGM2 model*. East Lansing, MI: Michigan State University.

Stynes, D. J., & Sun, Y.-Y. (2003). *Economic impacts of national park visitor spending on gateway communities: Systemwide estimates for 2001*. East Lansing, MI: Michigan State University.

Stynes, D. J., & White, E. M. (2006). Reflections on measuring recreation and travel spending. *Journal of Travel Research, 45*(1), 8–16.

White, E. M, Gooding, D. B., & Stynes, D. J. (2013). *Estimation of national forest visitor spending averages from national visitor use monitoring: Round 2* (Gen. Tech. Rep. No. PNW-GTR-883). Portland, OR: U.S. Department of Agriculture, Forest Service, Pacific Northwest Research Station. Retrieved from https://www.fs.fed.us/pnw/pubs/pnw_gtr883.pdf

White, E. M., & Stynes, D. J. (2008). National forest visitor spending averages and the influence of trip-type and recreation activity. *Journal of Forestry, 106*(1), 17–24.

Ziesler, P. (2017). *Statistical abstract: 2016* (Natural Resource Data Series NPS/NRSS/EQD/NRDS – 2017/1091). Fort Collins, CO: National Park Service.

Fringe stakeholder engagement in protected area tourism planning: inviting immigrants to the sustainability conversation

Anahita Khazaei, Statia Elliot and Marion Joppe

ABSTRACT

Effective and inclusive community participation is an essential and challenging component of sustainable tourism planning and development, especially as communities become increasingly diverse. The establishment of national parks and other protected areas closer to urban areas provides a unique opportunity for investigating community engagement in diverse contexts, as park agencies are mandated to connect with a broader range of community stakeholders. Historically, the engagement of immigrants and minorities with parks and protected areas has focused primarily on visitation, while their role as members of host communities has for the most part been overlooked. This qualitative study, conducted during the development of Canada's first National Urban Park, addresses this need by providing a deeper understanding of immigrants' engagement in planning. In-depth, semi-structured interviews are conducted with planners, politicians, community organizations, and first-generation immigrants who are now community leaders. The study draws upon, and expands on, earlier work by McCool and by Bramwell. It recommends five underlying principles for more inclusive public conversations: adopting an ongoing, long-term, and communicative approach; being open to new perspectives and willing to revisit assumptions; designing parallel strategies and customized tactics; collaborating with community leaders; and engaging in short-term and long-term learning.

Introduction

Sustainable tourism strategies have several general aims (Hunter, 1995), one of which is to meet the needs and wants of the local host community in terms of improved living standards and quality of life (Getz & Timur, 2005). This requires the full participation of all community stakeholders and the recognition that the community is not a homogeneous group (Okazaki, 2008) nor that all community subgroups have equal opportunities to participate in tourism planning. Yet, the lack of stakeholder participation has been found to be a major obstacle to the implementation of sustainable tourism strategies (Waligo, Clarke, & Hawkins, 2013). In traditional settlement countries such as Canada, but also increasingly in many other regions of the world, the resident profile is changing rapidly with a significant percentage of the population having immigrated. For example, as of 2011, 20.6% of Canada's population was foreign born, with recent immigrants (those who came since 2006) making up 17.2% of the total foreign-born population and 3.5% of the total population (Statistics Canada, 2013). Thus, it should be obvious that not only do these immigrants constitute an important component of today's host community, but they will be an increasingly significant one of future host communities, who will have to cope with any impacts from tourism development. Although governments often acknowledge that community engagement in the planning process enhances its "efficiency and legitimacy" (Schmidtke & Neumann, 2010, p. 6), it is rare that organizations are mandated to consult with host communities, and even less to specifically ensure that fringe stakeholders such as immigrants and minorities have equal opportunities to provide input into the process. More common are guidelines to ensure input is sought from a wide variety of stakeholders since public agencies generally act on behalf of the people in their jurisdiction. This is the case for the establishment of new national parks in Canada (Parks Canada, 2009), and the creation of the first national urban park in the Rouge Valley adjacent to Toronto presents a unique opportunity to observe the approach taken to engage fringe stakeholders. The goal of this research is to provide a deeper understanding of the current approach to immigrants' engagement in tourism-related public decision-making activities and to suggest underlying principles for designing more inclusive community engagement processes. Insights gained can easily be extrapolated to other tourism planning processes, especially those led by public and non-profit organizations.

Fringe stakeholders in community engagement processes

Although residents and community members have been the subject of numerous studies in tourism planning (e.g. Bornhorst, Ritchie, & Sheehan, 2010; Choi & Murray, 2010; Hung, Sirakaya-Turk, & Ingram, 2011; Jamal & Getz, 1995; Madrigal, 1995; Okazaki, 2008; Perdue, Long, & Allen, 1990; Ruiz-Balesteros, 2011), the heterogeneous nature of the host community has rarely been addressed in empirical works.

Immigrant communities are usually among the less engaged or *fringe* segments of a heterogeneous host community. Research shows that many immigrants, especially those that do not speak English at home, have lower levels of civic engagement compared to native Canadians (Baer, 2008). Yet, Hart and Sharma (2004) argue that a systematic approach to "identifying, exploring, and integrating the views of those on the periphery or at the *fringe*" is essential for "resolving the radical uncertainty of constantly evolving knowledge" (pp. 8–9). Recent immigrants may not be engaged in public planning and decision-making processes for a variety of reasons, ranging from personal and family factors to systematic issues related to planning and community engagement process. They may be highly educated but not have sufficient background information about the specific project or topic, especially in terms of tourism; not be familiar with the cultural aspects; experience language barriers; not be part of the active social groups; or even not be interested in the tourism related issues. They may not have a strong feeling of belonging to the new society yet, or perceive themselves as guests instead of hosts. They may even not be aware of their right to participate or not consider their input relevant or valuable. However, as mentioned by Reed, Graves, Dandy, Posthumus, Hubacek, and Morris (2009) stakeholders' level of interest or power changes over time. Today's newcomers will form the future host community and will influence the tourism industry in different ways: as residents who are part of the place identity and engaged in the co-creation of experience with tourists (Saraniemi & Kylänen, 2011), as employees and owners of tourism-related businesses, or as members of activist groups.

Sustainable tourism development requires planners to accommodate both current and future needs and expectations of host communities (Byrd, 2007). As there is not "a definable, single, generic interest for the future host community", incorporating their interests in tourism planning is very challenging for planners (p. 11). This issue is even more important and challenging for countries and cities that are open to new residents who are from different cultures and may have different attitudes and perception about the place and also about tourism. The "burkini" debates recently raging in some French coastal towns is illustrative ("Le Conseil d'Etat met un terme aux arrêtés « anti-burkini »", 2016). By engaging immigrants, planners can consider the future change factor in tourism planning.

They can also contribute to helping assimilate immigrants into their new country, helping them understand the concepts of protected area planning and the wider issues of conservation.

Immigrant engagement in parks planning

The literature on stakeholder theory shows increasing attention paid to the heterogeneity within stakeholder groups, with specific emphasis on marginalized and fringe segments (Burton & Dunn, 1996; Crane & Ruebottom, 2011; Hart & Sharma, 2004; Khazaei, Elliot, & Joppe, 2015; Roloff, 2008). Researchers are also moving toward suggesting more customized strategic alternatives for addressing diverse stakeholder claims and needs (e.g. Dunham, Freeman, & Liedtka, 2006). To date, neither of these advances have been reflected to any extent in tourism. Addressing specifically the research needs in park tourism, Eagles (2014) highlighted the need to understand "the methods that could be used to deal with new immigrants whose cultural background may not involve a focus on outdoor recreation" (p. 537) and that good governance implies that "all people [...] have a voice in decision making, either directly or through legitimate intermediate institutions that represent their interests" (p. 542).

In both research and practice, the engagement of immigrants and minorities with parks and protected areas is predominantly focused on their role as visitors. This is in response to the fact that visitation is lower among these groups (Floyd, 2001; Shultis & More, 2011; Weber & Sultana, 2013) and based on the assumption that visitation is the first step toward ensuring future support from increasingly diverse communities (Lovelock, Lovelock, Jellum, & Thompson, 2012). Many studies (mostly conducted in North America, New Zealand, and Australia) have been dedicated to understanding behaviours and preferences of immigrants and ethnic groups with regards to leisure activities in general and visiting national parks in particular (e.g. Carter, 2008; Deng, Walker, & Swinnerton, 2005; Floyd, 2001; Floyd, Shinew, McGuire, & Noe, 1994; Lovelock, Lovelock, Jellum & Thompson, 2011; Shinew et al., 2006; Stodolska, 2000). Yet, research on immigrants' role as members of host communities and their engagement in planning and decision-making activities is almost non-existent.

Studies on community engagement in planning and managing national parks are commonly focused on rural and Indigenous communities due to the fact that most of the studied sites are located in remote rural areas and/or near Indigenous communities. However, "a fundamental and pervasive shift away from nature-based recreation" noted in the United States and Japan (Pergams & Zaradic, 2008) but also in Canada (Bisby, 2015) has led to the establishment of national parks and other protected areas near population centres (e.g. Rouge National Urban Park in Canada and urban National Recreation Areas in the United States), in order to address decreasing visitation. Urban nature parks are different from traditional national parks in that they face new challenges with regard to conservation (Fortin & Gagnon, 1999) and incorporating "diverse demands" of an urban clientele (Gobster, 2002, p. 143). At the same time, urban national parks can benefit from unique opportunities for tapping into different cultures and taking advantage of local potential. The heterogeneity of host communities and its impact on planning for national parks and protected areas has remained an under-studied area.

Despite their lower visitation, it cannot be argued that immigrant and minority community members "perceive of or construct wilderness in culturally different terms" with regard to values they ascribe to it (Johnson, Bowker, Bergstrom, & Ken Cordell, 2004, p. 624). Research shows that "people from various ethnic backgrounds value being together in parks" (Peters, 2010). A study conducted by Peters, Elands, and Buijs (2010) on Dutch urban parks indicates that these parks are "more inclusive than non-urban green areas" and can "promote social cohesion" as immigrants and ethnic minorities are more likely to visit these sites (p. 93).

Agreement on the intrinsic value of nature and shared commitment to preserving it for future generations (Hester, Blazej, & Moore, 1999) as well as the fact that immigrants and members of ethnic communities are mostly living in urban and metropolitan areas and therefore more likely to visit urban parks (Peters et al., 2010) can provide a foundation for connecting with these fringe, yet important stakeholders through engaging them in planning activities. This would allow for their needs, expectations, and ideas to be incorporated in the design of parks from the outset. As a result, they would be able to "situate themselves in the nature" (Lovelock *et al.*, 2011, p. 252) and re-establish their relationship with it in the context of their new country. Visitation could be the natural outcome. Research already shows that local residents' participation in designing urban parks leads to increased visitation (Peters et al., 2010). A noteworthy finding of the current study, however, is that engaging immigrants and minorities in planning activities is not considered as an alternative but rather a complementary and long-term strategy to be adopted in addition to the current short-term initiatives focused on increasing visitation.

A national urban park case study

In 1995, the Province of Ontario officially announced the creation of Rouge Park to protect the Rouge River and surrounding land to the east of the City of Toronto, Canada's largest metropolitan area with almost 20% of the country's population. Over the years, the Province increased its size substantially. The federal Government's plan for creating a national urban park in the Rouge Valley was announced in the 2011 "Speech from the Throne" (a prepared speech to members of Parliament outlining the government's agenda for the coming session) and on 25 May 2012 $143.7 million was allocated to be spent over a 10-year period for its creation, and $7.6 million annually thereafter for its further development. In 2015, the federal government's *Rouge National Urban Park Act* came into effect, establishing the country's first National Urban Park. Once completed, it will be one of the largest and best protected urban parks of its kind in North America (Parks Canada, 2015).

Rouge National Urban Park, also referred to as the *People's Park* (Parks Canada, 2012), is particularly appropriate for studying immigrants' engagement as the communities living around the Park are among the most diverse in Canada with a high percentage of first-generation immigrants, defined as "persons who were born outside Canada. For the most part, these are people who are now, or have ever been, immigrants to Canada" by the National Household Survey. 40% of the population in the City of Toronto are immigrants with one-third of them having arrived in the last 10 years (City of Toronto, 2013); this percentage is even higher in some of the communities surrounding the Park.

Research approach and methodology

This study focuses on participants' perceptions about immigrants' engagement in tourism-related planning activities and their perspectives on what an inclusive community participation process would look like. The nature of the research resonates with the philosophical assumptions of the interpretivist paradigm, rooted in the pioneering German sociologist Max Weber's idea that social sciences are focused on "Verstehen" (understanding) rather than "Erklären" (explaining) (Crotty, 1998, p. 67). The focus of the interpretivist researcher is on "understanding the actual production of meanings and concepts used by social actors in real settings" rather than formulating and falsifying hypotheses (Gephart, 2004, p. 457). Participants' perceptions were deeply rooted in their personal experiences and therefore could be best studied in the context of a real project. Hence, the research was conducted in the form of an inductive qualitative case study. The case provided a platform that connected participants to each other through shared experiences, interests, or concerns. It also allowed researchers to take part in a conversation around an actual and ongoing project and to gain access to participants' broader experiences with regard to immigrants' engagement beyond the Rouge Park project. The qualitative approach was aligned with the interpretivist assumptions of the study and appropriate for addressing research objectives, as according to Gephart (2004) it "employs meanings in use by societal members to explain how they directly experience everyday life realities" (p. 455) and allows researchers to study the complexity of social systems (Patton & Appelbaum,

2003). In order to gain a comprehensive understanding of the current approach to immigrants' engagement, the phenomenon needed to be explored from different perspectives. Therefore, data were gathered from three groups of participants through in-depth, semi-structured interviews between July 2012 and April 2014:

- Planners (P): experts, senior managers, and politicians who were directly involved in designing and implementing the public consultation process for the Rouge National Urban Park and whose philosophies and visions were reflected in the process (17 interviews).
- Partner/community organizations (PO): intermediaries, mainly non-profit and community organizations that connect planners with immigrant communities (8 interviews).
- Community leaders (CL): first-generation immigrants who were actively involved in local and community projects (11 interviews).

Document analysis and observation served as secondary methods for gaining familiarity with the research context as well as identifying consistencies, inconsistencies, and potential gaps in the data.

Findings and discussion

Data analysis was informed by a constructivist grounded theory approach with three rounds of coding, namely, initial, focused, and axial coding (Charmaz, 2006). Over 40 hours of interview data, gathered from the three groups of participants, were first analysed as independent sets and then compared across groups. Transcripts were first coded phrase by phrase. Next, categories were formed based on initial codes, using NVivo software as a data management tool. Codes and themes were selected for entering into NVivo based on three criteria: frequency (number of times repeated), importance (offering a unique or deep insight into the phenomenon), and relevance to interview questions. During the axial coding process, categories and sub-categories were connected and integrated using the mind mapping technique. At this stage a second round of literature review, guided by research findings, was conducted as suggested by the constructivist grounded theory approach (Charmaz, 2006). The review (primarily focused on communicative planning and learning theories) was conducted after themes and categories emerged from the data in order to have a deeper understanding of the findings by situating them in related theoretical contexts. Through several rounds of coding and analysis, findings were elevated to higher levels of abstraction, moving from understanding the current engagement approach and the implicit and explicit assumptions contributing to it, to suggesting principles for more inclusive public consultation. During this process more than 300 codes were first distilled into 14, then 6, and finally 5 mind maps, capturing the 5 principles.

Specific strategies, informed by criteria introduced by Guba and Lincoln (1982) and Charmaz (2006), were incorporated in the research design to ensure rigour and trustworthiness. These strategies include (1) triangulation: data gathered from three groups of participants with a wide range of experiences and backgrounds, with observation and document analysis used as secondary data sources; (2) reflection on researchers' philosophical assumptions, perceptions, and expectations, as suggested by Phillimore and Goodson (2004); (3) detailed documentation of the emergent research process including important methodological decisions and details of the data analysis method in a reflective journal; (4) constant comparison between data gathered overtime and through different methods as well as emergent codes and categories; (5) theoretical sampling: additional data gathered after initial categories were formed in order to ensure the richness of themes and that "all voices are represented" (Tweed & Charmaz, 2011, pp. 132–133); and (6) participants' feedback received on Parks Canada's immigrant engagement approach.

Five principles of more inclusive community engagement processes are presented here. Each section includes specific references to immigrant communities, as the main focus of this research, and selected interview excerpts as supportive examples of the data.

Principle 1: Adopting an on-going, long-term, and communicative approach

The importance of establishing and maintaining an effective and ongoing dialogue between planners and community members was emphasized as one of the key building blocks of a more inclusive community engagement process. Participants suggested that planners should keep community members updated and informed even if they are not interested or ready to participate immediately. As explained in the following interview excerpts, although the relationship may start as a one-way distribution of information, it would provide community members the opportunity to participate when they feel comfortable and ready.

> You provide information today, but you will also provide information tomorrow and whenever they feel comfortable or they would like to participate, they have that invitation and a comfort level that they could do that and participate ... It takes time for people to hear about what is going on; and to understand what is going on; and how they might be able to input. (P5)

> I think first thing, they [first generation immigrants] are going to be very reserved in the process of observing. You are going to be reserved because you are not sure how is it similar to what you are accustomed to or not, and so once you are able to do that then you feel comfortable among the people around the table, you find that you are able to share your ideas and to share your perspective. (CL11)

Current engagement initiatives are focused on established immigrants. An example is offering one year of free access to all national parks, protected areas, and heritage sites as a gift to all new citizens at the Swearing-in Ceremony. However, community organizations argue that the outreach should start as soon as immigrants land in Canada, considering the fact that it takes at least four years for them to be eligible to apply for citizenship.

> If somebody waited 5 or 10 years to get engaged with their community, that's a long time to be disengaged and to not have a say in what's going on around you. If you live here, you care about the water; you care about highways; you care about schools. You might not know it. It might not be front and center in your priority list but it affects you and if it affects you, you should have your say. (PO6)

> The first five years of settlement [is] the most stressful and that is when you do need to have a space to go into and reflect and if you need to cry, if you need to laugh, if you want to run ... and humans feel safe in nature and it is an opportunity that is being missed because those memories of the first five years stay with that first generation settler. And if those memories are formed in the Rouge National Urban Park, guess who they are going to support: Rouge National Urban Park. Guess who they will want to volunteer with: Rouge National Urban Park. Guess who they will want to work diligently to grow and nurture and maybe even interact with on [a] multi-generational level. (PO3)

Adopting a short-term project-based approach, focused on collecting community input on one project, or even selected phase(s) of a project, may not be effective because "participatory capacity cannot be built like a road or dam; it must be developed" (Tosun, 2006, p. 503). This is captured in the following interview excerpts:

> It's all based on relationship and trust. If you just get a cold call or if you see somebody on the street with a flyer [that says] come to this program, it's great ... You are not going to have success. You have to invest the time and once you have even two or three people with good relationships they are your gateway into the community. (PO6)

> If your goal is to engage people then you are starting from a good base. If your goal is to get an agenda pushed through, you are not starting from a good place of understanding. But if you engage in a dialogue right from the beginning, even if it's a half baked idea [and ask] What do you think about this? and engage people right at the beginning you are going to have a much better success rate than being handed a flyer saying it's already done just come and give your input. (PO6)

> The biggest recommendation I have always had is, it's about relationships and it takes time to build a relationship. You can't just send out 5000 emails and expect something to happen. (PO4)

This approach is aligned with the concept of communicative planning, based on Habermas' theory of communicative action (see Habermas, 1983). It has gained interest among scholars and practitioners in the area of planning for many years now and is considered as specifically effective for planning in multicultural contexts and for engaging minority groups in decision making (Umemoto & Igarashi, 2009).[1] The following definition captures the essence of communicative planning:

> A method of civic engagement in decision making based on reasoned discourse where the conditions of interaction – mutually respectful, inclusive, transparent, honest, impartial – lead to empathy, social learning, and mutual transformation toward collaborative solutions to defined problems. (Umemoto & Igarashi, 2009, p. 39)

Communicative planning may not necessarily result in consensus in the short term or during the time that specific planning projects are being implemented. Planners are encouraged to "focus on the longue durée, and the lifeworld of lived experiences, where shared subjectivities over the built environment can develop" (Matthews, 2013, p. 139). The discourse can continue between community members and the environment even after formal community engagement projects are completed and in the absence of planners.

Principle 2: Being open to new perspectives and willing to revisit assumptions

Findings indicate that despite planners' attempts to adjust the consultation process to the needs and wants of a diverse urban population, it was still influenced by traditional assumptions and decision-making models, as expressed by research participants:

> Many years ago when there weren't as many immigrants, we were reaching out to a good number of the community [members]. More recently as more immigrants are moving into the area I think we are still relying on some of the old concepts. So our messaging, our communication, isn't that effective and we haven't developed the best strategy [for] communicating with all of the different immigrants and to make sure that we are getting their input. (P5)

> Systems seem to be slow to change and especially with government ... It's so old and entrenched in terms of structure ... I think we actually need to change our decision-making model. (PO4)

Current approaches may not necessarily resonate with and appeal to the increasingly diverse communities. Strong emphasis on traditionally accepted and preferred ways of engaging with parks may alienate new and less-engaged community members. According to research participants, general perceptions about the environmental movement as being predominantly white and middle class also contribute to this issue.

Different community subgroups (based on cultural, religious, philosophical, or other commonalities) may have different perceptions of an effective and inclusive planning process. According to Watson (2006), the possibility of reaching consensus during the communicative planning process is very low in diverse contexts because members of each subgroup engage in the discourse based on their own value systems and world views. Therefore, there is no one "universal set of deontological values" that can be used as the basis for designing the planning process. Although her focus is on values "shaped by the neoliberal tradition" the argument can also be made about any other dominant philosophy (p. 31). In the same vein, Bramwell (2004) emphasizes the importance of accommodating alternative "reasoning processes" and "the difficult translation between different world views" in the planning process (p. 546). Hence, it is suggested that planners consider adopting "moral philosophies which recognize the situated nature of knowledge" instead of approaching community participation from one specific point of view (Watson, 2006, p. 32). This view would allow for more flexibility to accommodate alternative interpretations of abstract concepts. The importance of flexibility and openness in understanding first-generation immigrants and adjusting planning processes accordingly is captured in the following interview excerpt:

> They [planners] are going to have to listen and watch and learn from these people. They can't come in with their kinds of assumptions and they have to be prepared to listen and adapt their concept, their proposal to the realities of this new group …We can't make assumptions about the way they are going to visit and use the resources. We are going to have to search for those new balance points between their realities in their situations and our offer … they have to come into that process with an open mind and flexible attitude and true integrity. (P4)

The neopragmatic approach to planning offers suggestions for enhancing flexibility and openness, specifically in the context of national parks and protected areas with a diverse group of stakeholders. According to Jamal, Stein, and Harper (2002), emphasis on reaching an agreement on definitions of "abstract principles" among stakeholders in the early stages can be detrimental to the planning process. They suggest that participants spend time on sharing their understandings and perceptions of key concepts instead of starting based on strict and taken-for-granted definitions. Focusing on meanings and core ideas as opposed to discussing them under certain labels can enhance the dialogue by helping participants to be more open to commonalities rather than differences. Ideas are considered and discussed in a less value-laden context and therefore are less likely to be rejected based on assumptions about the category they are associated with (Jamal et al., 2002). Issues with designing the process based on strict definitions of abstract concepts can become more prominent when working with recent immigrants that are not native English speakers. Although participants may be fluent in English and comfortable in routine interactions, when it comes to abstract concepts they may have different perceptions based on their native language and culture.

Principle 3: Designing parallel strategies and customized tactics

The study offers insights into differences between various community segments with regard to participation in public decision-making activities. First-generation immigrants, or new residents of an area more generally, may not be able to contribute to discussions because they do not have enough background information about the specific issue or project at hand.

> They don't have all that much background or history in the area. So I think something that is important is to make sure that we provide them with that background and understanding about some of the issues that we are dealing with so that they can become more informed. … They have expressed a lot of interest; they have expressed a lot of commitment; but at the same time they are holding back because they don't have the context; they don't have the background. (P5)

Access to knowledge, along with *trust* and *power*, is identified as one of the key conditions for designing successful tourism partnerships (McCool, 2009). Specifically, when topics are related to history, heritage, and culture, people who are raised in other countries may not feel confident to join the conversation. In addition, community segments and individuals are at different stages in terms of their readiness and willingness to participate and require different amounts of time to become familiar and comfortable with participation processes. Members of some cultural communities, especially during the first years of settlement, are less likely to participate in events outside their own communities and may take a backseat even if they attend public meetings. In contrast, some cultures encourage citizens to express their opinions. Immigrants from these countries are used to participating in community projects and attending public meetings, whereas in some other cultural contexts people may prefer not to get involved in order to avoid potential risks.

> You have a group who is from a culture where everything is dictated to them and you are going to find it difficult for that group to come out and share any ideas and be part of a community; be part of a group; or sit on a board. They are not going to do that because of the fear of their culture … [but] you [also] have a group that will be willing to go out and learn and participate and say what they have to say. (CL11)

When one or a few groups, specifically long-term residents of the area – in the case of the Rouge National Urban Park with strong environmental interests – dominate consultation sessions less-engaged community members might feel that they do not fit in.

They ask first generation Canadians, invite them as their neighbours or as [sic] their community neighbour friends to go but basically as far as talking it is not them who are talking. It is Canadians or the seniors who have [a] wide range of experience in this country, they talk. They are simply backbenchers or listeners. (CL9)

... [One of our community members] went to one part of ... those meetings and I think he was very fed up with the meeting because the meeting was very much about the environmentalists, and the species and those people taking up a lot of space and feeling entitled to take up a lot of space and ... [and I was wondering] was it really designed at all to incorporate people's feedback, who do not have environmental science degrees? (PO8)

Generally community [members] don't [participate] partly because they feel more comfortable within their community [sic] by language, by cultural experiences, and also by habits... Sometime they feel that they don't really fit in (CL10).

Holding separate meetings for "newer voices" and "traditional views" was recommended as a way to facilitate first-generation immigrants' participation in discussions, as reflected in the following interview excerpt:

I think what happens [is], if you bring them into the same room, it polarizes what the Canadian experience is to what the immigrant experience is and, you know, being a first generation Canadian you try not to stand out differently. ... [sic] I think you actually target ... and create a safe environment for them to express their opinions and not use the process to set up the challenge between the traditional view and newer view. (PO1)

Separate meetings might be helpful in addressing the power issue, specified as one of the main requirements of successful partnership (McCool, 2009). Critics of communicative planning argue that it does not consider the impact of power, politics and conflict on discourse, and decision-making (Dredge, 2006; Matthews, 2013; Umemoto & Igarashi, 2009; Watson, 2006). However, the downside of having separate meetings is that it may contribute to further isolation and the feeling of alienation among some first-generation immigrants. One suggestion would be to have parallel strategies with allocated time and resources focused on empowering and helping less-engaged community subgroups to develop the confidence and required skills to join mainstream decision-making processes.

This study emphasizes the importance of adopting a customized approach to community engagement with specific outreach plans for each segment based on characteristics such as cultural background, level of readiness and willingness to participate, information gaps, and empowerment and skill development needs – a case of equity rather than equality. A more targeted and customized approach to community engagement resonates with the progressive approaches identified in the stakeholder theory literature (Khazaei et al., 2015). According to research participants, a targeted approach would allow planners to adopt appropriate communication methods for each community; identify and work with facilitators from within the community; and use tailored discussion questions. These practices would help planners to create learning environments that foster participation for each community segment as suggested by Armitage, Marschke, and Plummer (2008).

Principle 4: Collaborating with community leaders

The complexity and heterogeneity of community subgroups as well as planners' limited experience in working with diverse communities require planners to engage communities in "shaping" and "defining" the decision-making process from the beginning. Otherwise, it would be almost impossible for them to gain an accurate and adequate understanding of experiences, perceptions, and preferences of all community segments.

I don't know that we have, to be honest, a lot of experience in targeting and reaching out to new immigrants to participate in the decision making. (P3)

I know there is not a lot of expertise anywhere in working specifically with immigrant communities to plan a specific park perhaps. (P8)

These limitations hinder planners from connecting with communities in the first place. Therefore, the key question is what would be the starting point? Connecting with community leaders can be the first step. Considering the fact that creating trust and building relationships happens over time, planners need to rely on community leaders that have already developed these relationships and are familiar with the needs and wants of their communities.

> You need to find the leaders that are in the community who are the connectors. You need to develop a relationship with them and so for me they are going to be the ones that are going to talk to the community first and invite them. (PO4)

> The only way is to have leadership from within that community to talk to, in the essence, their constituency... We couldn't engage those communities but we had leadership from within particular immigrant communities to engage their own people in the process. Because it's a question, not only of language but of listening to and of respect... So I think in engaging immigrant communities the most successful [approach] is to engage the leaders of the community. (P12)

Participants suggested "identifying" and "developing" local "champions" or "cheerleaders" and providing them with required resources for engaging their communities in decision-making activities. They can be "gateways" to communities and are connected to local networks. Therefore, they can help planners to distribute information and share their knowledge and experience with regard to appropriate methods for connecting with their communities. In addition, findings indicate that successful community leaders (like politicians) can be role models and inspire less-engaged members of the community, especially youth, to get involved.

Forming "advisory committees" or "boards of advisors" is suggested as a first step for engaging community leaders in designing public participation processes. It is a very limited form of participation, considered by Arnstein (1969) as part of the "manipulation" stage at the very first level of her Ladder of Citizen Participation, which is also referred to as "nonparticipation" (p. 218). However, participants mentioned it as an opportunity to start the conversation. These practices can be effective only if used as empowerment techniques and part of a comprehensive community engagement strategy that targets broader and higher levels of community participation. Becoming involved in local boards in general is an opportunity for first generation immigrants to become familiar with decision-making processes and to learn "how things work" in Canada.

Participation is, however, not just a Canadian problem in protected areas: it is worldwide and recognised as such. It is essential in dealing with "local-level complexities associated with multiple management objectives" (Becken & Job, 2014, p. 511).

Community leaders who participated in this research defined themselves as: agents of change trying to make tangible improvements in their communities; caring about the needs of the community and taking action to address them; willing to enhance community well-being through identifying problems in the neighbourhood; volunteering to be the voice of the community; and helping community members to integrate and feel comfortable. They actively seek opportunities to volunteer with community organizations on local projects. Therefore, planners can, with dedicated effort, connect with them through these organizations. Community organizations act as intermediaries and also provide familiar and safe environments for practicing community engagement in one's own cultural group as a first step toward engaging with the broader community. First-generation immigrants tend to feel more confident to participate in events when they are associated with strong community organizations.

> When people can associate with [a strong community organization] they won't feel as intimidated to go out. You can then participate in the activities organized by [the] main stream even if you are [a] visible minority; you don't feel bad if some other groups, other people ... are not as welcoming. (CL10)

According to Tosun (2006) "local Non-Governmental Organizations" can contribute to "a more participatory tourism development approach" through empowerment and "leading local people to take part in tourism development" (p. 502). In the same vein, Berkes (2009) emphasizes the importance of bridging organizations in the context of co-management of natural resources and considers them as specifically important when "local knowledge is based on a different epistemology and worldview than government science" (p. 1696). The current state of collaboration between planners and partner organizations, mainly limited to using their networks for distributing information and recruiting community members for public meetings, reveals opportunities for a more strategic relationship. Participants suggest that collaboration with community leaders and organizations can be most meaningful and effective if they become involved from the early stages, and in the design of consultation processes. This is where community organizations can play a key role by sharing their knowledge and experience about communities and techniques that work best for connecting with them. Moreover, a key factor in maintaining collaborative relationships with community leaders and organizations over time is providing feedback on how their input has influenced decisions. Research findings show that some of the community organizations and leaders feel that their comments and input will not affect final decisions partly because they do not receive any feedback after participating in meetings and consultation sessions.

> We bring 15-20 residents, we show up [at consultation meetings] and then that's it. A lot of times we don't even get the report back, we don't know, you tally all the information and you synthesize it and you analyze and you know you should be reporting back to the people that helped you gather that information. Most times that doesn't happen. We never see the reports. We never see follow-up. (PO6)

It is also important not to limit participation to community leaders but rather see them as a means to empower and engage community members.

Principle 5: Engaging in short-term and long-term learning

Sharing and learning are defining elements of community participation from community leaders' perspectives and happens mostly at the community level and among fellow community members:

> To me ... it is like a learning process because if I am engaged with other people and share [sic], you know, share the things ... what I learned. It is a transformation of my knowledge, my information, my skills to others and I want to get their knowledge, their expertise ... it's a journey of learning. (CL2)

Planners, on the other hand, see learning as the exchange of knowledge and information between themselves and communities mostly in the form of receiving input from community members, and in turn informing and educating them.

Transformative learning theory, rooted in the adult education field (see Mezirow, 1981), can be helpful in understanding and enhancing learning at the individual level, between different communities as well as planners and community members. Transformative learning has two components: instrumental learning, defined as "task-oriented problem solving actions to improve the performance of current activities" and communicative learning, which refers to "individuals' ability to examine and reinterpret meanings, intentions and values" (Armitage et al., 2008, p. 88). Instrumental learning can happen through participating in specific projects (Marschke & Sinclair, 2009). Planners' efforts to increase participation in visitation programs (e.g. tree planting, walking tours, bird watching, and camping) enhance instrumental learning. Visitation programs provide opportunities for learning through focused and short-term interactions between different communities; community members and planners, as well as the Park itself. This resonates with Jamal et al. (2002)'s suggestion that starting with "smaller, narrower, and more concrete" issues can result in enhanced collaboration and dialogue by avoiding potential conflicts over philosophical views and value systems. Research participants' suggestions for facilitating immigrants' participation in public consultations (e.g. sharing the agenda and background information in advance; holding preparatory meetings for community leaders before attending public consultation sessions; and holding smaller meetings) can facilitate instrumental learning.

Communicative learning, as defined below, can contribute to the establishment of more inclusive and pluralistic planning processes by enhancing mutual understanding and openness to different philosophies among individuals and groups involved in planning. This type of learning will result in more fundamental and long-term improvements in community engagement endeavours.

> Communicative learning involves learning that pertains to understanding what somebody means or the process by which others understand what you mean. … . Communicative competence refers to an individual's ability to negotiate meanings, intentions and values for oneself. (Marschke & Sinclair, 2009, p. 207)

Another theoretical lens that can be helpful in understanding how communication and dialogue can result in learning and change is social learning theory. Social learning happens through "sharing experiences, ideas and environments by others" at three levels:

> Single loop (correcting errors from routines); double loop (correcting errors by examining values and policies); and triple loop learning (designing governance norms and protocols). (Armitage et al., 2008, p. 89)

Tactical attempts for adjusting the public consultation process to the needs and characteristics of an urban population can be considered as single loop learning. Organizing youth workshops, conducting an online survey, and setting up information booths in downtown Toronto are among new community outreach initiatives undertaken in addition to traditional public consultation sessions. However, single loop learning "is not adequate" for an effective tourism protected area partnership, which happens in an increasingly complex and *messy* context (McCool, 2009, p. 141). Research findings indicate that planners are aware of the importance of, and the need for, reaching out to less-engaged community members and that it requires changes in the underlying assumptions and philosophies as well as developing a trust relationship with communities. "Trust building efforts and active engagement with civil society" are among the factors that foster double loop learning (Diduck, Bankes, Clark & Armitage, 2005 in Armitage et al., 2008, p. 89). Triple loop learning, focused on examining and revising systems, policies, and the broader context within which single- and double-loop learning are happening, can be considered as a long-term goal and potential outcome of the ongoing dialogue between different stakeholders (Armitage et al., 2008).

Deconstructing learning and examining it as instrumental versus communicative or single, double, and triple loop learning helps to highlight the practical aspect of community engagement and the fact that planners need to balance short-term and long-term learning goals with regard to first generation immigrants' engagement. Attempts should be made to enhance participation in current decision-making opportunities like the Rouge National Urban Park. At the same time, it is important for all stakeholders to collaborate and move toward more inclusive community engagement processes in the future through capacity building and empowerment. Instrumental and single-loop learning contribute to achievement of short-term objectives, whereas communicative and multiple loop learning goes beyond the scope of one single project and institution.

Conclusion

A critical review of the tourism planning literature through the lens of stakeholder theory shows that immigrants' participation has remained an understudied subject (Khazaei et al., 2015). A closer examination of the literature on immigrants' engagement in parks planning, as the context selected for this research, emphasizes the gap by revealing that despite increasing attention to immigrants as visitors, their participation in planning as community members has not been studied. This research contributes to the literature on community participation in tourism planning by providing recommendations for enhancing fringe stakeholders' engagement, with specific focus on immigrants. In addition, the inductive and constructivist grounded theory approach allowed for insights from other disciplines (primarily learning theories, communicative planning, and management) to be brought to the tourism literature. Following Darbellay and Stock (2012)'s definition of interdisciplinary research, this study attempted to "mobilize different institutionalised disciplines through dynamic interaction in order to describe, analyse, and understand" the complexity of the phenomenon (p. 453).

In a conceptual paper on successful tourism planning partnerships in the context of protected areas, McCool (2009) suggests four criteria for developing and maintaining successful partnerships in such contexts: *representing* different and even conflicting views; creating a sense of *ownership* for the protected area as well as the plan; providing opportunities for *double-loop learning* to deal with complex and non-linear issues; and building strong *relationships* based on openness, honesty, mutual respect, and clear roles and responsibilities. This research is specifically concerned with enhancing representativeness and inclusiveness of public decision-making processes by focusing on fringe stakeholders, specifically first-generation immigrants, as an under-represented group. In line with McCool's (2009) criteria, relationship is considered as a key success factor and the required foundation for the other four principles of more inclusive community engagement processes, as illustrated in Figure 1. These principles highlight the importance of establishing a long-term relationship, initiated by planners, as soon as immigrants enter the new country. The relationship should be viewed at the institutional level and beyond the scope of one project to account for different degrees of readiness and to allow for continuous learning, capacity building, and revising dominant assumptions.

Establishing and maintaining an inclusive dialogue requires all players to be open to new perspectives and worldviews and to be willing to revise their assumptions as communities become more diverse. This is especially important for planners and participants who have more experience and a longer history of community engagement, in other words traditionally salient stakeholders. Designing engagement processes based on dominant values and insisting on traditional ways of participation discourages less-engaged and newer community members from participating in the first place, and from expressing their opinions if they attend meetings. It is recommended that planners and facilitators use neutral language and spend time on developing shared definitions for key concepts instead of using value laden terminology (Jamal et al., 2002).

Research findings reveal that learning and sharing are considered as key elements of community participation from community leaders' point of view. The importance of double-loop learning, as suggested by McCool (2009) is supported. In addition, the study provides examples of how participation in specific projects can enhance single-loop and instrumental learning, whereas the ongoing and long-term dialogue can result in double-loop and communicative learning. Aligned with the long-term approach and essential for an inclusive planning philosophy, we also raise triple-loop learning for consideration. It can lead to improvements in (or at least a better understanding of) the broader socio-economic, cultural, and political environment that influence immigrants' engagement. Learning extends beyond the scope of one project or one institution. It is an on-going process that happens within and between park agencies, partner organizations, and communities.

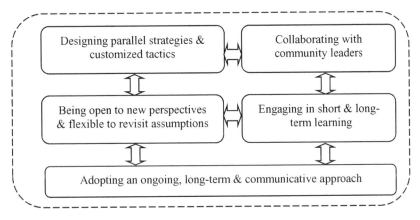

Figure 1. Underlying principles of more inclusive community engagement processes.

This study takes a step back and looks at the very first phase of community engagement: initiating the relationship and making first contacts with communities, identified by research participants as one of the biggest challenges (if not *the* biggest challenge) for planners. It may not be as significant in small and homogenous communities. However, differences between segments within diverse urban communities, coupled with limited resources (including knowledge and expertise), create a serious barrier for planners.

Bramwell (2004) suggests "capacity building among parties outside partnerships" (p. 547) as a strategy for addressing barriers to stakeholder participation. He emphasizes that specific attention needs to be paid to marginalized groups and that capacity building should happen regardless of "whether or not they might subsequently join in multi-party initiatives" (p. 547). This study supports Bramwell's point by indicating that different sub-groups or even individuals within the broader immigrant community are at various stages of readiness and willingness to participate. Therefore, capacity building should be a key priority in the long-term relationship and dialogue among planners, immigrant communities, and partner organizations. This approach would give less-engaged community members the opportunity to develop their skills and join the conversation at the time and in a way that suites them. Close collaboration with community leaders and adopting a targeted approach would enable planners to have a better understanding of the characteristics and needs of each community.

Inclusiveness in terms of diversity among residents attending consultation meetings, as well as their effective engagement in discussions is key. This is aligned with Bramwell and Sharman (1999)'s criteria for assessing the inclusiveness of local collaborative tourism policymaking projects, namely *scope and intensity of collaboration* as well as the *degree to which consensus emerges* (p. 393). Findings indicate low participation in both domains. However, an interesting observation is that for community members with limited or no previous community engagement experience, especially recent immigrants, observation and listening can be considered as the first stage of participation. Although it may seem that they are not contributing to discussions, this silent observation is part of a learning process that helps them to become comfortable and ready for more active forms of participation.

Studies on communicative planning in general (e.g. Matthews, 2013; Umemoto & Igarashi, 2009; Watson, 2006) and in the tourism context in particular (e.g. Bramwell, 2004; Dredge, 2006; Jamal et al., 2002) discuss the impact of power structures on participation, especially among marginalized groups. This research provides insights into how traditional assumptions, mental models, and priorities shape decision-making processes and prevail in discussions. It is partly the result of limited diversity among planners as well as active and engaged community members. This homogenous image can lead to further alienation of disengaged groups, despite being provided the opportunity to participate and to voice their opinion. Research findings suggest that there is potential in implementing parallel strategies, in collaboration with community leaders and local organizations, for addressing this problem. These mechanisms can help disengaged community sub-groups to experience community engagement in smaller and friendlier settings that are customized based on their needs and preferences. Parallel strategies should be aimed at facilitating and accelerating immigrants' participation in the mainstream decision-making process.

As shown in Figure 1, the long-term approach and on-going dialogue are closely related to openness and flexibility to new perspectives. Maintaining a long-term dialogue with a diverse group of participants is not possible without the willingness and ability to understand and accommodate alternative perspectives. At the same time, communication and dialogue enhance flexibility and openness through learning. A long-term approach is essential for communicative, double and triple-loop learning. It also contributes to instrumental and single-loop learning by allowing planners to have access to a broader range of community engagement skills and experiences gained over time. These conditions then provide a good foundation for more practical steps to be taken in terms of collaborating with community leaders and designing parallel strategies and customized tactics, typically in the context of specific projects.

For the purpose of this study, community is defined based on the geographic region and looks at neighbourhoods adjacent to the Rouge National Urban Park. This approach was in line with Parks Canada's definition and allowed the research to reflect the real situation and account for the diversity within the resident community. Further comparative studies could contribute to a better understanding of similarities and differences between specific cultural communities. It would also be interesting to compare the experience of immigrants from English speaking and other countries.

The five principles are intended to provide guidelines at the strategic level. However, each theme also provides specific insights into barriers contributing to immigrants' low participation and offers suggestions for addressing them (e.g. the need to initiate communication and trust building at the point of entry rather than focusing on established immigrants; alienation of immigrants and persistence of traditional assumptions due to lack of diversity among planners and engaged community members; the risk of creating tension between "newer voices" and "traditional views" by adopting a general strategy). Although the research is conducted in the context of the Rouge National Urban Park, many of the research participants had extensive experience in working with cultural communities on different projects. Therefore, their insights and reflections were not limited solely to this specific case. It is hoped that the principles presented in this study will be useful in enhancing fringe stakeholders' engagement in similar settings to support greater diversity in future park planning processes.

Note

1. The scholarly debate on applications and limitations of communicative planning provides valuable insights and an appropriate foundation for presenting and discussing research findings. However, an in-depth and critical discussion of the theory is not intended here and is beyond the scope of this paper.

Acknowledgments

The authors would like to thank Dr. Elizabeth C. Kurucz, Dr. Heather Mair, and Dr. Chris Choi for their valuable insights and contributions to this research and Parks Canada for its genuine support of this study through access to information, time, and open willingness to advance planning processes.

Disclosure statement

No potential conflict of interest was reported by the authors.

ORCID

Marion Joppe (iD) http://orcid.org/0000-0002-6360-3218

References

Armitage, D., Marschke, M., & Plummer, R. (2008). Adaptive co-management and the paradox of learning. *Global Environmental Change, 18*(1), 86–98.

Arnstein, S. R. (1969). A ladder of citizen participation. *Journal of the American Institute of Planners, 35*(4), 216–224.

Baer, D. (2008). Community context and civic participation in immigrant communities: A multilevel study of 137 Canadian communities. Retrieved May 4, 2015, from Metropolis British Columbia: http://mbc.metropolis.net/assets/uploads/files/wp/2008/WP08-03.pdf

Becken, S., & Job, H. (2014). Protected areas in an era of global–local change. *Journal of Sustainable Tourism, 22*(4), 507–527.

Berkes, F. (2009). Evolution of co-management: Role of knowledge generation, bridging organizations and social learning. *Journal of Environmental Management, 90*(5), 1692–1702.

Bisby, A. (2015, February 9). Why aren't Canadians camping anymore? *The Globe and Mail.* Retrieved April 10, 2017 from http://www.theglobeandmail.com/life/travel/travel-news/why-arent-canadians-camping-any-more/article22860024/

Bornhorst, T., Ritchie, J. R. B., & Sheehan, L. (2010). Determinants of tourism success for DMOs & destinations: An empirical examination of stakeholders' perspectives. *Tourism Management, 31*(5), 572–589.

Bramwell, B. (2004). Partnerships, participation, and social science research in tourism planning. In A. A. Lew, C. M. Hall, & A. M. Williams (Eds.), *A companion to tourism* (pp. 541–554). Malden, MA: Blackwell.

Bramwell, B., & Sharman, A. (1999). Collaboration in local tourism policy making. *Annals of Tourism Research, 26*(2), 392–415.

Burton, B. K., & Dunn, C. P. (1996). Feminist ethics as moral grounding for stakeholder theory. *Business Ethics Quarterly, 6*(2), 133–147.

Byrd, E. T. (2007). Stakeholders in sustainable tourism development and their roles: Applying stakeholder theory to sustainable tourism development. *Tourism Review, 62*(2), 6–13.

Carter, P. L. (2008). Coloured places and pigmented holidays: Racialized leisure travel. *Tourism Geographies, 10*(3), 265–284.

Charmaz, K. (2006). *Constructing grounded theory: A practical guide through qualitative analysis.* Thousand Oaks, CA: Sage.

City of Toronto. (2013). 2011 National Household Survey: Immigration, citizenship, place of birth, ethnicity, visible minorities, religion and aboriginal peoples. Retrieved August 20, 2016, from https://www1.toronto.ca/city_of_toronto/social_development_finance__administration/files/pdf/nhs_backgrounder.pdf

Choi, H. C., & Murray, I. (2010). Resident attitudes toward sustainable community tourism. *Journal of Sustainable Tourism, 18*(4), 575–594.

Crane, A., & Ruebottom, T. (2011). Stakeholder theory and social identity: Rethinking stakeholder identification. *Journal of Business Ethics, 102*(1), 77–87.

Crotty, M. (1998). *The foundations of social research: Meaning and perspective in the research process.* Thousand Oaks, CA: Sage.

Darbellay, F., & Stock, M. (2012). Tourism as complex interdisciplinary research object. *Annals of Tourism Research, 39*(1), 441–458.

Deng, J., Walker, G. J., & Swinnerton, G. S. (2005). A comparison of attitudes toward appropriate use of national parks between Chinese in Canada and Anglo-Canadians. *World Leisure Journal, 47*(3), 28–41.

Dredge, D. (2006). Networks, conflict and collaborative communities. *Journal of Sustainable Tourism, 14*(6), 562–581.

Dunham, L., Freeman, R. E., & Liedtka, J. (2006). Enhancing stakeholder practice: A particularized exploration of community. *Business Ethics Quarterly, 16*, 23–42.

Eagles, P. F. J. (2014). Research priorities in park tourism. *Journal of Sustainable Tourism, 22*(4), 528–549.

Floyd, M. (2001). Managing national parks in a multicultural society: Searching for common ground. *George Wright Forum, 18*(3), 41–51.

Floyd, M. F., Shinew, K. J., McGuire, F. A., & Noe, F. P. (1994). Race, class, and leisure activity preferences: Marginality and ethnicity revisited. *Journal of Leisure Research, 26*(2), 158–173.

Fortin, M. J., & Gagnon, C. (1999). An assessment of social impacts of national parks on communities in Quebec, Canada. *Environmental Conservation, 26*(03), 200–211.

Gephart, R. P. (2004). Qualitative research and the *Academy of Management Journal. Academy of Management Journal, 47*(4), 454–462.

Getz, D., & Timur, S. (2005). Stakeholder involvement in sustainable tourism: Balancing the voices. In W. F. Theobald (Ed.), *Global tourism* (pp. 230–247, 3rd ed.). Burlington, MA: Elsevier.

Gobster, Paul H. (2002). Managing urban parks for a racially and ethnically diverse clientele. *Leisure Sciences, 24*(2), 143–159.

Guba, E. G., & Lincoln, Y. S. (1982). Epistemological and methodological bases of naturalistic inquiry. *Educational Communications and Technology Journal*, *30*(4), 233–252.

Habermas, J. (1983). *The theory of communicative action,* vol.1. Cambridge, MA: MIT Press.

Hart, S. L., & Sharma, S. (2004). Engaging fringe stakeholders for competitive imagination. *Academy of Management Executive*, *18*(1), 7–18.

Hester, R. T., Blazej, N. J., & Moore, I. S. (1999). Whose wild? Resolving cultural and biological diversity conflicts in urban wilderness. *Landscape Journal*, *18*(2), 137–146.

Hung, K., Sirakaya-Turk, E., & Ingram, L. (2011). Testing the efficacy of an integrative model for community participation. *Journal of Travel Research*, *50*(3), 276–288.

Hunter, C. J. (1995). On the need to re-conceptualize sustainable tourism development. *Journal of Sustainable Tourism*, *3*(3), 155–165.

Jamal, T. B., & Getz, D. (1995). Collaboration theory and community tourism planning. *Annals of Tourism Research*, *22*(1), 186–204.

Jamal, T. B., Stein, S. M., & Harper, T. L. (2002). Beyond labels pragmatic planning in multi-stakeholder tourism-environmental conflicts. *Journal of Planning Education and Research*, *22*(2), 164–177.

Johnson, C. Y., Bowker, J. M., Bergstrom, J. C., & Ken Cordell, H. (2004). Wilderness values in America: Does immigrant status or ethnicity matter? *Society and Natural Resources*, *17*(7), 611–628.

Khazaei, A., Elliot, S., & Joppe, M. (2015). An application of stakeholder theory to advance community participation in tourism planning: The case for engaging immigrants as fringe stakeholders. *Journal of Sustainable Tourism*, *23*(7), 1049–1062.

"Le Conseil d'Etat met un terme aux arrêtés « anti-burkini »" [The Council of State ends the decrees against the burkini], *Le Monde*, Retrieved March 20, 2017, from http://www.lemonde.fr/societe/article/2016/08/26/le-conseil-d-etat-suspend-l-arrete-anti-burkini-de-villeneuve-loubet_4988472_3224.html#xb1PTXCLApHeJUJ1.99

Lovelock, K., Lovelock, B., Jellum, C., & Thompson, A., (2011). In search of belonging: Immigrant experiences of outdoor nature-based settings in New Zealand. *Leisure Studies*, *30*(4), 513–529.

Lovelock, B., Lovelock, K., Jellum, C., & Thompson, A. (2012). Immigrants' experiences of nature-based recreation in New Zealand. *Annals of Leisure Research*, *15*(3), 204–226.

Madrigal, R. (1995). Residents' perceptions and the role of government. *Annals of Tourism Research*, *22*(1), 86–102.

Marschke, M., & Sinclair, A. J. (2009). Learning for sustainability: Participatory resource management in Cambodian fishing villages. *Journal of Environmental Management*, *90*(1), 206–216.

Matthews, P. (2013). The longue durée of community engagement: New applications of critical theory in planning research. *Planning Theory*, *12*(2), 139–157.

McCool, S. F. (2009). Constructing partnerships for protected area tourism planning in an era of change and messiness. *Journal of Sustainable Tourism*, *17*(2), 133–148.

Mezirow, J. (1981). A critical theory of adult learning and education. *Adult Education Quarterly*, *32*(1), 3–24.

Okazaki, E. (2008). A community-based tourism model: Its conception and use. *Journal of Sustainable Tourism*, *16*(5), 511–529.

Parks Canada. (2009). Parks Canada Guiding Principles and Operational Policies - 8. Public Involvement. Retrieved April 7, 2017, from http://www.pc.gc.ca/en/docs/pc/poli/princip/sec1/part1d#a8

Parks Canada. (2012). Rouge National Urban Park Concept. Retrieved 8 June 2017, from http://www.pc.gc.ca/en/progs/np-pn/cnpn-cnnp/rouge/~/media/progs/np-pn/cnpn-cnnp/rouge/PDF/concept_rouge_national_urban_park.ashx

Parks Canada. (2015). Rouge National Urban Park. Retrieved August 20, 2016, from http://www.pc.gc.ca/eng/pn-np/on/rouge/nouvelles-news/agrandissement-expansion.aspx

Patton, E., & Appelbaum, S. H. (2003). The case for case studies in management research. *Management Research News*, *26*(5), 60–71.

Perdue, R. R., Long, P. T., & Allen, L. (1990). Resident support for tourism development. *Annals of Tourism Research*, *17*(4), 586–599.

Pergams, O. R., & Zaradic, P. A. (2008). Evidence for a fundamental and pervasive shift away from nature-based recreation. *Proceedings of the National Academy of Sciences*, *105*(7), 2295–2300.

Peters, K. (2010). Being together in urban parks: Connecting public space, leisure, and diversity. *Leisure Sciences*, *32*(5), 418–433.

Peters, K., Elands, B., & Buijs, A. (2010). Social interactions in urban parks: Stimulating social cohesion? *Urban Forestry & Urban Greening*, *9*(2), 93–100.

Phillimore, J., & Goodson, L. (2004). The inquiry paradigm in qualitative tourism research. In J. Phillimore & L. Goodson (Eds.), *Qualitative research in tourism ontologies, epistemologies and methodologies*. London: Routledge.

Reed, M. S., Graves, A., Dandy, N., Posthumus, H., Hubacek, K., Morris, J., ... Stringer, L. C. (2009). Who's in and why? A typology of stakeholder analysis methods for natural resource management. *Journal of Environmental Management*, *90*(5), 1933–1949.

Roloff, J. (2008). Learning from multi-stakeholder networks: Issue-focussed stakeholder management. *Journal of Business Ethics*, *82*(1), 233–250.

Ruiz-Ballesteros, E. (2011). Social-ecological resilience and community-based tourism: An approach from Agua Blanca, Ecuador. *Tourism Management, 32*(3), 655–666.

Saraniemi, S., & Kylänen, M. (2011). Problematizing the concept of tourism destination: An analysis of different theoretical approaches. *Journal of Travel Research, 50*(2), 133–143.

Schmidtke, O., & Neumann, S. (2010). *Engaging the migrant community outside of Canada's main metropolitan centers: Community engagement – The Welcoming Community Initiative and the case of Greater Victoria* (Working Paper No. 10-13). Metropolis British Columbia, Centre of Excellence for Research on Immigration and Diversity. Retrieved March, 15, 2017, from http://mbc.metropolis.net/assets/uploads/files/wp/2010/WP10-13.pdf

Shinew, K. J., Stodolska, M., Floyd, M., Hibbler, D., Allison, M., Johnson, C., & Santos, C. (2006). Race and ethnicity in leisure behaviour: Where have we been and where do we need to go? *Leisure Sciences, 28*(4), 403–408.

Shultis, J., & More, T. (2011). American and Canadian national park agency responses to declining visitation. *Journal of Leisure Research, 43*(1), 110–132.

Statistics Canada. (2013). Immigration and ethnocultural diversity in Canada. Catalogue no. 99-010-X2011001. Retrieved August 20, 2016, from http://www12.statcan.gc.ca/census-recensement/index-eng.cfm

Stodolska, M., (2000). Changes in leisure participation patterns after immigration. *Leisure Sciences, 22*(1), 39–63.

Tosun, C. (2006). Expected nature of community participation in tourism development. *Tourism Management, 27*(3), 493–504.

Tweed, A., & Charmaz, K. (2011). Grounded theory methods for mental health practitioners. In D. Harper & A. R. Thompson (Eds.), *Qualitative research methods in mental health and psychotherapy: A guide for students and practitioners*. Chichester: John Wiley & Sons.

Umemoto, K., & Igarashi, H. (2009). Deliberative planning in a multicultural milieu. *Journal of Planning Education and Research, 29*(1), 39–53.

Waligo, V. M., Clarke, J., & Hawkins, R. (2013). Implementing sustainable tourism: A multi-stakeholder involvement management framework. *Tourism Management, 36*, 342–353.

Watson, V. (2006). Deep difference: Diversity, planning and ethics. *Planning Theory, 5*(1), 31–50.

Weber, J., & Sultana, S. (2013). Why do so few minority people visit National Parks? Visitation and the accessibility of "America's Best Idea". *Annals of the Association of American Geographers, 103*(3), 437–464.

An interview with a protected area insider

Barbara Engels, Susanne Becken, Hubert Job and Bernard Lane

Susanne Becken, Hubert Job, and Bernard Lane talk to Barbara Engels, an experienced officer at the German Agency for Nature Conservation, specializing in Protected Area management, and a member of the German Delegation to the UNESCO Word Heritage Committee.

Susanne Becken, Hubert Job, and Bernard Lane (SB, HJ, and BL): How does UNESCO manage the work load created by the 1073 World Heritage Sites (WHSs) in 167 countries and 669 Biosphere Reserves (BRs) in 120 countries around the world?

Barbara Engels (BE): As the Secretariat of the World Heritage Convention, UNESCO ensures the day-to-day management of the Convention, the World Heritage Centre in Paris, organizes the annual sessions of the World Heritage Committee and its Bureau, provides advice to States Parties on the preparation of site nominations, organizes international assistance from the World Heritage Fund upon request, and coordinates both the reporting on the condition of sites and the emergency action undertaken when a site is threatened. The Centre also organizes technical seminars and workshops, updates the World Heritage List and database, develops teaching materials to raise awareness among young people of the need for heritage preservation, and keeps the public informed of World Heritage issues. At present, around 30 fixed term staff are posted to the World Heritage Centre. For BRs, some 10 fixed term staff are posted to UNESCO headquarters. Given the different procedures under both the WH Convention and the Man and Biosphere (MAB) programme, a comparison of staff and workload is not easily possible.

However, the implementation of the Convention and the programme are not only handled within the UNESCO. On a national level the WHSs and BRs are supported by national ministries, protected area agencies, civil society organizations, National UNESCO Commissions, and National MAB Committees. Furthermore, UNESCO offices as well as International Union for the Conservation of Nature (IUCN) offices all around the world provide support to both WHSs and BRs.

It is also to be noted that the World Heritage Convention has three international Advisory bodies: IUCN, ICOMOS (International Council of Monuments and Sites), and ICCROM (International Centre for the Study of the Preservation and Restoration of Cultural Property). These organizations play an important role in the evaluation of nominations and the monitoring processes of the Convention.

SB, HJ, and BL: Is there a planned limit to the number of WHSs or BRs, and how many might there be one day?

BE: There is no limit to the possible overall number of either WHSs or BRs. However, the number of new WHSs to be evaluated and considered for selection each year is currently limited to 35 and one nomination per State Party, with priority given to countries with no or very few sites inscribed. This takes into account that there is still a global imbalance between regions and states on the WHS list. For BRs, no limit currently applies. No overall limit has been discussed so far, but the concept of

"Outstanding Universal Value" (OUV) as the requirement for World Heritage status somehow implies that at some point all sites with OUV might be included in the WHS list.

SB, HJ, and BL: Do you think that the role of protected areas in general has changed over the last two decades, and, if it has, what does that mean for Governments and also for international organizations like UNESCO?

BE: The role of protected areas has changed considerably over the last two decades. Protected areas today are considered much more than "only" areas for species, habitat, or biodiversity conservation. They are expected to fulfil a multitude of other tasks: foster sustainable development, provide education for sustainable development, provide space, adequate infrastructure, and opportunities for nature-based tourism, deliver nature-based solutions to global environmental problems such as climate change, establish research and monitoring programmes, and evaluate their management effectiveness – to name just a few examples. For governments and international organizations that means somehow finding additional resources, and employing staff with new skills.

The MAB programme has undergone an interesting and major shift from a science focused programme in the 1970s/1980s to a true sustainable development focussed programme since the mid-1990s. The World Heritage Convention has also seen this shift, although the focus of the Convention remains on conservation. However, over the past few years, there have been a number of policy milestones that could have an impact on protected areas in the future including the policy adopted in 2015 by the General Assembly of States Parties to the World Heritage Convention on the integration of a sustainable development perspective into the processes of the World Heritage Convention, which allows States Parties, practitioners, institutions, communities, and networks to harness the potential of World Heritage properties, and heritage in general, to contribute to sustainable development.

The Convention has adopted a policy on World Heritage and sustainable development and runs a sustainable Tourism Programme. For governments, this means translating these different and changing concepts into national legislation and implementation. The World Heritage Committee adopted the World Heritage and Sustainable Tourism Programme in 2013 (see http://whc.unesco.org/archive/2012/whc12-36com-5E-en.pdf). A key objective of the programme is to strengthen the enabling environment by advocating policies, strategies, frameworks, and tools that support sustainable tourism as an important vehicle for protecting and managing cultural and natural heritage of OUV. This, coupled with the objective of integrating sustainable tourism principles into the mechanisms of the World Heritage Convention, provides the framework for action on sustainable tourism in the public sector for international organizations.

SB, HJ, and BL: What special roles do UNESCO enlisted sites, WHSs and BRs, play in implementing the global PA mandate?

BE: They play a crucial role as the UNESCO labelled sites receive special attention – both on the political level and from the broader public. As both systems – the WH Convention and the MAB programme – put emphasis on effective and adequate management, WHSs and BRs can serve as models for other protected areas.

SB, HJ, and BL: The opening paper of this special issue considers that there could be a case for bringing together the concepts of BRs and WHSs. Do you have any thoughts about that?

BE: The need to integrate sustainable development with and in WHSs clearly links the two concepts nowadays. Nevertheless, the World Heritage Convention's primary goal is protection and safeguarding the World's Natural and Cultural Heritage, whilst BRs are meant to serve as model regions for sustainable development. To achieve their goals, BRs very often include much larger landscapes and seascapes where sustainable land use shapes these areas – especially in their so-called development zones. But integration of the two concepts can happen, when, for example, a zone of strict nature conservation is formed by a WHS surrounded by a buffer zone (also a requirement under the WH

Convention) and a much larger development zone. A WH site as the core of a BR reserve can serve both concepts and allow for both: proper conservation of the core values and sustainable development in the surrounding landscape.

SB, HJ, and BL: How important is the consistent economic monitoring of tourism in all protected areas worldwide?

BE: This is indeed very important. In many protected areas, tourism is considered as potentially one of the most adequate forms of sustainable land use (as it can be non-consumptive). And many protected areas use the argument of the creation of income though tourism in the protected area to convince politicians or local stakeholders of their value. Political and local acceptance of a protected area is very important. But what is needed is comparable, reliable, and easily comprehensible data to prove this argument. Without consistent economic monitoring, protected areas cannot deliver that data.

SB, HJ, and BL: How serious is the risk of the commodification of nature (and maybe of the commodification of culture and communities too?) in WHSs and BRs and how do organizations at both the international level and locally manage this risk?

BE: Again, proper monitoring not only of the economic but also of the ecological and social impacts of tourism is essential to evaluate the impacts tourism might have on a protected area. This is even more important as the marketing of sites as WH and BR might considerably increase not only the numbers of tourists but also the demand for specific tourism products.

Although the World Heritage Convention does not specifically reference tourism, it does require in Article 4 for each State Party to ensure the identification, protection, conservation, presentation, and transmission to future generations of the cultural and natural heritage. The presentation and transmission of heritage using interpretation, communication, and marketing is the foundation of UNESCO's work on tourism and it is in this paradigm that the risk of commodification rests. A stakeholder approach to ensure authentic and heritage based tourism is needed to manage this risk.

SB, HJ, and BL: Is tourism a main driver in the designation of new UNESCO sites, and how are the UNESCO WHS, and BR governance committees shaped by tourism influence/interests?

BE: The driving force for UNESCO designation often comes from the local level. And while the possibility of increasing tourism can influence their motivation to prepare a nomination, there are other factors that inspire inscription. For example, a study by James Rebanks[1] found that for some WHs and BRs, tourism was not the only reason for inscription. The study found that UNESCO WHSs fall into four conceptual categories:

(1) "Celebration" Designation – many areas treat it as a celebration or reward designation for heritage already preserved.
(2) Heritage "SOS" Designation – many areas treat it as an emergency designation to safeguard outstanding universal heritage at risk.
(3) Marketing/Quality Logo/Brand – a growing number of sites see the value as a marketing or quality brand for historic places.
(4) A "Place Making" Catalyst – this view treats WHS status as a powerful catalyst for economic development using heritage as a tool to develop powerful new identities for places, and to create powerful programmes of actions to change places fundamentally.

SB, HJ, and BL: It would be interesting to know how the "committee" that chooses new sites (and determines the future of wayward old sites) is chosen. Is it a secret?

BE: No, this is not a secret. The decision to nominate sites as WH sites or BR reserves is a sovereign national decision by the States Parties. The WH Committee which finally decides which sites become

WHSs is made up of 21 States Parties elected by the General Assembly of the World Heritage Convention. The same is valid for the BRs where the MAB International Coordinating Council (elected by the UNESCO General Assembly) decides on the new BRs.

SB, HJ, and BL: And finally, is there an overall vision for UNESCO's programme linking tourism and heritage?

BE: The UNESCO World Heritage and Sustainable Tourism Programme (WH+ST Programme) provides a global platform to catalyse positive change to protect and conserve the WHSs while enriching the lives of local communities and at the same time enhancing the experience of travellers.

The vision of the UNESCO WH+ST Programme is for World Heritage and tourism stakeholders to share responsibility for the conservation of our common cultural and natural heritage of OUV and for sustainable development through appropriate tourism management. It is part of a growing movement to make WHSs places which lead the world in sustainable tourism.

Note

1. See http://www.lakedistrict.gov.uk/__data/assets/pdf_file/0009/393966/WHSEconomicGainSupplement.pdf.

Acknowledgments

The interviewers, and the interviewee, wish to thank Peter Debrine for his help with information for this interview, and with the meetings that have taken place that led to the publication of the Special Issue of the *Journal of Sustainable Tourism* in which this interview is published. Peter Debrine is the Senior Project Officer at UNESCO's World Heritage and Sustainable Tourism Programme.

Disclosure statement

No potential conflict of interest was reported by the authors.

Index